Springer Series in Reliability Engineering

Series Editor

Professor Hoang Pham
Department of Industrial and Systems Engineering
Rutgers, The State University of New Jersey
96 Frelinghuysen Road
Piscataway, NJ 08854-8018
USA

Other titles in this series

The Universal Generating Function in Reliability Analysis and Optimization
Gregory Levitin

Warranty Management and Product Manufacture
D.N.P. Murthy and Wallace R. Blischke

Maintenance Theory of Reliability
Toshio Nakagawa

System Software Reliability
Hoang Pham

Reliability and Optimal Maintenance
Hongzhou Wang and Hoang Pham

Applied Reliability and Quality
B.S. Dhillon

Shock and Damage Models in Reliability Theory
Toshio Nakagawa

Risk Management
Terje Aven and Jan Erik Vinnem

Satisfying Safety Goals by Probabilistic Risk Assessment
Hiromitsu Kumamoto

Offshore Risk Assessment (2nd Edition)
Jan Erik Vinnem

The Maintenance Management Framework
Adolfo Crespo Márquez

Human Reliability and Error in Transportation Systems
B.S. Dhillon

Complex System Maintenance Handbook
D.N.P. Murthy and Khairy A.H. Kobbacy

Recent Advances in Reliability and Quality in Design
Hoang Pham

Product Reliability
D.N.P. Murthy, Marvin Rausand and Trond Østerås

Mining Equipment Reliability, Maintainability, and Safety
B.S. Dhillon

Advanced Reliability Models and Maintenance Policies
Toshio Nakagawa

Justifying the Dependability of Computer-based Systems
Pierre-Jacques Courtois

Reliability and Risk Issues in Large Scale Safety-critical Digital Control Systems
Poong Hyun Seong

Maxim Finkelstein

Failure Rate Modelling for Reliability and Risk

Springer

Maxim Finkelstein, PhD, DSc
Department of Mathematical Statistics
University of the Free State
Bloemfontein
South Africa
and
Max Planck Institute for Demographic Research
Rostock
Germany

ISBN 978-1-84996-810-2 e-ISBN 978-1-84800-986-8

DOI 10.1007978-1-84800-986-8

Springer Series in Reliability Engineering ISSN 1614-7839

A catalogue record for this book is available from the British Library

Cover design: deblik, Berlin, Germany

Printed on acid-free paper

9 8 7 6 5 4 3 2 1

springer.com

To my wife Olga

Preface

In the early 1970s, after obtaining a degree in mathematical physics, I started working as a researcher in the Department of Reliability of the Saint Petersburg Elektropribor Institute. Founded in 1958, it was the first reliability department in the former Soviet Union. At first, for various reasons, I did not feel a strong inclination towards the topic. Everything changed when two books were placed on my desk: Barlow and Proshcan (1965) and Gnedenko *et al.* (1964). On the one hand, they showed how mathematical methods could be applied to various reliability engineering problems; on the other hand, these books described reliability theory as an interesting field in applied mathematics/probability and statistics. And this was the turning point for me. I found myself interested–and still am after more than 30 years of working in this field.

This book is about reliability and reliability-related stochastics. It focuses on failure rate modelling in reliability analysis and other disciplines with similar settings. Various applications of risk analysis in engineering and biological systems are considered in the last three chapters. Although the emphasis is on the failure rate, one cannot describe this topic without considering other reliability measures. The mean remaining lifetime is the first in this list, and we pay considerable attention to describing and discussing its properties.

The presentation combines classical results and recent results of other authors with our research over the last 10 to15 years. The recent excellent encyclopaedic books by Lai and Xie (2006) and Marshall and Olkin (2007) give a broad picture of the modern mathematical reliability theory and also present an up-to-date source of references. Along with the classical text by Barlow and Proschan (1975), the excellent textbook by Rausand and Hoyland (2004) and a mathematically oriented reliability monograph by Aven and Jensen (1999), these books can be considered as complementary or further reading. I hope that our text will be useful for reliability researchers and practitioners and to graduate students in reliability or applied probability.

I acknowledge the support of the University of the Free State, the National Research Foundation (South Africa) and the Max Planck Institute for Demographic Research (Germany).

I thank those with whom I had the pleasure of working and (or) discussing reliability-related problems: Frank Beichelt, Ji Cha, Pieter van Gelder, Waltraud

Kahle, Michail Nikulin, Jan van Noortwijk, Michail Revjakov, Michail Rosenhaus, Fabio Spizzichino, Jef Teugels, Igor Ushakov, James Vaupel, Daan de Waal, Tertius de Wet, Anatoly Yashin, Vladimir Zarudnij. Chapters 6 and 7 are written in co-authorship with my daughter Veronica Esaulova on the basis of her PhD thesis (Esaulova, 2006). Many thanks to her for this valuable contribution.

I would like to express my gratitude and appreciation to my colleagues in the department of mathematical statistics of the University of the Free State. Annual visits (since 2003) to the Max Planck Institute for Demographic Research (Germany) also contributed significantly to this project, especially to Chapter 10, which is devoted to demographic and biological applications.

Special thanks to Justin Harvey and Lieketseng Masenyetse for numerous suggestions for improving the presentation of this book. Finally, I am indebted to Simon Rees, Anthony Doyle and the Springer staff for their editorial work.

University of the Free State
South Africa
July 2008

Maxim Finkelstein

Contents

1 Introduction ... 1
1.1 Aim and Scope of the Book ... 1
1.2 Brief Overview.. ... 5

2 Failure Rate and Mean Remaining Lifetime ... 9
2.1 Failure Rate Basics ... 10
2.2 Mean Remaining Lifetime Basics ... 13
2.3 Lifetime Distributions and Their Failure Rates ... 19
2.3.1 Exponential Distribution ... 19
2.3.2 Gamma Distribution ... 20
2.3.3 Exponential Distribution with a Resilience Parameter ... 22
2.3.4 Weibull Distribution ... 23
2.3.5 Pareto Distribution ... 24
2.3.6 Lognormal Distribution ... 25
2.3.7 Truncated Normal Distribution ... 26
2.3.8 Inverse Gaussian Distribution ... 27
2.3.9 Gompertz and Makeham–Gompertz Distributions ... 27
2.4 Shape of the Failure Rate and the MRL Function ... 28
2.4.1 Some Definitions and Notation ... 28
2.4.2 Glaser's Approach ... 30
2.4.3 Limiting Behaviour of the Failure Rate and the MRL Function... 36
2.5 Reversed Failure Rate ... 39
2.5.1 Definitions ... 39
2.5.2 Waiting Time ... 42
2.6 Chapter Summary ... 43

3 More on Exponential Representation ... 45
3.1 Exponential Representation in Random Environment ... 45
3.1.1 Conditional Exponential Representation ... 45
3.1.2 Unconditional Exponential Representation ... 47
3.1.3 Examples ... 48
3.2 Bivariate Failure Rates and Exponential Representation ... 52

3.2.1 Bivariate Failure Rates ... 52
3.2.2 Exponential Representation of Bivariate Distributions ... 54
3.3 Competing Risks and Bivariate Ageing ... 59
3.3.1 Exponential Representation for Competing Risks ... 59
3.3.2 Ageing in Competing Risks Setting ... 60
3.4 Chapter Summary ... 65

4 Point Processes and Minimal Repair ... 67
4.1 Introduction – Imperfect Repair ... 67
4.2 Characterization of Point Processes ... 70
4.3 Point Processes for Repairable Systems ... 72
4.3.1 Poisson Process ... 72
4.3.2 Renewal Process ... 73
4.3.3 Geometric Process ... 76
4.3.4 Modulated Renewal-type Processes ... 79
4.4 Minimal Repair. ... 81
4.4.1 Definition and Interpretation ... 81
4.4.2 Information-based Minimal Repair ... 83
4.5 Brown–Proschan Model ... 84
4.6 Performance Quality of Repairable Systems ... 85
4.6.1 Perfect Restoration of Quality ... 86
4.6.2 Imperfect Restoration of Quality ... 88
4.7 Minimal Repair in Heterogeneous Populations ... 89
4.8 Chapter Summary ... 92

5 Virtual Age and Imperfect Repair ... 93
5.1 Introduction – Virtual Age ... 93
5.2 Virtual Age for Non-repairable Objects ... 95
5.2.1 Statistical Virtual Age ... 95
5.2.2 Recalculated Virtual Age ... 98
5.2.3 Information-based Virtual Age ... 102
5.2.4 Virtual Age in a Series System ... 105
5.3 Age Reduction Models for Repairable Systems ... 107
5.3.1 G-renewal Process ... 107
5.3.2 'Sliding' Along the Failure Rate Curve ... 109
5.4 Ageing and Monotonicity Properties ... 115
5.5 Renewal Equations ... 123
5.6 Failure Rate Reduction Models ... 125
5.7 Imperfect Repair *via* Direct Degradation ... 127
5.8 Chapter Summary ... 130

6 Mixture Failure Rate Modelling ... 133
6.1 Introduction – Random Failure Rate ... 133
6.2 Failure Rate of Discrete Mixtures ... 138
6.3 Conditional Characteristics and Simplest Models ... 139
6.3.1 Additive Model ... 141
6.3.2 Multiplicative Model ... 143

6.4 Laplace Transform and Inverse Problem ... 144
6.5 Mixture Failure Rate Ordering ... 149
6.5.1 Comparison with Unconditional Characteristic ... 149
6.5.2 Likelihood Ordering of Mixing Distributions ... 152
6.5.3 Mixing Distributions with Different Variances ... 157
6.6 Bounds for the Mixture Failure Rate ... 159
6.7 Further Examples and Applications ... 163
6.7.1 Shocks in Heterogeneous Populations ... 163
6.7.2 Random Scales and Random Usage ... 164
6.7.3 Random Change Point ... 165
6.7.4 MRL of Mixtures ... 167
6.8 Chapter Summary ... 168

7 Limiting Behaviour of Mixture Failure Rates ... 171
7.1 Introduction ... 171
7.2 Discrete Mixtures ... 172
7.3 Survival Models ... 175
7.4 Main Asymptotic Results ... 177
7.5 Specific Models ... 179
7.5.1 Multiplicative Model ... 179
7.5.2 Accelerated Life Model ... 182
7.5.3 Proportional Hazards and Other Possible Models ... 183
7.6 Asymptotic Mixture Failure Rates for Multivariate Frailty ... 184
7.6.1 Introduction ... 184
7.6.2 Competing Risks for Mixtures ... 185
7.6.3 Limiting Behaviour for Competing Risks ... 187
7.6.4 Bivariate Frailty Model ... 189
7.7 Sketches of the Proofs ... 192
7.8 Chapter Summary ... 196

8 'Constructing' the Failure Rate ... 197
8.1 Terminating Poisson and Renewal Processes ... 197
8.2 Weaker Criteria of Failure ... 201
8.2.1 Fatal and Non-fatal Shocks ... 201
8.2.2 Fatal and Non-fatal Failures ... 205
8.3 Failure Rate for Spatial Survival ... 207
8.3.1 Obstacles with Fixed Coordinates ... 207
8.3.2 Crossing the Line Process ... 210
8.4 Multiple Availability on Demand ... 213
8.4.1 Introduction ... 213
8.4.2 Simple Criterion of Failure ... 215
8.4.3 Two Consecutive Non-serviced Demands ... 218
8.4.4 Other Weaker Criteria of Failure ... 221
8.5 Acceptable Risk and Thinning of the Poisson Process ... 222
8.6 Chapter Summary ... 223

9 Failure Rate of Software 225
9.1 Introduction 225
9.2 Several Empirical Models for Software Reliability 226
9.2.1 The Jelinski–Moranda Model 227
9.2.2 The Moranda Model 228
9.2.3 The Schick and Wolverton Model 229
9.2.4 Models Based on the Number of Failures 230
9.3 Time-dependant Operational Profile 231
9.3.1 General Setting 231
9.3.2 Special Cases 233
9.4 Chapter Summary 235

10 Demographic and Biological Applications 237
10.1 Introduction 237
10.2 Unobserved Overall Resource 242
10.3 Mortality Model with Anti-ageing 246
10.4 Mortality Rate and Lifesaving 250
10.5 The Strehler–Mildvan Model and Generalizations 252
10.6 'Quality-of-life Transformation' 253
10.7 Stochastic Ordering for Mortality Rates 255
10.7.1 Specific Population Modelling 256
10.7.2 Definitions of Life Expectancy 260
10.7.3 Comparison of Life Expectancies 263
10.7.4 Further Inequalities 265
10.8 Tail of Longevity 268
10.9 Chapter Summary 273

References 275

Index 287

1

Introduction

1.1 Aim and Scope of the Book

As the title suggests, this book is devoted to failure rate modelling for reliability analysis and other disciplines that employ the notion of the failure rate or its equivalents. The conditional hazard in risk analysis and the mortality rate in demography are the relevant examples of these equivalent concepts. Although the main focus in the text is on this crucial characteristic, our presentation cannot be restricted to failure rate analysis alone; other important reliability measures are studied as well.

We consider non-negative random variables, which are called *lifetimes*. The time to failure of an engineering component or a system is a lifetime, as is the time to death of an organism. The number of casualties after an accident and the wear accumulated by a degrading system are also positive random variables. Although we deal here mostly with engineering applications, the reliability-based approach to lifetime modelling for organisms is one of the important topics discussed in the last chapter of this book. Obviously, the human organism is not a machine, but nothing prevents us from using stochastic reasoning developed in reliability theory for lifespan modelling of organisms.

The presented models focus on reliability applications. However, some of the considered methods are already formulated in terms of risk and safety assessment (*e.g.*, Chapters 8 and 10); most of the others can also be used for this purpose after a suitable adjustment.

It is well known that the failure rate function can be interpreted as the probability (risk) of failure in an infinitesimal unit interval of time. Owing to this interpretation and some other properties, its importance in reliability, survival analysis, risk analysis and other disciplines is hard to overestimate. For example, the increasing failure rate of an object is an indication of its deterioration or ageing of some kind, which is an important property in various applications. Many engineering (especially mechanical) items are characterized by the processes of "wear and tear", and therefore their lifetimes are described by an increasing failure rate. The failure (mortality) rate of humans at adult ages is also increasing. The empirical Gompertz law of human mortality (Gompertz, 1825) defines the exponentially increasing mortality rate. On the other hand, the constant failure rate is usually an indication of a non-ageing property, whereas a decreasing failure rate can describe, *e.g.*, a period of "infant mortality" when early failures, bugs, *etc.*, are eliminated or corrected. Therefore, the shape of the failure rate plays an important role in reliability analysis. Figure 1.1 shows probably the most popular graph in reliability applications: a typical life cycle failure rate function (bathtub shape) of an engineering object. Note that, the usage period with a near-constant failure rate is mostly typical for various electronic items, whereas mechanical and electro-mechanical devices are usually subject to processes of wear. When the lifetime distribution function $F(t)$ is absolutely continuous, the failure rate $\lambda(t)$ can be defined as $F'(t)/(1-F(t))$. In this case, there exists a simple, well-known exponential representation for $F(t)$ (Section 2.1). It defines an important characterization of the distribution function *via* the failure rate $\lambda(t)$. Moreover, the failure rate contains information on the chances of failure of an operating object in the next sufficiently small interval of time. Therefore, the shape of $\lambda(t)$ is often much more informative in the described sense than, for example, the shapes of the distribution function or of the probability density function.

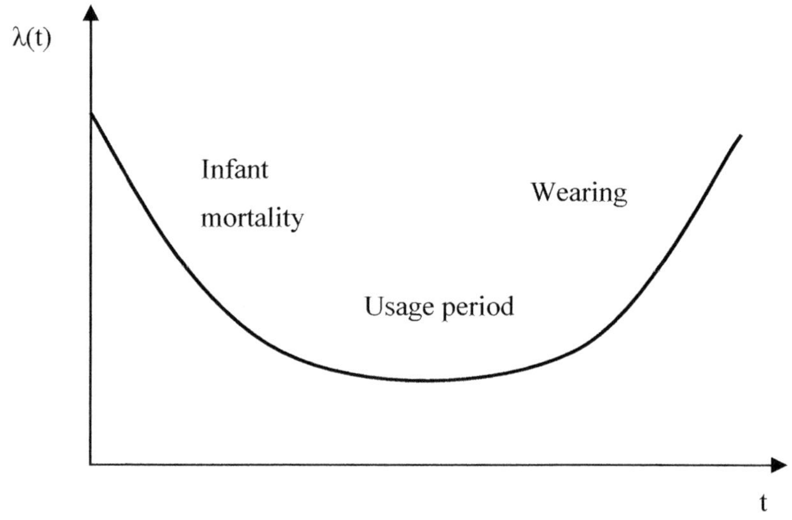

Figure 1.1. The bathtub curve

Many tools and approaches developed in reliability engineering are naturally formulated *via* the failure rate concept. For example, a well-known proportional hazards model that is widely used in reliability and survival analysis is defined directly in terms of the failure rate; the hazard (failure) rate ordering used in stochastic comparisons is the ordering of the failure rates; many software reliability models are directly formulated by means of the corresponding failure rates (see various models of Chapter 9). For example, each 'bug', in accordance with the Jelinski–Moranda model (Jelinski and Moranda, 1972), has an independent input of a fixed size into the failure rate of the software.

Although the emphasis in this book is on the failure rate, one cannot describe this topic without considering other reliability characteristics. The mean remaining lifetime is the first on this list, and we pay considerable attention to describing and discussing its properties. In many applications, the stochastic description of ageing by means of the mean remaining lifetime function that is decreasing with time is more appropriate than the description of ageing *via* the corresponding increasing failure rate.

In this text, we consider several generalizations of the 'classical' notion of the failure rate $\lambda(t)$. One of them is the *random failure rate.* Engineering and biological objects usually operate in a random environment. This random environment can be described by a stochastic process $Z_t, t \geq 0$ or by a random variable Z as a special case. Therefore, the failure rate, which corresponds to a lifetime T, can also be considered as a stochastic process $\lambda(t, Z_t)$ or a random variable $\lambda(t, Z)$. These functions should be understood conditionally on realizations $\lambda(t \mid z(u), 0 \leq u \leq t)$ and $\lambda(t \mid Z = z)$, respectively. Similar considerations are valid for the corresponding distribution functions $F(t, Z_t)$ and $F(t, Z)$. What happens when we try to average these characteristics and obtain the marginal (observed) distribution functions and failure rates? The following is obviously true for the distribution functions:

$$F(t) = E[F(t, Z_t)], \quad F(t) = E[F(t, Z)],$$

where the expectations should be obtained with respect to $Z_t, t \geq 0$ and Z, respectively. Note that explicit computations in accordance with these formulas are usually cumbersome and can be performed only for some special cases. On the other hand, it is clear that as the failure rate $\lambda(t)$ is a conditional characteristic (on the condition that an object did not fail up to t), the corresponding conditioning should be performed, *i.e.*,

$$\lambda(t) = E[\lambda(t, Z_t) \mid T > t], \quad \lambda(t) = E[\lambda(t, Z) \mid T > t].$$

This 'slight' difference can be decisive, as it not only complicates the computational part of the problem but often changes the important monotonicity properties of $\lambda(t)$ (compared with the monotonicity properties of the family of conditional failure rates $\lambda(t \mid Z = z)$). For example, when $\lambda(t \mid Z = z)$ is an increasing power function for each z (the Weibull law) and Z is a gamma-distributed random variable, $\lambda(t)$ appears to have an upside-down bathtub shape: this function is equal to 0 at $t = 0$, then increases to reach a maximum at some point in time and eventually monotonically decreases to 0 as $t \to \infty$. Another relevant example is when the conditional failure rate $\lambda(t \mid Z = z)$ is an exponentially increasing function (the Gompertz law). Assuming again that Z is gamma-distributed, it is easy to derive (Chapter 6) that $\lambda(t)$ tends to a constant as $t \to \infty$. The dramatic changes in the shapes of failure rates in these examples and in many other instances should be taken into account in theoretical analysis and in practical applications. Note that the second example provides a possible explanation for the mortality rate plateau of humans observed recently for the 'oldest-old' populations in developed countries (Thatcher, 1999). According to these results, the mortality rate of centenarians is either increasing very slowly or not increasing at all, which contradicts the Gompertz law of human mortality.

Another important generalization of the conventional failure rate $\lambda(t)$ deals with repairable systems and considers the failure rate of a repairable component as an intensity process (stochastic intensity) $\lambda_t, t \geq 0$. The 'randomness' of the failure rate in this case is due to random times of repair. This approach is in line with the modern description of point processes (see, *e.g.*, Daley and Vere–Jones, 1988, and Aven and Jensen, 1999). Assume for simplicity that the repair action is perfect and instantaneous. This means that after each repair a component is 'as good as new'. Let the governing failure rate for this component be $\lambda(t)$. Then the intensity process at time t for this simplest case of perfect repair is defined as

$$\lambda_t = \lambda(t - T_-),$$

where T_- denotes the random time of the last repair (renewal) before t. Therefore, the probability of a failure in $[t, t + dt)$ is $\lambda(t - T_-)dt$, which should also be understood conditionally on realizations of T_-. The main focus in Chapters 4 and 5 is on considering the intensity processes for the case of imperfect (general) repair when a component after the repair action is not *as good as new*. Various models of imperfect repair and of imperfect maintenance can be found in the literature (see, for example, the recent book by Wang and Pham, 2006, and references therein). We investigate only the most popular models of this kind and also discuss our recent findings in this field.

This book provides a comprehensive treatment of different reliability models focused on properties of the failure rate and other relevant reliability characteristics. Our presentation combines classical and recent results of other authors with our research findings of the last 10 to 15 years. We discuss the subject mostly using necessary tools and approaches and do not intend to present a self-sufficient textbook on reliability theory. The choice of topics is driven by the research interests of the author. The recent excellent encyclopaedic books by Lai and Xie (2006) and Marshall and Olkin (2007) give a broad picture of modern mathematical reliability theory and also present up-to-date reference sources. Along with the classical text by Barlow and Proschan (1975), an excellent textbook by Rausand and Hoyland (2004) and a mathematically oriented reliability monograph by Aven and Jensen (1999), these books can be considered the first-choice complementary or further reading.

In this book, we understand risk (hazard) as a chance (probability) of failure or of another undesirable, harmful event. The consequences of these events (Chapter 8) can also be taken into account to comply with the classical definition of risk (Bedford and Cooke, 2001).

The book is mostly targeted at researchers and 'quantitative engineers'. The first two chapters, however, can be used by undergraduate students as a supplement to a basic course in reliability. This means that the reader should be familiar with the basics of reliability theory. The other parts can form a basis for graduate courses on imperfect (general) repair and on mixture failure rate modelling for students in probability, statistics and engineering. The last chapter presents a collection of stochastic, reliability-based approaches to lifespan modelling and ageing concepts of organisms and can be useful to mathematical biologists and demographers.

We follow a general convention regarding the monotonicity properties of a function. We say that a function is increasing (decreasing) if it is not decreasing (increasing). We also prefer the term "failure rate" to the equivalent "hazard rate", although many authors use the second term. Among other considerations, this choice is supported by the fact that the most popular nonparametric classes of distributions in applications are the increasing failure rate (IFR) and the decreasing failure rate (DFR) classes.

Note that all necessary acronyms and nomenclatures are defined below in the appropriate parts of the text, when the corresponding symbol or abbreviation is used for the first time. For convenience, where appropriate, these explanations are often repeated later on in the text as well. This means that each section is self-sufficient in terms of notation.

1.2 Brief Overview

Chapter 2 is devoted to reliability basics and can be viewed as a brief introduction to some reliability notions and results. We pay considerable attention to the shapes of the failure rate and of the mean remaining life function as these topics are crucial for the rest of the book. The properties of the reversed failure rate have recently attracted noticeable interest. In the last section, definitions and the main properties for the reversed failure rate and related characteristics are considered. Note that, in this chapter, we consider only those facts, definitions and properties that are necessary for further presentation and do not aim at a general introduction to reliability theory.

Chapter 3 deals with two meaningful generalizations of the main exponential formula of reliability and survival analysis: the exponential representation of lifetime distributions with covariates and an analogue of the exponential representation for the multivariate (bivariate) case. The first meaningful generalization is used in Chapter 6 on mixture modelling and in the last chapter on applications to demography and biological ageing. Other chapters do not directly rely on this material and therefore can be read independently. The bivariate setting is studied in Chapter 7 only, where the competing risks model of Chapter 3 is generalized to the case of correlated covariates.

In **Chapter 4**, we present a brief introduction to the theory of point processes that is necessary for considering models of repairable systems. We define the stochastic intensity (intensity process) and the equivalent complete intensity function for the point processes that usually describe the operation of repairable systems. It is well known that renewal processes and alternating renewal processes are used for this purpose. Therefore, a repair action in these models is considered to be perfect, *i.e.*, returning a system to the *as good as new* state. This assumption is not always true, as repair in real life is usually imperfect. Minimal repair is the simplest case of imperfect repair, and therefore we consider this topic in detail. Specifically, information-based minimal repair is studied using some meaningful practical examples. The simplest models for minimal repair in heterogeneous populations are also considered.

Chapter 5 is devoted to repairable systems with imperfect (general) repair. When repair is perfect, the age of an item is just the time elapsed since the last repair, which is modelled by a renewal process. If it is minimal, then the age is equal to the time since a repairable item started operating. The point process of minimal repairs is the non-homogeneous Poisson process. When the repair is imperfect in a more general sense than minimal, the corresponding equivalent or *virtual* age should be defined. We describe the concept of virtual age for different settings and apply it to reliability modelling of repairable systems. An important feature of this concept is the assumption that the repair does not change the shape of the baseline failure rate and only the 'starting age' changes after each repair. We develop the renewal theory for this setting and also consider the asymptotic properties of the corresponding imperfect repair process. We prove that, as $t \to \infty$, this process converges to an ordinary renewal process.

Chapter 6 provides a comprehensive treatment of mixture failure rate modelling in reliability analysis. We present the relevant theory and discuss various applications. It is well known that mixtures of distributions with decreasing failure rate always have a decreasing failure rate. On the other hand, mixtures of increasing failure rate distributions can decrease at least in some intervals of time. As the latter distributions usually model lifetimes governed by ageing processes, this means that the operation of mixing can dramatically change the pattern of ageing, *e.g.*, from 'positive ageing' to 'negative ageing'. We prove that the mixture failure rate is 'bent down' due to "the weakest populations are dying out first" effect. Among other results, it is shown that if mixing random variables are ordered in the sense of likelihood ratio ordering, the mixture failure rates are ordered accordingly. We also define the operation of mixing for the mean remaining lifetime function and study its properties.

In **Chapter 7**, we present the asymptotic theory for mixture failure rates. It is mostly based on Finkelstein and Esaulova (2006, 2008). The chapter is rather technical and can be omitted by a less mathematically oriented reader. We obtain explicit asymptotic results for the mixture failure rate as $t \to \infty$. A general class of distributions is suggested that contains as specific cases the additive, multiplicative and accelerated life models that are widely used in practice. The most surprising is the result for the accelerated life model: when the support of the mixing distribution is $[0, \infty)$, the mixture failure rate for this model converges to 0 as $t \to \infty$ and does not depend on the baseline distribution. The ultimate behaviour of $\lambda(t)$ for other models, however, depends on a number of factors, specifically the baseline distribution. The univariate approach developed in this chapter is applied to the bivariate competing risks model. The components in the corresponding series system are dependent *via* a shared frailty parameter. An interesting feature of this model is that this dependence 'vanishes' as $t \to \infty$. This result may have an analogue in the life sciences, *e.g.*, for statistical analysis of correlated life spans of twins.

Chapter 8 deals with several specific problems where the failure rate can be obtained (constructed) directly as an exact or approximate relationship. Along with meaningful heuristic considerations, exact solutions and approaches are also discussed. Most examples are based on the operation of thinning of the Poisson process (Cox and Isham, 1980) or on equivalent reasoning. Among other settings, we apply the developed approach to obtaining the survival probability of an object moving in a plane and encountering moving or (and) fixed obstacles. In the 'safety at sea' application terminology, each foundering or collision results in a failure (accident) with a predetermined probability. It is shown that this setting can be reduced to the one-dimensional case. We assume that the field of fixed obstacles in the plane is described by the spatial non-homogeneous Poisson process. A spatial-temporal process is used for modelling moving obstacles. As another example, we also introduce the notion of *multiple availability* when an object must be available at all (random) instants of demand. We obtain the relevant probabilities using the thinning of the corresponding Poisson process and consider various generalizations.

Chapter 9 is devoted to software reliability modelling, and specifically to a discussion of some of the software failure rate models. It should be considered not as a comprehensive study of the subject, but rather a brief illustration of methods and approaches developed in the previous chapters. We consider several well-known empirical models for software failure rates, which can be described in terms of the corresponding stochastic intensity processes. Note that most of the models of this kind considered in the literature are based on very strong assumptions. A different approach, based on our stochastic model, which is similar to the model used for constructing the failure rate for spatial survival, is also discussed.

Chapter 10 is focused on another application of reliability-based reasoning. Reliability theory possesses the well-developed 'machinery' for stochastic modelling of ageing and failures in technical objects, which can be successfully applied to lifespan modelling of humans and other organisms. Thus, not only the final event (*e.g.*, death) can be considered, but the process, which eventually results in this event, as well. Several simple stochastic approaches to this modelling are described in this chapter. We revise the original Strehler–Mildvan (1960) model that was widely applied to human mortality data and show that from a mathematical point of view it is valid only under the assumption of the Poisson property of the point process of shocks (demands for energy). It also turns out that the thinning of the Poisson process described in Chapter 8 can be used for the probabilistic explanation of the lifesaving procedure, which results in decrease in mortality rates of contemporary human populations. We apply the concept of stochastic ordering to stochastic comparisons of different populations. An important feature of this modelling is that the mortality rate in demographic studies is usually not only a function of age (as in reliability) but of calendar time as well. Finally, in the last section, the tail of longevity for human populations is discussed. This notion is somehow close to the notion of the mean remaining lifetime, but the corresponding definition is based on two population distributions: on an 'ordinary' lifetime distribution and on the distribution of time to death of the last survivor.

2

Failure Rate and Mean Remaining Lifetime

Reliability engineering, survival analysis and other disciplines mostly deal with positive random variables, which are often called *lifetimes*. As a random variable, a lifetime is completely characterized by its distribution function. A realization of a lifetime is usually manifested by a failure, death or some other 'end event'. Therefore, for example, information on the probability of failure of an operating item in the next (usually sufficiently small) interval of time is really important in reliability analysis. The failure (hazard) rate function $\lambda(t)$ defines this probability of interest. If this function is increasing, then our object is usually degrading in some suitable probabilistic sense, as the conditional probability of failure in the corresponding infinitesimal interval of time increases with time. For example, it is well known that the failure (mortality) rate of adult humans increases exponentially with time; the failure rate of many mechanically wearing devices is also increasing. Thus, understanding and analysing the shape of the failure rate is an essential part of reliability and lifetime data analysis. Similar to the distribution function $F(t)$, the failure rate also completely characterizes the corresponding random variable. It is well known that there exists a simple, meaningful exponential representation for the absolutely continuous distribution function in terms of the corresponding failure rate (Section 2.1).

The study of the failure rate function, the main topic of this book, is impossible without considering other reliability measures. The mean remaining (residual) lifetime function is probably first among these; it also plays a crucial role in the aforementioned disciplines. These functions complement each other nicely: the failure rate gives a description of the random variable in an infinitesimal interval of time, whereas the mean remaining lifetime describes it in the whole remaining interval of time. Moreover, these two functions are connected *via* the corresponding differential equation and asymptotically, as time approaches infinity, one tends to the reciprocal of the other (Section 2.4.3).

In this introductory chapter, we consider only some basic facts, definitions and properties. We will use well-known results and approaches to the extent sufficient for the presentation of other chapters. The topic of reversed failure rate, which has attracted considerable interest recently, and the rather specific Section 2.4.3 on the limiting behaviour of the mean remaining life function can be skipped at first reading.

This chapter is, in fact, a mathematically oriented introduction to some of the main reliability notions and approaches. Recent books by Lai and Xie (2006), Marshall and Olkin (2007), a classic monograph by Barlow and Proschan (1975) and a useful textbook by Rausand and Hoyland (2004) can be used for further reading and as sources of numerous reliability-related results and facts.

2.1 Failure Rate Basics

Let $T \geq 0$ be a continuous lifetime random variable with a cumulative distribution function (Cdf)

$$F(t) = \begin{cases} \Pr[T \leq t], & t \geq 0, \\ 0, & t < 0. \end{cases}$$

Unless stated specifically, we will implicitly assume that this distribution is 'proper', *i.e.*, $F^{-1}(1) = \infty$, and that $F(0) = 0$. The support of $F(t)$ will usually be $[0, \infty)$, although other intervals of $\Re_+ = [0, \infty)$ will also be used. We can view T as some time to failure (death) of a technical device (organism), but other interpretations and parameterizations are possible as well. Inter-arrival times in a sequence of ordered events or the amount of monotonically accumulated damage on the failure of a mechanical item are also relevant examples of lifetimes.

Denote the expectation of the lifetime variable $E[T]$ by m and assume that it is finite, *i.e.*, $m < \infty$. Assume also that $F(t)$ is absolutely continuous, and therefore the probability density function (pdf) $f(t) = F'(t)$ exists (almost everywhere). Recall that a function $g(t)$ is absolutely continuous in some interval $[a, b]$, $0 \leq a < b \leq \infty$, if for every positive number ε, no matter how small, there is a positive number δ such that whenever a sequence of disjoint subintervals $[x_k, y_k]$, $k = 1, 2, ..., n$ satisfies

$$\sum_1^n | y_k - x_k | < \delta,$$

the following sum is bounded by ε:

$$\sum_1^n | g(y_k) - g(x_k) | < \varepsilon.$$

Owing to this definition, the uniform continuity in $[a, b]$, and therefore the 'ordinary' continuity of the function $g(t)$ in this interval, immediately follows.

In accordance with the definition of $E[T]$ and integrating by parts:

$$\begin{aligned} m &= \lim_{t \to \infty} \int_0^t x f(x) dx \\ &= \lim_{t \to \infty} \left[tF(t) - \int_0^t F(x) dx \right] \end{aligned}$$

$$= \lim_{t \to \infty} \left[-t\bar{F}(t) + \int_0^t \bar{F}(x)dx \right],$$

where

$$\bar{F}(t) = 1 - F(t) = \Pr[T > t]$$

denotes the corresponding survival (reliability) function. As $0 < m < \infty$, it is easy to conclude that

$$m = \int_0^\infty \bar{F}(x)dx, \tag{2.1}$$

which is a well-known fact for lifetime distributions. Thus, the area under the survival curve defines the mean of T.

Let an item with a lifetime T and a Cdf $F(t)$ start operating at $t = 0$ and let it be operable (alive) at time $t = x$. The remaining (residual) lifetime is of significant interest in reliability and survival analysis. Denote the corresponding random variable by T_x. The Cdf $F_x(t)$ is obtained using the law of conditional probability (on the condition that an item is operable at $t = x$), *i.e.*,

$$\begin{aligned} F_x(t) = \Pr[T_x \le t] &= \frac{\Pr[x < T \le x + t]}{\Pr[T > x]} \\ &= \frac{F(x+t) - F(x)}{\bar{F}(x)}. \end{aligned} \tag{2.2}$$

The corresponding conditional survival probability is given by

$$\bar{F}_x(t) = \Pr[T_x > t] = \frac{\bar{F}(x+t)}{\bar{F}(x)}. \tag{2.3}$$

Although the main focus of this book is on failure rate modelling, analysis of the remaining lifetime, and especially of the mean remaining lifetime (MRL), is often almost as important. We will use Equations (2.2) and (2.3) for definitions of the next section.

Now we are able to define the notion of failure rate, which is crucial for reliability analysis and other disciplines. Consider an interval of time $(t, t + \Delta t]$. We are interested in the probability of failure in this interval given that it did not occur before in $[0, t]$. This probability can be interpreted as the risk of failure (or of some other harmful event) in $(t, t + \Delta t]$ given the stated condition. Using a relationship similar to (2.2), *i.e.*,

$$\begin{aligned} \Pr[t < T \le t + \Delta t \mid T > t] &= \frac{\Pr[t < T \le t + \Delta t]}{\Pr[T > t]} \\ &= \frac{F(t + \Delta t) - F(t)}{\bar{F}(t)}. \end{aligned}$$

Consider the following quotient:

$$\lambda_{\Delta t}(t) = \frac{F(t+\Delta t) - F(t)}{\overline{F}(t)\Delta t}$$

and define the failure rate $\lambda(t)$ as its limit when $\Delta t \to 0$. As the pdf $f(t)$ exists,

$$\begin{aligned}\lambda(t) &= \lim_{\Delta t \to 0} \frac{\Pr[t < T \le t + \Delta t \mid T > t]}{\Delta t} \\ &= \lim_{\Delta t \to 0} \frac{F(t+\Delta t) - F(t)}{\overline{F}(t)\Delta t} = \frac{f(t)}{\overline{F}(t)} \ .\end{aligned} \tag{2.4}$$

Therefore, when $\Delta(t)$ is sufficiently small,

$$\Pr[t < T \le t + \Delta t \mid T > t] \approx \lambda(t)\Delta t \ ,$$

which gives a very popular and important interpretation of $\lambda(t)\Delta t$ as an approximate conditional probability of a failure in $(t, t + \Delta t]$. Note that $f(t)\Delta t$ defines the corresponding approximate unconditional probability of a failure in $(t, t + \Delta t]$. It is very likely that, owing to this interpretation, failure rate plays a pivotal role in reliability analysis, survival analysis and other fields. In actuarial and demographic disciplines, it is usually called the *force of mortality* or the *mortality rate*. To be precise, the force of mortality in demographic literature is usually the infinitesimal version ($\Delta t \to 0$), whereas the term *mortality rate* more often describes the discrete version when Δt is set equal to a calendar year. For convenience, we will always use the term *mortality rate* as an equivalent of *failure rate* when discussing demographic applications. Chapter 10 will be devoted entirely to some aspects of mortality rate modelling. Note that, when considering real populations, the mortality rate becomes a function of two variables: age t and calendar time x. This creates many interesting problems in the corresponding stochastic analysis. We will briefly discuss some of them in this chapter. For a general introduction to mathematical demography, where the mortality rate also plays a pivotal role, the interested reader is referred to Keyfitz and Casewell (2005).

Definition 2.1. The failure rate $\lambda(t)$, which corresponds to the absolutely continuous Cdf $F(t)$, is defined by Equation (2.4) and is approximately equal to the probability of a failure in a small unit interval of time $(t, t + \Delta t]$ given that no failure has occurred in $[0, t]$.

The following theorem shows that the failure rate uniquely defines the absolutely continuous lifetime Cdf:

Theorem 2.1. Exponential Representation of $F(t)$ by Means of the Failure Rate
Let T be a lifetime random variable with the Cdf $F(t)$ and the pdf $f(t)$.

Then

$$F(t)=1-\exp\left(-\int_0^t \lambda(u)du\right). \tag{2.5}$$

Proof. As $f(t)=F'(t)$, we can view Equation (2.4) as an elementary first-order differential equation with the initial condition $F(0)=0$. Integration of this equation results in the main *exponential formula of reliability and survival analysis* (2.5). ■

The importance of this formula is hard to overestimate as it presents a simple characterization of $F(t)$ *via* the failure rate. Therefore, along with the Cdf $F(t)$ and the pdf $f(t)$, the failure rate $\lambda(t)$ uniquely describes a lifetime T. At many instances, however, this characterization is more convenient, which is often due to the meaningful probabilistic interpretation of $\lambda(t)\Delta t$ and the simplicity of Equation (2.5).

Equation (2.5) has been derived for an absolutely continuous Cdf. Does the probability of failure in a small unit interval of time (which always exists) define the corresponding distribution function of a random variable under weaker assumptions? This question will be addressed in the next chapter.

Remark 2.1 Equation (2.4) can be used for defining the simplest empirical estimator for the failure rate. Assume that there are $N>>1$ independent, statistically identical items (*i.e.*, having the same Cdf) that started operating in a common environment at $t=0$. A population of this kind in the life sciences is often called a *cohort*. Failure times of items are recorded, and therefore the number of operating items $N(t), N(0)=N$ at each instant of time $t\geq 0$ is known. Thus, for $N\to\infty$, Equation (2.4) is equivalent to

$$\lambda(t)=\lim_{\Delta t\to 0}\frac{N(t+\Delta t)-N(t)}{N(t)\Delta t}, \tag{2.6}$$

which can be used as an estimate for the failure rate for finite N and Δt, whereas $(N(t+\Delta t)-N(t))/N(t)$ is an estimate for the probability of failure in $(t,t+\Delta t]$.

2.2 Mean Remaining Lifetime Basics

How much longer will an item of age x live? This question is vital for reliability analysis, survival analysis, actuarial applications and other disciplines. For example, how much time does an average person aged 65 (which is the typical retirement age in most countries) have left to live? The distribution of this remaining lifetime T_x, $T_0\equiv T$ is given by Equation (2.2). Note that this equation defines a conditional probability, *i.e.*, the probability on condition that the item is operating at time $t=x$.

Assume, as previously, that $E[T]\equiv m<\infty$. Denote $E[T_t]\equiv m(t)$, $m(0)=m$, where, for the sake of notation, the variable x in Equation (2.2) has been interchanged with the variable t. The function $m(t)$ is called the mean remaining (residual) life (MRL) function. It defines the mean lifetime left for an item of age t.

Along with the failure rate, it plays a crucial role in reliability analysis, survival analysis, demography and other disciplines. In demography, for example, this important population characteristic is called the "life expectancy at time t" and in risk analysis the term "mean excess time" is often used.

Whereas the failure rate function at t provides information on a random variable T about a small interval after t, the MRL function at t considers information about the whole remaining interval (t,∞) (Guess and Proschan, 1988). Therefore, these two characteristics complement each other, and reliability analysis of, *e.g.*, engineering systems is often carried out with respect to both of them. It will be shown in this section that, similar to the failure rate, the MRL function also uniquely defines the Cdf of T and that the corresponding exponential representation is also valid.

In accordance with Equations (2.1) and (2.3),

$$\begin{aligned} m(t) &= E[T_t] = E[T - t \mid T > t] \\ &= \int_0^\infty \bar{F}_t(u)du \\ &= \frac{\int_t^\infty \bar{F}(u)du}{\bar{F}(t)}. \end{aligned} \tag{2.7}$$

Assuming that the failure rate exists and using Equation (2.5), Equation (2.7) can be transformed into

$$m(t) = \int_0^\infty \exp\left\{-\int_t^{t+u} \lambda(x)dx\right\}du .$$

It easily follows from these equations that the MRL function, which corresponds to the constant failure rate λ, is also constant and is equal to $1/\lambda$.

Definition 2.2. The MRL function $m(t) = E[T_t]$, $m(0) \equiv m < \infty$, is defined by Equation (2.7), obtained by integrating the survival function of the remaining lifetime T_t.

Alternatively, integrating by parts, similar to (2.1),

$$\int_t^\infty uf(u)du = \int_t^\infty \bar{F}(u)du + t\bar{F}(t) .$$

Therefore, the last integral in (2.7) can be obtained from this equation, which results in the equivalent expression

$$m(t) = \frac{\int_t^\infty uf(u)du}{\bar{F}(t)} - t . \tag{2.8}$$

Equation (2.8) can be sometimes helpful in reliability analysis.

Assume that $m(t)$ is differentiable. Differentiation in (2.7) yields

$$m'(t)=\frac{\lambda(t)\int_t^{\infty}\bar{F}(u)du-\bar{F}(t)}{\bar{F}(t)}$$
$$=\lambda(t)m(t)-1. \tag{2.9}$$

From Equation (2.9) the following relationship between the failure rate and the MRL function is obtained:

$$\lambda(t)=\frac{m'(t)+1}{m(t)}. \tag{2.10}$$

This simple but meaningful equation plays an important role in analysing the shapes of the MRL and failure rate functions.

Consider now the following lifetime distribution function:

$$F_e(t)=\frac{\int_0^t \bar{F}(u)du}{m}, \tag{2.11}$$

where, as usual, $m(0)\equiv m$. The right-hand side of Equation (2.11) defines an equilibrium distribution, which plays an important role in renewal theory (Ross, 1996). This distribution will help us to prove the following simple but meaningful theorem. An elegant idea of the proof belongs to Meilijson (1972).

Theorem 2.2. Exponential Representation of $F(t)$ by Means of the MRL Function Let T be a lifetime random variable with the Cdf $F(t)$, the pdf $f(t)$ and with finite first moment: $m=m(0)<\infty$.

Then

$$\bar{F}(t)=\frac{m}{m(t)}\exp\left\{-\int_0^t\frac{1}{m(u)}du\right\}. \tag{2.12}$$

Proof. It follows from Equation (2.11) that

$$\bar{F}_e(t)=1-\frac{\int_0^t\bar{F}(u)du}{\int_0^{\infty}\bar{F}(u)du}=\frac{\int_t^{\infty}\bar{F}(u)du}{m}$$

and that $f_e(t) = \overline{F}(t)/m$. Therefore, the failure rate, which corresponds to the equilibrium distribution $F_e(t)$, is

$$\lambda_e(t) = \frac{f_e(t)}{\overline{F}_e(t)} = \frac{1}{m(t)}. \tag{2.13}$$

Applying Theorem 2.1 to $\overline{F}_e(t)$ results in

$$\overline{F}_e(t) = \exp\left(-\int_0^t \frac{1}{m(u)} du\right). \tag{2.14}$$

Therefore, the corresponding pdf is

$$f_e(t) = \frac{1}{m(t)} \exp\left(-\int_0^t \frac{1}{m(u)} du\right).$$

Finally, substitution of this density into the equation $\overline{F}(t) = mf_e(t)$ results in Equation (2.12). ■

On differentiating Equation (2.12), we obtain the pdf $f(t)$ that is also expressed in terms of the MRL function $m(t)$ (Lai and Xie, 2006), *i.e.*,

$$f(t) = \frac{m(m'(t)+1)}{m^2(t)} \exp\left(-\int_0^t \frac{1}{m(u)} du\right).$$

Theorem 2.2 has meaningful implications. Firstly, it defines another useful exponential representation of the absolutely continuous distribution $F(t)$. Whereas (2.5) is obtained in terms of the failure rate $\lambda(t)$, Equation (2.12) is expressed in terms of the MRL function $m(t)$. Secondly, it shows that, under certain assumptions, $\lambda(t)$ and $1/m(t)$ could be close, at least in some sense to be properly defined. This topic will be discussed in the next section, where the shapes of the failure rate and the MRL functions will be studied.

Equation (2.12) can be used for 'constructing' distribution functions when $m(t)$ is specified. Zahedi (1991) shows that in this case, differentiable functions $m(t)$ should satisfy the following conditions:

- $m(t) > 0,\ t \in [0,\infty)$;
- $m(0) < \infty$;
- $m'(t) > -1,\ t \in (0,\infty)$;
- $\int_0^\infty \frac{1}{m(u)} du = \infty$;

The first two conditions are obvious. The third condition is obtained from Equation (2.10) and states that $\lambda(t)m(t)$ is strictly positive for $t > 0$. Note that, $m(0)\lambda(0) = 0$ when $\lambda(0) = 0$. The last condition states that the cumulative failure rate

$$\int_0^t \lambda_e(u)du = \int_0^\infty \frac{1}{m(u)} du$$

of equilibrium distribution (2.11) should tend to infinity as $t \to \infty$. This condition ensures a proper Cdf, as $\lim_{t \to \infty} \bar{F}_e(t) = 0$ in this case.

In accordance with Equation (2.3) and exponential representation (2.5), the survival function for T_t can be written as

$$\bar{F}_t(x) = \Pr[T_t > x] = \exp\left\{ -\int_t^{t+x} \lambda(u)du \right\}. \tag{2.15}$$

This equation means that the failure rate, which corresponds to the remaining lifetime T_t, is a shift of the baseline failure rate, namely

$$\lambda_t(x) = \lambda(t + x). \tag{2.16}$$

Assume that $\lambda(t)$ is an increasing (decreasing) function. Note that, in this book, as usual, by *increasing* (*decreasing*) we actually mean *non-decreasing* (*non-increasing*). The first simple observation based on Equation (2.15) tells us that in this case, for each fixed $x > 0$, the function $\bar{F}_t(x)$ is decreasing (increasing), and therefore, in accordance with (2.7), the MRL function $m(t)$ is decreasing (increasing). The inverse is generally not true, *i.e.*, a decreasing $m(t)$ does not necessarily lead to an increasing $\lambda(t)$. This topic will be addressed in Section 2.4.

The operation of conditioning in the definition of the MRL function is performed with respect to the event that states that an item is operating at time t. In this approach, an item is considered as a 'black box' without any additional information on its state. Alternatively, we can define the information-based MRL function, which makes sense in many situations when this information is available. The following example (Finkelstein, 2001) illustrates this approach.

Example 2.1 Information-based MRL
Consider a parallel system of two components with independent, identically distributed (i.i.d.) exponential lifetimes defined by the failure rate λ. The survival function of this structure is

$$\bar{F}(t) = 2\exp\{-\lambda t\} - \exp\{-2\lambda t\},$$

and therefore, the corresponding failure rate is defined by

$$\lambda(t) = \frac{2\lambda \exp\{-\lambda t\} - 2\lambda \exp\{-2\lambda t\}}{2\exp\{-\lambda t\} - \exp\{-2\lambda t\}}.$$

It can easily be seen that $\lambda(t)$ monotonically increases from $\lambda(0)=0$ to λ as $t \to \infty$. The corresponding MRL function, in accordance with (2.7), is

$$m(t) = \frac{1}{\lambda}\frac{(4-\exp\{-\lambda t\})}{(4-2\exp\{-\lambda t\})}.$$

This function decreases from $3/2\lambda$ to $1/\lambda$ as $t \to \infty$. Therefore, the following bounds are obvious for $t \in (0,\infty)$:

$$\frac{1}{\lambda} < m(t) < \frac{3}{2\lambda} = m(0). \tag{2.17}$$

These inequalities can be interpreted in the following way. The left-hand side defines the information-based MRL when observation of the system confirms that only one component is operating at $t \in (0,\infty)$, whereas the right-hand side is the information-based MRL when observation confirms that both components are operating. Thus the values of the information-based MRL are the bounds for $m(t)$ in this simple case.

For the case of independent components with different failure rates λ_1, λ_2 ($\lambda_1 < \lambda_2$), the result of the comparison appears to be dependent on the time of observation. The corresponding survival function is defined as

$$\bar{F}(t) = \exp\{-\lambda_1 t\} + \exp\{-\lambda_2 t\} - \exp\{-(\lambda_1+\lambda_2)t\},$$

and the system's failure rate is

$$\lambda(t) = \frac{\lambda_1 \exp\{-\lambda_1 t\} + \lambda_2 \exp\{-\lambda_2 t\} - (\lambda_1+\lambda_2)\exp\{-(\lambda_1+\lambda_2)t\}}{\exp\{-\lambda_1 t\} + \exp\{-\lambda_2 t\} - \exp\{-(\lambda_1+\lambda_2)t\}}.$$

It can be shown that the function $\lambda(t)$ ($\lambda(0)=0$) is monotonically increasing in $[0, t_{\max}]$ and monotonically decreasing in $(t_{\max}, \infty)$, asymptotically approaching λ_1 from above as $t \to \infty$, as stated in Barlow and Proschan (1975). It crosses the line $y = \lambda_1$ at $t = t_c < t_{\max}$. The value of $t_{\max}$ is uniquely obtained from the equation

$$\lambda_2^2 \exp\{-\lambda_1 t\} + \lambda_1^2 \exp\{-\lambda_2 t\} = (\lambda_1 - \lambda_2)^2; \quad \lambda_1 \neq \lambda_2.$$

As in the previous case, the MRL function can be explicitly obtained, but we are more interested in discussing the information-based bounds. When both components are operating at $t > 0$, then, similar to the right-hand inequality in (2.17), the MRL function $m(t)$ is bounded from above by $m(0)$:

$$m(t) < \frac{\lambda_1}{\lambda_1+\lambda_2}\frac{1}{\lambda_2} + \frac{\lambda_2}{\lambda_1+\lambda_2}\frac{1}{\lambda_1}.$$

Now, let only the second component be operating at the time of observation. As this component is the worst one $(\lambda_2 < \lambda_1)$, the system's MRL should be better: $m(t) > 1/\lambda_2$. On the other hand, if only the first component is operable at time t, then

$$m(t) \le \frac{1}{\lambda_1}, \quad t \in [t_c, \infty). \tag{2.18}$$

This inequality immediately follows by combining the shape of the failure rate (*i.e.*, $\lambda(t)$ is larger than λ_1 for $t > t_c$), Equation (2.15) and the definition of the MRL function in (2.7). It is also clear that $m(t) > 1/\lambda_1$ for sufficiently small values of t, as two components are 'better' than one component in this case. This fact suggests that there should be some equilibrium point $\tilde{t}$ in $(0, t_c)$, where $m(\tilde{t}) = 1/\lambda_1$.

2.3 Lifetime Distributions and Their Failure Rates

There are many lifetime distributions used in reliability theory and in practice. In this section, we briefly discuss the important properties of several important lifetime distributions that we will use in this book. Complete information on the subject can be found in Johnson *et al.* (1994, 1995). A recent book by Marshall and Olkin (2007) also presents a thorough analysis of statistical distributions with an emphasis on reliability theory.

2.3.1 Exponential Distribution

The exponential distribution (or negative exponential), owing to its simplicity and relevance in many applications, is still probably the most popular distribution in practical reliability analysis. Many engineering devices (especially electronic) have a constant failure rate $\lambda > 0$ during the usage period. The Cdf and the pdf of the exponential distribution are given by

$$F(t) = \Pr[T \le t] = 1 - \exp\{-\lambda t\} \tag{2.19}$$

and

$$f(t) = \lambda \exp\{-\lambda t\},$$

respectively.

The expected value and variance are respectively given by

$$E[T] = \frac{1}{\lambda}, \quad \mathrm{var}(T) = \frac{1}{\lambda^2}.$$

The MRL function is also a constant, *i.e.*,

$$m(t) \equiv m = E[T].$$

The exponential distribution is the only distribution that possesses the memoryless property:

$$F(t \mid x) = F(t), \forall x, t \geq 0,$$

and therefore, it is the only non-trivial solution of the functional equation

$$\bar{F}(t+x) = \bar{F}(t)\bar{F}(x).$$

As the failure rate λ is constant, the items described by the exponential distribution do not age in the sense to be defined in Section 2.4.1. The exponential distribution has many characterizations (Marshall and Olkin, 2007). The simplest is *via* the constant failure rate. Another natural characterization is as follows: a distribution is exponential if and only if its mean remaining lifetime is a constant. The memoryless property can also be used as a characterization for this distribution.

2.3.2 Gamma Distribution

Consider the sum of n i.i.d. exponential random variables:

$$T = X_1 + X_2 + \ldots + X_n.$$

The corresponding $(n-1)$-fold convolution of Cdf (2.19) with itself results in the following Cdf for this sum:

$$F(t) = 1 - \sum_{k=0}^{n-1} \frac{(\lambda t)^k}{k!} \exp\{-\lambda t\}, \tag{2.20}$$

whereas the pdf is

$$f(t) = \frac{\lambda^n t^{n-1}}{(n-1)!} \exp\{-\lambda t\}.$$

For $n = 1$, this distribution reduces to the exponential one. Therefore, (2.20) can be considered a generalization of the exponential distribution. The mean and variance are respectively

$$E[T] = \frac{n}{\lambda}, \quad \operatorname{var}(T) = \frac{n}{\lambda^2},$$

and the failure rate is given by the following equation:

$$\lambda(t) = \frac{\lambda^n t^{n-1}}{(n-1)! \sum_{k=0}^{n-1} \frac{(\lambda t)^k}{k!}}. \tag{2.21}$$

It can easily be seen from this formula that $\lambda(t)$ ($\lambda(0) = 0$) is an increasing function asymptotically approaching λ from below, *i.e.*,

$$\lim_{t\to\infty} \lambda(t) = \lambda .$$

This distribution, which is a special case of the gamma distribution for integer n, is often called the *Erlangian* distribution. It plays an important role in reliability engineering. For example, the distribution function of the time to failure of a 'cold' standby system, where the lifetimes of components are exponentially distributed, follows this rule. As $\lambda(t)$ increases, this system ages.

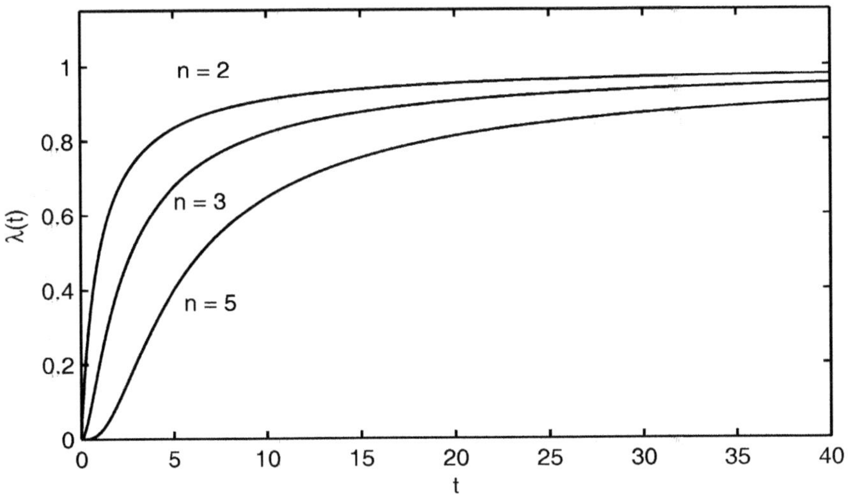

Figure 2.1. The failure rate of the Erlangian distribution ($\lambda = 1$)

We will use this graph for deterioration curve modelling in Chapter 5.

The probability density function for a non-integer n, which for the sake of notation is denoted by α, is

$$f(t) = \frac{\lambda^{\alpha} t^{\alpha-1}}{\Gamma(\alpha)} \exp\{-\lambda t\}, \tag{2.22}$$

where the gamma function is defined in the usual way as

$$\Gamma(\alpha) = \int_0^{\infty} u^{\alpha-1} \exp\{-u\} du$$

and the scale parameter λ and the shape parameter α are positive. For non-integer α, the corresponding Cdf does not have a 'closed form' as in the integer case (2.20). Equation (2.22) defines a standard two-parameter gamma distribution that is very popular in various applications. The gamma distribution naturally appears in statistical analyses as the distribution of the sum of squares of independent normal variables.

It can be shown (Lai and Xie, 2006) that the failure rate of the gamma distribution can be represented in the following way:

$$\frac{1}{\lambda(t)} = \int_0^\infty \left(1+\frac{u}{t}\right)^{\alpha-1} \exp\{-\lambda u\} du \,.$$

It follows from this equation that $\lambda(t)$ is an increasing function for $\alpha \geq 1$ and is decreasing for $0 < \alpha \leq 1$. When $\alpha = 1$, we arrive at the exponential distribution, which has a failure rate 'that is increasing and decreasing at the same time'.

As we stated in the previous section, it follows from Equations (2.15) and (2.7) that for increasing (decreasing) $\lambda(t)$, the MRL function $m(t)$ is decreasing (increasing). This is a general fact, which means in the case of the gamma distribution that $m(t)$ is a decreasing function for $\alpha \geq 1$ and is increasing for $0 < \alpha \leq 1$. Govil and Agraval (1983) have shown that

$$m(t) = \frac{\lambda^{\alpha-1} t^{\alpha} \exp\{-\lambda t\}}{\Gamma(\alpha)\bar{F}(t)} + \frac{\alpha}{\lambda} - t \,,$$

where $\bar{F}(t)$ is the survival function for the gamma distribution. It can be verified by direct differentiation that the monotonicity properties of $m(t)$ defined by this equation comply with those obtained from general considerations. As the corresponding integrals can usually be calculated explicitly, the gamma distribution is often used in stochastic and statistical modelling. For example, it is a prime candidate for a mixing distribution in mixture models (Chapters 6 and 7).

2.3.3 Exponential Distribution with a Resilience Parameter

The two-parameter distribution obtained from the exponential distribution by introducing a resilience parameter r has not received much attention in the literature (Marshall and Olkin, 2007). However, when r is an integer, similar to the Erlangian distribution, it plays an important role in reliability, as it defines the time-to-failure distribution of a parallel system of r exponentially distributed components. Therefore, the Cdf and the pdf are defined respectively as

$$F(t) = (1-\exp\{-\lambda t\})^r \,,\ \lambda, r > 0 \,,$$

$$f(t) = \lambda r \exp\{-\lambda t\}(1-\exp\{-\lambda t\})^{r-1} \,,\ \lambda, r > 0 \,.$$

The failure rate is

$$\lambda(t) = \frac{\lambda r \exp\{-\lambda t\}(1-\exp\{-\lambda t\})^{r-1}}{1-(1-\exp\{-\lambda t\})^r} \,. \tag{2.23}$$

It is easy to show by direct computation that $\lambda(t)$ is increasing for $r > 1$. Therefore, the described parallel system is ageing. Using L'Hospital's rule, it can also be shown that for $r > 0$,

$$\lim_{t \to \infty} \lambda(t) = \lambda ,$$

which, similar to the case of the Erlangian distribution, also follows from the definition of the failure rate as a conditional characteristic. Also: $\lambda(0) = 0$ for $r > 1$ and $\lambda(t) \to \infty$ as $t \to 0$ for $0 < r < 1$.

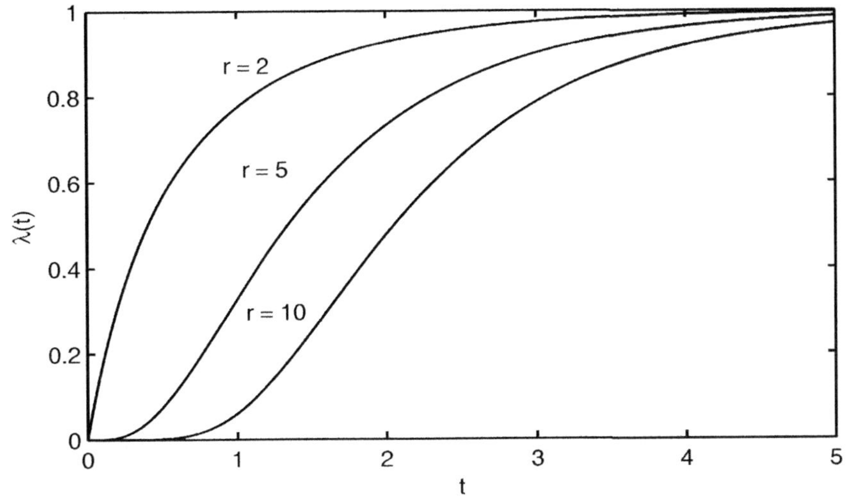

Figure 2.2. The failure rate of the exponential distribution ($\lambda = 1$) with a resilience parameter

2.3.4 Weibull Distribution

The Weibull distribution is one of the most popular distributions for modelling stochastic deterioration. It has been widely used in reliability analysis of ball bearings, engines, semiconductors, various mechanical devices and in modelling human mortality as well. It also appears as a limiting distribution for the smallest of a large number of the i.i.d. positive random variables. If, for example, a series system of n i.i.d. components is considered, then the time to failure of this system is asymptotically distributed ($n \to \infty$) as the Weibull distribution. The monograph by Murthy *et al.* (2003) covers practically all topics on the theory and practical usage of this distribution.

The standard two-parameter Weibull distribution is defined by the following survival function:

$$\bar{F}(t) = \exp\{-(\lambda t)^{\alpha}\}, \quad \lambda, \alpha > 0 . \tag{2.24}$$

The failure rate is

$$\lambda(t) = \alpha\lambda(\lambda t)^{\alpha-1} . \tag{2.25}$$

For $\alpha \geq 1$, it is an increasing function and therefore is suitable for deterioration modelling. When $0 < \alpha \leq 1$, this function is decreasing and can be used, *e.g.*, for infant-mortality modelling. The corresponding expectation is given by

$$m(0) = \frac{1}{\lambda}\Gamma\left(\frac{1}{\alpha}+1\right).$$

In general, $m(t)$ has a rather complex form, but for some specific cases (Lai and Xie, 2006) it can be reasonably simple. On the other hand, as $\lambda(t)$ is monotone, $m(t)$ is also monotone: it is increasing for $0 < \alpha \leq 1$ and is decreasing for $\alpha \geq 1$.

2.3.5 Pareto Distribution

The Pareto distribution can be viewed as another interesting generalization of the exponential distribution. We will derive it using mixtures of distributions, which is a topic of Chapters 6 and 7 of this book. Therefore, the following can be considered as a meaningful example illustrating the operation of mixing.

Assume that the failure rate in (2.19) is random, *i.e.*,

$$\lambda = Z ,$$

where Z is a gamma-distributed random variable with parameters α (shape) and β (scale). When considering mixing distributions, we will usually use the notation β for the scale parameter and not λ as in (2.23). Thus, if $Z = z$, the pdf of the random variable T is given by

$$f(t \mid Z = z) \equiv f(t,z) = z\exp\{-zt\} .$$

Denote the pdf of Z by $\pi(z)$. The marginal (or observed) pdf of T is

$$f(t) = \int_0^\infty f(t,z)\pi(z)dz = \frac{\alpha\beta^\alpha}{(\beta+t)^{\alpha+1}}$$

and the corresponding survival function is given by

$$\bar{F}(t) = \left(1+\frac{t}{\beta}\right)^{-\alpha}, \alpha,\beta > 0 . \tag{2.26}$$

Equation (2.26) defines the Pareto distribution of the second kind (the Lomax distribution) for $t \geq 0$. Note that the survival function of the Pareto distribution of the first kind is usually given by $\bar{F}(t) = t^{-c}$, where $c > 0$ is the corresponding shape parameter. Therefore, this distribution has a support in $[1,\infty)$, whereas (2.26) is defined in $[0,\infty)$, which is usually more convenient in applications.

The failure rate is given by a very simple relationship:

$$\lambda(t) = \frac{f(t)}{\overline{F}(t)} = \frac{\alpha}{(\beta + t)}, \tag{2.27}$$

which is a decreasing function. Therefore, the MRL function $m(t)$ is increasing. Oakes and Dasu (1990) show that it can be a linear function for some specific values of parameters α and β. The expectation is

$$m(0) = \frac{\beta}{\alpha - 1}, \quad \alpha > 1.$$

Unlike exponentially decreasing functions, survival function (2.26) is a 'slowly decreasing' function. This property makes the Pareto distribution useful for modelling of extreme events.

2.3.6 Lognormal Distribution

The most popular statistical distribution is the normal distribution. However, it is not a lifetime distribution, as its support is $(-\infty, +\infty)$. Therefore, usually two 'modifications' of the normal distribution are considered in practice for positive random variables: the lognormal distribution and the truncated normal distribution.

A random variable $T \geq 0$ follows the lognormal distribution if $Y = \ln T$ is normally distributed. Therefore, we assume that Y is $N(\alpha, \sigma^2)$, where α and σ^2 are the mean and the variance of Y, respectively. The Cdf in this case is given by

$$F(t) = \Phi\left\{\frac{\ln t - \alpha}{\sigma}\right\}, t \geq 0, \tag{2.28}$$

where, as usual, $\Phi(\cdot)$ denotes the standard normal distribution function. The pdf is given by

$$f(t) = \frac{\exp\left\{-\frac{(\ln t - \alpha)^2}{2\sigma^2}\right\}}{(t\sqrt{2\pi}\sigma)},$$

and it can be shown (Lai and Xie, 2006) that the failure rate is

$$\lambda(t) = \frac{1}{t\sqrt{2\pi}\sigma} \frac{\exp\left\{-\frac{(\ln at)^2}{2\sigma^2}\right\}}{1 - \Phi\left\{\frac{\ln at}{\sigma}\right\}}, \quad a \equiv \exp\{-\alpha\}. \tag{2.29}$$

The expected value of T is

$$m(0) = \exp\left\{\alpha + \frac{\sigma^2}{2}\right\}.$$

The MRL function for this distribution will be discussed in the next section. The shape of the failure rate for $\alpha = 0$ is illustrated by Figure 2.3. Sweet (1990) showed that the failure rate has the upside-down bathtub shape (see the next section) and that $\lim_{t\to\infty} \lambda(t) = 0$, $\lim_{t\to 0} \lambda(t) = 0$.

It is worth noting that, along with the Weibull distribution, the lognormal distribution is often used for fatigue analysis, although it models different dynamics of deterioration than the dynamics described by the Weibull law. It is also considered as a good candidate for modelling the repair time in engineering systems.

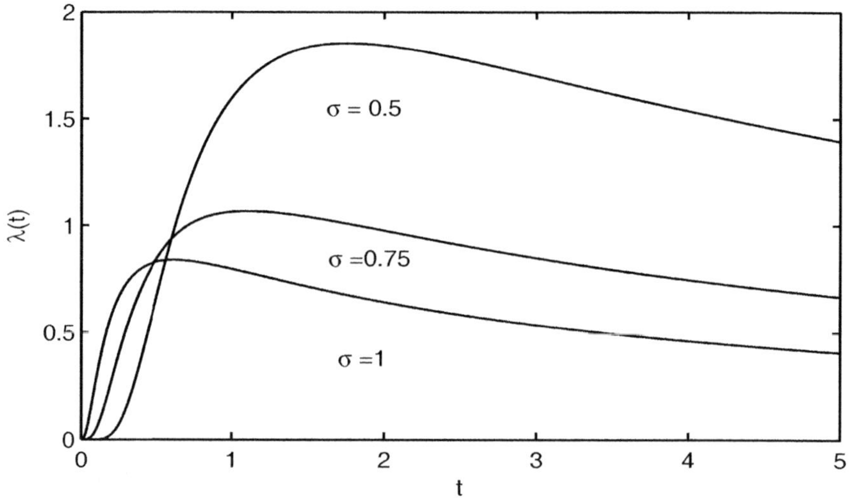

Figure 2.3. The failure rate of the lognormal distribution

2.3.7 Truncated Normal Distribution

The density of the truncated normal distribution is given by

$$f(t) = c\exp\left\{-\frac{(t-\mu)^2}{2\sigma^2}\right\}, \sigma > 0, -\infty < \mu < \infty,\ t \geq 0,$$

where

$$c = \frac{1}{\sqrt{2\pi\sigma^2}}\frac{1}{\Phi(\mu/\sigma)}.$$

The corresponding failure rate then follows as

$$\lambda(t)=\frac{1}{\sqrt{2\pi\sigma^2}}\left(1-\Phi\left(\frac{t-\mu}{\sigma}\right)\right)^{-1}\exp\left\{-\frac{(t-\mu)^2}{2\sigma^2}\right\}.$$

It can be shown that this failure rate is increasing and asymptotically approaches the straight line, as defined by (Navarro and Hernandez, 2004):

$$\lim_{t\to\infty}\lambda(t)=\sigma^{-2}.$$

If $\mu+3\sigma>>0$, then the truncated normal distribution practically coincides for $t\geq 0$ with the corresponding standard normal distribution, which is known to have an increasing failure rate.

2.3.8 Inverse Gaussian Distribution

This distribution is popular in reliability, as it defines the first passage time probability for the Wiener process with drift. Although realizations of this process are not monotone, it is widely used for modelling deterioration. The distribution function of the inverse Gaussian distribution is defined by the following equation:

$$F(t)=\Phi\left\{\sqrt{\frac{\lambda}{t}}\left(\frac{t}{\mu}-1\right)\right\}+\exp\left\{\frac{2\lambda}{\mu}\right\}\Phi\left\{-\sqrt{\frac{\lambda}{t}}\left(\frac{t}{\mu}+1\right)\right\},\ t\geq 0, \tag{2.30}$$

where λ and μ are parameters. The pdf of the inverse Gaussian distribution is

$$f(t)=\sqrt{\frac{\lambda}{2\pi t^3}}\exp\left\{-\frac{\lambda}{2\mu^2 t}(t-\mu)^2\right\}.$$

The mean and the variance are respectively

$$E[T]=\mu,\ \mathrm{var}(T)=\frac{\mu^3}{\lambda}.$$

We will show in Section 2.4 that its failure rate has an upside-down bathtub shape. The MRL function will also be analysed.

2.3.9 Gompertz and Makeham–Gompertz Distributions

These distributions have their origin in demography and describe the mortality of human populations.

Gompertz (1825) was the first to suggest the following exponential form for the mortality (failure) rate of humans (see Chapter 10 for more details):

$$\lambda(t)=a\exp\{bt\},\ a,b>0. \tag{2.31}$$

The data on human mortality in various populations are in good agreement with this curve. In Section 10.1, we will present a simple original 'justification' of this model, but in fact, there is no suitable biological explanation of exponentiality in (2.31) so far. Therefore, this distribution should only be considered as an empirical law. Note that this is the first distribution in this section that is defined directly *via* the failure (mortality) rate. The corresponding survival function is

$$\bar{F}(t) = \exp\left\{-\int_0^t \lambda(u)du\right\} = \exp\left\{-\frac{a}{b}(\exp\{bt\}-1)\right\}. \tag{2.32}$$

The mortality rate (2.31) is increasing, therefore the corresponding MRL function is decreasing.

The Makeham–Gompertz distribution is a slight generalization of (2.32). It takes into account the initial period, where the mortality is approximately constant and is mostly due to external causes (accidents, suicides, *etc.*). This distribution was also defined in Makeham (1867) directly *via* the mortality rate, although the equation-based explanation was also provided by this author (Chapter 10):

$$\lambda(t) = A + a\exp\{bt\},\ A, a, b > 0\,.$$

The corresponding survival function in this case is

$$\bar{F}(t) = \exp\left\{-At + \frac{a}{b}(\exp\{bt\}-1\right\}. \tag{2.33}$$

Both of these distributions are still widely used in demography. Numerous generalizations and alterations have been suggested in the literature and applied in practice.

2.4 Shape of the Failure Rate and the MRL Function

2.4.1 Some Definitions and Notation

Understanding the shape of the failure rate is important in reliability, risk analysis and other disciplines. The conditional probability of failure in $(t, t+dt]$ describes the ageing properties of the corresponding distributions, which are crucial for modelling in many applications. A qualitative description of the monotonicity properties of the failure rate can be very helpful in the stochastic analysis of failures, deaths, disasters, *etc.* As the failure rate of the exponential distribution is constant (as is the corresponding MRL function), this distribution describes stochastically non-ageing lifetimes.

Survival and failure data are frequently modelled by monotone failure rates. This may be inappropriate when, *e.g.*, the course of a disease is such that the mortality reaches a peak after some finite interval of time and then declines (Gupta, 2001). In such a case, the failure rate has an *upside-down bathtub shape* and the data should be analysed with the help of, *e.g.*, lognormal or inverse Gaussian distributions. On the other hand, many engineering devices possess a period of 'infant mortality' when the failure rate declines in an initial time interval, reaches a minimum and then increases. In such a case, the failure rate has a *bathtub shape* and can be modelled, *e.g.*, by mixtures of distributions. Navarro and Hernandez (2004) show how to obtain the bathtub-shaped failure rates from the mixtures of truncated normal distributions. Many other relevant examples can be found in Section 2.8 of Lai and Xie (2006) and in references therein. We will consider in this section only some basic facts, which will be helpful for obtaining and discussing the results in the rest of this book.

Most often, the Cdf and the failure rate of a lifetime are modelled or estimated only on the basis of the corresponding failures (deaths). However, one can also use information (if available) on the process of a 'failure development'. If, *e.g.*, a failure occurs when the accumulated random damage or wear exceeds a predetermined level, then the failure rate can be derived analytically for some simple stochastic processes of wear. The shape of the failure rate in this case can also be analysed using properties of underlying stochastic processes (Aalen and Gjeissing, 2001). These underlying processes are largely unknown. However, this does not imply that they should be ignored. Some simple models of this kind will be discussed in Chapter 10.

As we saw in the previous section, many popular parametric lifetime models are described by monotone failure rates. If $\lambda(t)$ increases (decreases) in time, then we say that the corresponding distribution belongs to the increasing (decreasing) failure rate (IFR (DFR)) class. These are the simplest nonparametric classes of ageing distributions. A natural generalization on the non-monotone failure rates is when

$$\frac{\int_0^t \lambda(u)du}{t} \tag{2.34}$$

is increasing (decreasing) in t. These classes are called IFRA (DFRA), where "A" stands for "average".

We say that the Cdf $F(x)$ belongs to the decreasing (increasing) mean remaining lifetime (DMRL (IMRL)) class if the corresponding MRL function $m(t)$ is decreasing (increasing). These classes are in some way dual to IFR (DFR) classes. See Section 3.3.2 for formal definitions of IFR (DFR) and DMRL (IMRL) classes.

The Cdf $F(x)$ is said to be new better (worse) than used (NBU (NWU)) if

$$\bar{F}(x \mid t) \le (\ge) \bar{F}(x), \forall x, t \ge 0\,. \tag{2.35}$$

This definition means that an item of age t has a stochastically smaller (larger) remaining lifetime (Definition 3.4) than a new item at age $t = 0$.

The described classes will usually be sufficient for presentation in this book. Each of them has a clear, simple 'physical' meaning describing some kind of deterioration. A variety of other ageing classes of distributions can be found in the literature (Barlow and Proschan, 1975; Rausand and Hoyland, 2004; Lai and Xie, 2006; Marshall and Olkin, 2007, to name a few). Many of them do not have this clear interpretation and are of mathematical interest only.

Note that IFR (DFR) and DMRL (IMRL) classes are defined directly by the shape of the failure rate and the MRL function, respectively. If $\lambda(t)$ is monotonically (strictly) increasing (decreasing) in time, we say that it is I (D) shaped and for brevity write $\lambda(t) \in \mathrm{I}$ (D). A similar notation will be used for the DMRL (IMRL) classes, *i.e.*, $m(t) \in \mathrm{D}$ (I).

Figure 1.1 of Chapter 1 gives an illustration of the bathtub shape of a failure rate with a useful period, where it is approximately constant. This can be the case in practical life-cycle applications, but formally we will define the bathtub shape without a useful period plateau of this kind.

Definition 2.3. The differentiable failure rate $\lambda(t)$ has a bathtub shape if

$$\lambda'(t) < 0 \text{ for } t \in [0, t_0),\ \lambda'(t_0) = 0,\ \lambda'(t) > 0 \text{ for } t \in (t_0, \infty),$$

and it has an upside-down bathtub shape if

$$\lambda'(t) > 0 \text{ for } t \in [0, t_0),\ \lambda'(t_0) = 0,\ \lambda'(t) < 0 \text{ for } t \in (t_0, \infty).$$

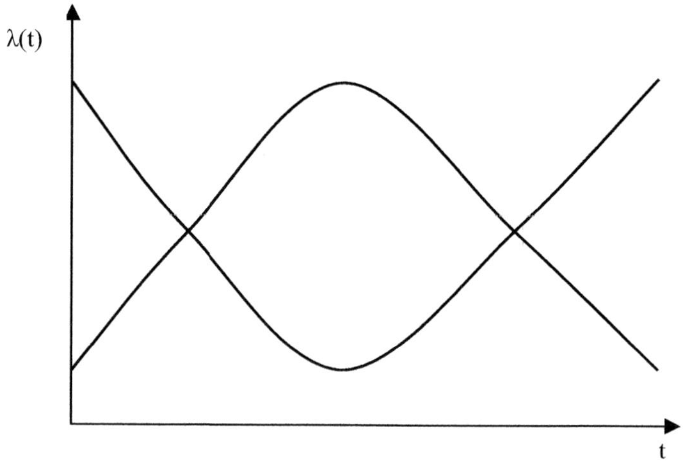

Figure 2.4. The BT and the UBT shapes of the failure rate

We will use the notation $\lambda(t) \in \mathrm{BT}$ and $\lambda(t) \in \mathrm{UBT}$, respectively. There can be modifications and generalizations of these shapes (*e.g.*, when there is more than one minimum or maximum for the function $\lambda(t)$), but for simplicity, only BT and UBT shapes will be considered.

2.4.2 Glaser's Approach

As we have already stated, the lognormal and the inverse Gaussian distributions have a UBT failure rate. We will see in Chapter 6 that many mixing models with an increasing baseline failure rate result in a UBT shape of the mixture (observed) failure rate. For example, mixing in a family of increasing (as a power function) failure rates (the Weibull law) 'produces' the UBT shape of the observed failure rate. From this point of view, the BT shape is 'less natural' and often results as a combination of different standard distributions defined for different time intervals. For example, infant mortality in $[0, t_0]$ is usually described by some DFR distribution in this interval, whereas the wear-out in (t_0, ∞) is modelled by an IFR distribution. However, mixing of specific distributions can also result in the BT shape of the failure rate as, *e.g.*, in Navarro and Fernandez (2004). Note that the infant mortality curve can also be explained *via* the concept of mixing, as, *e.g.*, mixtures of exponential distributions are always DFR (Chapter 6).

The function

$$\eta(t) = -\frac{f'(t)}{f(t)} \tag{2.36}$$

appears to be extremely helpful in the study of the shape of the failure rate $\lambda(t) = f(t)/\bar{F}(t)$. This function contains useful information about $\lambda(t)$ and is simpler because it does not involve $\bar{F}(t)$. In particular, the shape of $\eta(t)$ often defines the shape of $\lambda(t)$ (Gupta, 2001).

Assume that the pdf $f(t)$ is a twice differentiable, positive function in $(0, \infty)$. Define a function $g(t)$ as the reciprocal of the failure rate, *i.e.*,

$$g(t) = \frac{1}{\lambda(t)} = \frac{\bar{F}(t)}{f(t)}. \tag{2.37}$$

Then

$$g'(t) = g(t)\eta(t) - 1, \tag{2.38}$$

which means that the turning point of $\lambda(t)$ is the solution of the equation $\lambda(t) = \eta(t)$ (compare with Equation (2.9)). It can also be verified that (Gupta, 2001)

$$\lim_{t\to\infty} \lambda(t) = \lim_{t\to\infty} \eta(t).$$

Using Equations (2.37) and (2.38):

$$\begin{aligned} g'(t) &= \int_t^\infty \left[\frac{f(y)}{f(t)}\right] \eta(t)dy - 1 \\ &= \int_t^\infty \left[\frac{f(y)}{f(t)}\right] [\eta(t) - \eta(y)]dy + \int_t^\infty \left[\frac{f(y)}{f(t)}\right] \eta(y)dy - 1. \end{aligned}$$

Taking into account that

$$\int_t^{\infty}\left[\frac{f(y)}{f(t)}\right]\eta(y)dy = -\frac{1}{f(t)}\int_t^{\infty} f'(y)dy = 1,$$

we arrive eventually at

$$g'(t) = \int_t^{\infty}\left[\frac{f(y)}{f(t)}\right][\eta(t)-\eta(y)]dy. \qquad (2.39)$$

Using (2.39) as a supplementary result, we are now able to prove Glaser's theorem, which is crucial for the analysis of the shape of the failure rate function (Glaser, 1980).

Theorem 2.3.

- If $\eta(t)\in \mathrm{I}$, then also $\lambda(t)\in \mathrm{I}$;
- If $\eta(t)\in \mathrm{D}$, then also $\lambda(t)\in \mathrm{D}$;
- If $\eta(t)\in \mathrm{BT}$ and there exists y_0 such that $g'(y_0)=0$, then $\lambda(t)\in \mathrm{BT}$, otherwise $\lambda(t)\in \mathrm{I}$;
- If $\eta(t)\in \mathrm{UBT}$ and there exists y_0 such that $g'(y_0)=0$, then $\lambda(t)\in \mathrm{UBT}$, otherwise $\lambda(t)\in \mathrm{D}$.

Proof. If $\eta(t)\in \mathrm{I}$, then $g'(t)$, as follows from Equation (2.39), is negative for all $t>0$. Therefore, $g(t)\in \mathrm{D}$ and $\lambda(t)\in \mathrm{I}$. The proof of the second statement is similar.

Let us prove the first part of the third statement. This proof follows the original proof in Glaser (1980). Another proof, which is obtained using more general considerations, can be found in Marshall and Olkin (2007). It follows from the definition of the BT shape that $\eta(t)\in \mathrm{BT}$ if

$$\eta'(t)<0 \text{ for } t\in[0,t_0),\, \eta'(t_0)=0,\ \eta'(t)>0 \text{ for } t\in(t_0,\infty). \qquad (2.40)$$

Assume that $g''(y_0)<0$. Since $g'(y_0)=0$ in accordance with the conditions of the theorem, it follows from the differentiation of (2.38) that

$$g''(y_0)=g(y_0)\eta'(y_0).$$

Therefore,

$$g''(y_0)<0 \Leftrightarrow \eta'(y_0)<0 \Leftrightarrow y_0<t_0.$$

Thus, if our assumption is true, then $y_0<t_0$. Suppose the opposite: $y_0\ge t_0$. From Equations (2.39) and (2.40) it follows that $g'(t)<0$ for $t\ge t_0$. Therefore, $g'(y_0)<0$, which contradicts the condition of the theorem stating that $g'(y_0)=0$. Hence $y_0<t_0$ and $g''(y_0)<0$. On the other hand, it is clear that $y=y_0$ is the only root of equation $g'(y)=0$ and that $g(t)$ attains its maximum at this point.

The proof of the second part is simpler: indeed, either $g'(t)>0$ for all $t>0$ or $g'(t)<0$. It follows from Equation (2.39) that $g'(t)<0$ for all $t\ge t_0$. Therefore, $g'(t)<0$ for all $t>0$ and $\lambda(t)\in \mathrm{I}$.

The proof of the last statement is similar. ■

This important theorem states that the monotonicity properties of $\lambda(t)$ are defined by those of $\eta(t)$, and because $\eta(t)$ is often much simpler than $\lambda(t)$, its analysis is more convenient. The simplest meaningful example is the standard normal distribution. Although it is not a lifetime distribution, the application of Glaser's theorem is very impressive in this case. Indeed, the failure rate of the normal distribution does not have an explicit expression, whereas the function $\eta(t)$, as can be easily verified, is very simple:

$$\eta(t) = (t - \mu)/\sigma^2 .$$

Therefore, as $\eta(t) \in \mathrm{I}$, the failure rate is also increasing, which is a well-known fact for the normal distribution.

Note that Gupta and Warren (2001) generalized Glaser's theorem to the case where $\lambda(t)$ has two or more turning points.

Example 2.2 Failure Rate Shape of the Truncated Normal Distribution
The function $\eta(t)$ in this case is the same as for the normal distribution, and therefore the failure rate is also increasing. Navarro and Hernandez (2004) also show that

$$\lambda(t) > (t - \mu)/\sigma^2, t \geq 0 .$$

Example 2.3 Failure Rate Shapes of Lognormal and Inverse Gaussian Distributions
The function $\eta(t)$ for the lognormal distribution is

$$\eta(t) = -\frac{f'(t)}{f(t)} = \frac{1}{\sigma^2 t}(\sigma^2 + \ln t - \alpha) . \tag{2.41}$$

It can be shown that $n(t) \in \mathrm{UBT}$ (Lai and Xie, 2006) and that the second condition in the last statement of Theorem 2.3 is also satisfied, since, in accordance with Equation (2.29),

$$\lim_{t \to 0} \lambda(t) = 0 , \lim_{t \to \infty} \lambda(t) = 0 .$$

Therefore, $\lambda(t) \in \mathrm{UBT}$, and this is illustrated by Figure 2.2.

The $\eta(t)$ function for the inverse Gaussian distribution (2.30) is

$$\eta(t) = \frac{3\mu^2 t + \lambda(t^2 - \mu^2)}{2\mu^2 t^2} . \tag{2.42}$$

Using arguments similar to those used in the case of the lognormal distribution, it can be shown (Lai and Xie, 2006) that $\lambda(t) \in \mathrm{UBT}$. The exact MRL function for this distribution (Gupta, 2001) is very cumbersome to derive.

Glaser's approach was generalized by Block *et al.* (2002) by considering the ratio of two functions

$$G(t) = \frac{N(t)}{D(t)}, \tag{2.43}$$

where the functions on the right-hand side are continuously differentiable and $D(t)$ is positive and strictly monotone. As with (2.36), where the numerator is the derivative of $f(t)$ and the denominator is the derivative of $\bar{F}(t)$, we define the function $\eta(t)$ as

$$\eta(t) = \frac{N'(t)}{D'(t)}. \tag{2.44}$$

These authors show that the monotonicity properties of $G(t)$ are 'close' to those of $\eta(t)$, as in the case where $\eta(t)$ is defined by (2.36). Consider, for example, the MRL function

$$m(t) = \frac{\int_t^{\infty} \bar{F}(u)du}{\bar{F}(t)}.$$

We can use it as $G(t)$. It is remarkable that $\eta(t)$ in this case is simply the reciprocal of the failure rate, *i.e.*,

$$\eta(t) = \frac{\bar{F}(t)}{f(t)} = \frac{1}{\lambda(t)}.$$

Therefore, the functions $m(t)$ and $1/\lambda(t)$ can be close in some suitable sense; this will be discussed in Section 2.4.3.

Glaser's theorem defines sufficient conditions for monotonic or BT (UBT) shapes of the failure rate. The next three theorems establish relationships between the shapes of $\lambda(t)$ and $m(t)$. The first one is obvious and in fact has already been used several times.

Theorem 2.4. If $\lambda(t) \in \mathrm{I}$ (or $(\lambda(t))^{-1} \in D$), then $m(t) \in D$.

Proof. The result follows immediately from Equations (2.7) and (2.15). The symmetrical result is also evident: if $\lambda(t) \in \mathrm{D}$, then $m(t) \in \mathrm{I}$. ■

Thus, a monotone failure rate always corresponds to a monotone MRL function. The inverse is true only under additional conditions.

Theorem 2.5. Let the MRL function $m(t)$ be twice differentiable and the failure rate $\lambda(t)$ be differentiable in $(0,\infty)$. If $m(t) \in \mathrm{D}$ (I) and is a convex (concave) function, then $\lambda(t) \in \mathrm{I}$ (D).

Proof. Differentiation of both sides of Equation (2.9) gives

$$m''(t) = m'(t)\lambda(t) + m(t)\lambda'(t).$$

If $m(t)$ is strictly decreasing, then its derivative is negative for all $t \in (0,\infty)$. Owing to convexity defined by $m''(t) \geq 0$ and taking into account that the functions $\lambda(t)$ and $m(t)$ are positive in $(0,\infty)$, $\lambda'(t)$ should be positive as well. This means that $\lambda(t) \in$ I. The 'symmetrical' result is proved in a similar way. ■

Gupta and Kirmani (2000) state that if $\lambda(t)$ is concave, then $m(t)$ is a convex function. Theorem 2.5 gives the sufficient conditions for the monotonicity of the failure rate in terms of the monotonicity of $m(t)$. The following theorem generalizes the foregoing results to a non-monotone case (Gupta and Akman, 1995; Mi, 1995; Finkelstein, 2002a). It states that the BT (UBT) failure rate under certain assumptions can correspond to a monotone MRL function (compare with Theorem 2.4, which gives a simpler correspondence rule).

Theorem 2.6. Let $\lambda(t)$ be a differentiable BT failure rate in $[0,\infty)$.

- If
$$m'(0) = \lambda(0)m(0) - 1 \leq 0, \tag{2.45}$$
then $m(t) \in$ D;
- If $m'(0) > 0$, then $m(t) \in$ UBT.

Let $\lambda(t)$ be a differentiable UBT failure rate in $[0,\infty)$.

- If $m'(0) \geq 0$, then $m(t) \in$ I;
- If $m'(0) < 0$, then $m(t) \in$ BT.

Proof. We will prove only the first statement. Other results follow in the same manner. Denote the numerator in (2.9) by $d(t)$, *i.e.*,

$$d(t) = \lambda(t)\int_t^{\infty}\bar{F}(u)du - \bar{F}(t). \tag{2.46}$$

The sign of $d(t)$ in (2.9) defines the sign of $m'(t)$. On the other hand,

$$d'(t) = \lambda'(t)\int_t^{\infty}\bar{F}(u)du, \tag{2.47}$$

and the monotonicity properties of $\lambda(t)$ are the same as for $d(t)$. Recall that t_0 is the change (turning) point for the BT failure rate. Therefore,

$$\lambda'(t_0) = d'(t_0) = 0\,;\, \lambda(t) > \lambda(t_0) \text{ for } t > t_0$$

and

$$d(t_0) = \lambda(t_0)\int_{t_b}^{\infty}\overline{F}(u)du - \overline{F}(t_0)$$

$$< \int_{t_b}^{\infty}\lambda(u)\overline{F}(u)du - \overline{F}(t_0) = 0. \quad (2.48)$$

Owing to the assumption $m'(0) \le 0$ and to Equation (2.9), the function $d(t)$ is negative at $t = 0$. It then follows from (2.47) that $d(t)$ decreases to $d(t_0)$ and then increases in (t_0, ∞), being negative. The latter can be seen from Inequality (2.48), where t_0 can be substituted by any $t > t_0$. Therefore, in accordance with (2.9), $m'(t) < 0$ in $(0, \infty)$, which completes the proof. ■

Corollary 2.1. Let $\lambda(0) = 0$. If $\lambda(t)$ is a differentiable UBT failure rate, then $m(t)$ has a bathtub shape.

Proof. This statement immediately follows from Theorem 2.6, as Equation (2.45) reads $m'(0) = \lambda(0)m(0) - 1 = -1 \le 0$ in this case. ■

Example 2.4 (Gupta and Akman, 1995)
Consider a lifetime distribution with $\lambda(t) \in \mathrm{BT}$, $t \in [0, \infty)$ of the following specific form:

$$\lambda(t) = \frac{(1 + 2.3t^2) - 4.6t}{1 + 2.3t^2}.$$

It can easily be obtained using Equation (2.22) that the corresponding MRL is

$$m(t) = \frac{1}{1 + 2.3t^2},$$

which is a decreasing function. Obviously, the condition $\lambda(0) \le 1/m(0)$ is satisfied.

2.4.3 Limiting Behaviour of the Failure Rate and the MRL Function

In this section, we will discuss and compare the simplest asymptotic (as $t \to \infty$) properties of $\lambda(t)$ and $1/m(t)$. When a lifetime T has an exponential distribution, these functions are equal to the same constant. It has already been mentioned that Block *et al.* (2001) stated that the monotonicity properties of the function $G(t)$ defined by Equation (2.43) are 'close' to those of the function $\eta(t)$ defined by Equation (2.44). When we choose $G(t) = m(t)$, the function $\eta(t)$ is equal to $1/\lambda(t)$, and therefore the monotonicity properties of these functions are similar. Moreover, we will show now that they are asymptotically equivalent.

Denote $r(t) \equiv 1/m(t)$ and, as in Finkelstein (2002a), rewrite Equation (2.10) in form that connects the failure rate and the reciprocal of the MRL function

$$\lambda(t) = -\frac{r'(t)}{r(t)} + r(t). \quad (2.49)$$

The following obvious result is a direct consequence of Equation (2.49).

Theorem 2.7. Let $\lim_{t\to\infty} r(t) = c, 0 < c \le \infty$.
Then $r(t)$ is asymptotically equivalent to $\lambda(t)$ in the following sense:

$$\lim_{t\to\infty} \left|\lambda(t) - r(t)\right| = 0, \tag{2.50}$$

if and only if

$$\left|\frac{r'(t)}{r(t)}\right| = \left|\frac{m'(t)}{m(t)}\right| \to 0 \text{ as } t \to \infty. \tag{2.51}$$

Let, *e.g.*, $r(t) = t^{\beta}$; $\beta > 0$. Then Theorem 2.7 holds and the reciprocal of the MRL function for the Weibull distribution with an increasing failure rate can be approximated as $t \to \infty$ by this failure rate. The exact formula for the MRL function in this case is rather cumbersome, and thus this result can be helpful for asymptotic analysis. Note that Relationship (2.51) does not hold for sharply increasing functions $r(t)$, such as, *e.g.*, $r(t) = \exp\{t\}$ or $r(t) = \exp\{t^2\}$.

Remark 2.2 Applying L'Hopital's rule to the right-hand side of (2.7), the following asymptotic relation can be obtained (Calabra and Pulchini, 1987; Bradley and Gupta, 2003):

$$\lim_{t\to\infty} m(t) = \lim_{t\to\infty} \frac{1}{\lambda(t)},$$

provided the latter limit exists and is finite. It is clear that this statement differs from the stronger one (2.50) only when $\lim_{t\to\infty} \lambda(t) = \infty$.

The asymptotic equivalence in (2.50) is a very strong one, especially when $\lim_{t\to\infty} r(t) = \infty$ and $\lim_{t\to\infty} \lambda(t) = \infty$. Therefore, it is reasonable to consider the following relative distance between $\lambda(t)$ and $r(t)$:

$$\frac{|\lambda(t) - r(t)|}{r(t)} = m'(t).$$

This distance tends to zero when

$$\lim_{t\to\infty} |m'(t)| = \lim_{t\to\infty} \left|\frac{r'(t)}{r^2(t)}\right| = 0, \tag{2.52}$$

which, in fact, is equivalent to the following asymptotic relationship:

$$\lambda(t) = r(t)(1 + o(1)) \text{ as } t \to \infty, \tag{2.53}$$

where, as usual, the notation $o(1)$ means $\lim_{t\to\infty} o(1) = 0$. Asymptotic relationships of this kind are also often written as $\lambda(t) \sim r(t)$, meaning that

$$\lim_{t\to\infty} \frac{r(t)}{\lambda(t)} = 1. \tag{2.54}$$

We will use both types of asymptotic notation.

It can easily be verified that $|m'(t)| \to 0$, *e.g.*, for functions $r(t) = \exp\{t\}$ or $r(t) = \exp\{t^2\}$, for which (2.51) does not hold.

When $\lim_{t\to\infty} r(t) = 0$ ($\lim_{t\to\infty} m(t) = \infty$), which corresponds to $\lambda(t) \to 0$ as $t \to \infty$, the reasoning should be slightly different. Relationships (2.50) and (2.52) do not make much sense in this case. Therefore, the corresponding reciprocal values should be considered. From Equation (2.10):

$$\frac{1}{\lambda(t)} = \frac{m(t)}{m'(t)+1}$$

and

$$\frac{1}{\lambda(t)} - m(t) = -\frac{m'(t)m(t)}{m'(t)+1}.$$

The relative distance in this case is

$$\frac{1}{\lambda(t)m(t)} - 1 = -\frac{m'(t)}{m'(t)+1}.$$

Therefore, Relationship (2.52) is also valid if

$$\lim_{t\to\infty} |m'(t)| - 0.$$

Example 2.5 (Bradley and Gupta, 2003)
Consider the linear MRL function

$$m(t) = a + bt,\ a, b > 0.$$

The corresponding failure rate is

$$\lambda(t) = \frac{1+b}{a+bt}.$$

Thus, Condition (2.52) is not satisfied, and therefore (2.53) does not hold.

Remark 2.3 Assume that $r(t)$ is ultimately (*i.e.*, for large t) increasing. It is easy to see from (2.49) that $\lambda(t)$ is also ultimately increasing if $r'(t)/r(t)$ is ultimately decreasing, which holds, *e.g.*, for the power law.

Many of the standard distributions have failure rates that are polynomials or ratios of polynomials. The same is true for the MRL function. Theorem 2.7 can be generalized to these rather general classes of functions by assuming that $r(t)$ is a regularly varying function (Bingham *et al.*, 1987). A regularly varying function is defined as a function with the following structure:

$$r(t) = t^{\beta} l(t)(1+o(1)),\ t \to \infty\,; -\infty < \beta < \infty,\ \beta \neq 0,$$

where $l(t)$ is a slowly varying function: $l(kt)/l(t) \to 1$ for all $k > 0$. Therefore, as $t \to \infty$, it is asymptotically equivalent to the product of a power function and a function, which, *e.g.*, increases slower than any increasing power function (for example, $\ln t$).

Theorem 2.8. Let the function $r(t)$ in Theorem 2.7 be a regularly varying function with $\beta > 0$. Assume that $r'(t)$ is ultimately monotone.

Then Relationship (2.51) holds, and therefore (2.50) is also true.

Proof (Finkelstein, 2002a). In accordance with the Monotone Density Theorem (Bingham *et al.*, 1987), the ultimately monotone $r'(t)$ can be written in the following way:

$$r'(t) = t^{\beta-1}\tilde{l}(t)(1+o(1)) \text{ as } t \to \infty,$$

where $\tilde{l}(t)$ is a slowly varying function. Using expressions for regularly varying $r(t)$ and $r'(t)$:

$$\left|\frac{r'(t)}{r(t)}\right| = t^{-1}\hat{l}(t)(1+o(1)) \text{ as } t \to \infty,$$

where $\hat{l}(t)$ is another slowly varying function. Owing to the definition of the slowly varying function, $t^{-1}\hat{l}(t) \to 0$ as $t \to \infty$, and therefore Relationship (2.51) holds.

2.5 Reversed Failure Rate

2.5.1 Definitions

As stated earlier, the failure rate plays a crucial role in reliability and survival analysis. The interpretation of $\lambda(t)dt$ as the conditional probability of failure of an item in $(t, t+dt]$ given that it did not fail before in $[0,t]$ is meaningful. It describes the chances of failure of an operable object in the next infinitesimal interval of time.

The reversed failure (hazard) rate (RFR) function was introduced by von Mises in 1936 (von Mises, 1964). It has been largely ignored in the literature primarily because it was believed that this function did not have the strong intuitive probabilistic content of the failure rate (Marshall and Olkin, 2007). In the next section, we will show that it still has an interesting probabilistic meaning, although not similar to that of the 'ordinary' failure rate. Most likely owing to this meaning, the properties of the reversed failure rate have attracted considerable interest among researchers (Block *et al.*, 1998; Chandra, and Roy, 2001; Gupta and Nanda, 2001; Finkelstein, 2002, to name a few). Here we will only consider definitions and some of the simplest general properties. For more details, the reader is referred to the above-mentioned papers and references therein.

Definition 2.4. The RFR $\rho(t)$ is defined by the following equation:

$$\rho(t) = \frac{f(t)}{F(t)}. \tag{2.55}$$

Thus, $\rho(t)dt$ can be interpreted as an approximate probability of a failure in $(t-dt,t]$ given that the failure had occurred in $[0,t]$.

Similar to exponential representation (2.5), it can be easily shown solving, for instance, the elementary differential equation $F'(t) = \rho(t)F(t)$ with the initial condition $F(0) = 0$ that the following analogue of (2.5) holds:

$$F(t) = \exp\left\{-\int_t^\infty \rho(u)du\right\} \tag{2.56}$$

and that the corresponding pdf is given by

$$f(t) = \rho(t)\exp\left\{-\int_t^\infty \rho(u)du\right\}.$$

Therefore, $\rho(t)$ defines another characterization for the absolutely continuous Cdf $F(t)$. Note that for proper lifetime distributions,

$$\int_0^\infty \rho(u)du = \infty, \quad \int_t^\infty \rho(u)du \neq \infty, \forall t > 0, \tag{2.57}$$

which means that

$$\lim_{t\to 0} \rho(t) = \infty,$$

and $F(0) = 0$ should also be understood as the corresponding limit.

Unlike $\lambda(t)$, the RFR $\rho(t)$ cannot be a constant or an increasing function in $(a,\infty), a \geq 0$. It is easy to verify that (2.57) holds, *e.g.*, for the power function $\rho(t) = t^{-\alpha}, \alpha > 1$.

After a simple transformation, the following relationship between $\rho(t)$ and $\lambda(t)$ can be obtained:

$$\rho(t) = \frac{\lambda(t)\bar{F}(t)}{1-\bar{F}(t)} = \lambda(t)\frac{1}{(\bar{F}(t))^{-1}-1} \tag{2.58}$$
$$= \frac{\lambda(t)}{\exp\left\{\int_0^t \lambda(u)du\right\}-1}.$$

Let, *e.g.*, $\lambda(t)$ be a constant: $\lambda(t) = \lambda$. In accordance with Equation (2.58),

$$\rho(t) = \frac{\lambda}{\exp\{\lambda t\} - 1},$$

and therefore, $\rho(t)$ decreases exponentially as $t \to \infty$, whereas its behaviour for $t \to 0$ is defined by the function t^{-1}.

It follows from Equation (2.58) that if $\lambda(t)$ is decreasing, then $\rho(t)$ is also decreasing. For $t \to \infty$, Equation (2.55) can be written asymptotically as

$$\rho(t) = f(t)(1 + o(1)).$$

Thus $\rho(t)$ and $f(t)$ are asymptotically equivalent, which means that the study of the RFR function is relevant only for finite time.

Example 2.6 Consider a series system of two independent components with survival functions $\bar{F}_1(t), \bar{F}_2(t)$, failure rates $\lambda_1(t), \lambda_2(t)$ and RFRs $\rho_1(t), \rho_2(t)$, respectively. As the survival function of the system in this case is the product of the components' survival functions $\bar{F}_s(t) = \bar{F}_1(t)\bar{F}_2(t)$, it follows from (2.5) that $\lambda_s(t) = \lambda_1(t) + \lambda_2(t)$, where $\lambda_s(t)$ denotes the failure rate of the system. On the other hand, $F_s(t)$ can be written in terms of the RFRs as

$$\begin{aligned} F_s(t) &= 1 - \bar{F}_1(t)\bar{F}_2(t) \\ &= 1 - \left(1 - \exp\left\{-\int_t^\infty \rho_1(u)du\right\}\right)\left(1 - \exp\left\{-\int_t^\infty \rho_2(u)du\right\}\right), \end{aligned} \tag{2.59}$$

and the system's RFR can be obtained using Definition 2.4. This will be a much more cumbersome expression than the self-explanatory $\lambda_1(t) + \lambda_2(t)$.

Using the same notation, consider now a parallel system of two independent components. The failure rate of this system is defined by the distribution $F_i(t)F_2(t)$ which, similar to (2.59), does not give a 'nice' expression for $\lambda_s(t)$. The RFR for this system, however, is simply the sum of individual reversed failure rates, *i.e.*,

$$\rho_s(t) = \rho_1(t) + \rho_2(t),$$

which can be seen by substituting (2.56) into the product $F_1(t)F_2(t)$. A similar result is obviously valid for more than two independent components in parallel.

Remark 2.4 It is well known that the probability that the i th component is the cause of the failure of the series system described in Example 2.6 (given that this failure had occurred in $(t, t+dt]$) is $\lambda_i(t)/\lambda_s(t),\ i = 1,2$. It can easily be seen, however (Cha and Mi, 2008), that a similar relationship holds for the probability that the i th component is the last to fail in the described parallel system (given that the failure of a system had occurred in $(t, t+dt]$) and that probability is $\rho_i(t)/\rho_s(t), i = 1,2$.

The foregoing reasoning indicates that some characteristics of parallel systems can be better described *via* the RFR than *via* the 'ordinary' failure rate.

2.5.2 Waiting Time

It turns out that the RFR is closely related to another important lifetime characteristic: the waiting time since failure. Indeed, as the condition of a failure in $[0,t]$ is already imposed in the definition of the RFR, it is of interest in different applications (reliability, actuarial science, survival analysis) to describe the time that has elapsed since the failure time T to the current time t. Denote this random variable by $T_{w,t}$. Similar to (2.3), the corresponding survival function with support in $[0,t]$ (Finkelstein, 2002b) is

$$\bar{F}_{w,t}(x) = P\{t - T > x \mid T \le t\}$$

$$= \frac{F(t-x)}{F(t)},\ x \in [0,t], \tag{2.60}$$

and the corresponding pdf is

$$f_{w,t}(x) = \frac{f(t-x)}{F(t)},\ x \in [0,t],$$

which, taking into account (2.55), leads to an intuitively evident relationship

$$\rho(t) = f_{w,t}(0).$$

Similar to Equation (2.7):

Definition 2.5. The mean waiting time (MWT) function $m_w(t)$ for an item that had failed in the interval $[0,t]$ is

$$m_w(t) \equiv E[T_{w,t}] = \int_0^t \bar{F}_{w,t}(u)du$$

$$= \frac{\int_0^t F(u)du}{F(t)}. \tag{2.61}$$

Assume that $m_w(t)$ is differentiable. Differentiating (2.61) and similar to (2.9), the following equation is obtained:

$$m_w'(t) = 1 - \rho(t)m_w(t). \tag{2.62}$$

Equivalently,

$$\rho(t) = \frac{1 - m_w'(t)}{m_w(t)}. \tag{2.63}$$

Substituting the RFR defined by Equation (2.63) into the right-hand side of Equation (2.56), we arrive at the exponential representation for the Cdf $F(t)$, which can also be considered as another characterization of the absolutely continuous distribution function *via* the MWT function $m_w(t)$:

$$F(t) = \exp\left\{ -\int_t^{\infty} \frac{1 - m'_w(u)}{m_w(u)} du \right\}. \tag{2.64}$$

Remark 2.5 Sufficient conditions for the function $m_w(t)$ to be a MWT function for some proper lifetime distribution are similar to the corresponding conditions for the MRL function in Section 2.2.

Note that the properties of $m_w(x)$ and $m(x)$ differ significantly, which can be illustrated by the following example.

Example 2.7 Let $\lambda(t) = \lambda$. Then $m(t) = \lambda^{-1}$, whereas

$$m_w(t) = \frac{\int_0^t F(u)du}{F(t)} = \frac{t + \lambda^{-1}(\exp\{-\lambda t\} - 1)}{1 - \exp\{-\lambda t\}}.$$

It can be shown that

$$sign(m'_w(t)) = sign(\exp\{-\lambda t\} - 1 - \lambda t) > 0,$$

and therefore $m_w(t)$ is increasing in $t \in [0, \infty)$.

Transform (2.61) in the following way:

$$m_w(t) = \frac{\int_0^t F(u)du}{F(t)} = \frac{t - \int_0^t \bar{F}(u)du}{1 - \bar{F}(t)}, \tag{2.65}$$

and, as usual, assume that $E[T] = m(0) < \infty$. Then (2.65) results in the following asymptotic relationship:

$$m_w(t) = (t - m(0))(1 + o(1)), \quad t \to \infty.$$

As $m(0) = m$ is the mean time to failure, this relationship means that for t sufficiently large, $m_w(t)$ is approximately equal to the corresponding unconditional mean waiting time, when the condition that the failure had occurred in $[0, t]$ is not imposed. This result is intuitively evident.

2.6 Chapter Summary

In this chapter, we have discussed the definitions and basic properties of the failure rate, the mean remaining lifetime function and of the reversed failure rate. These facts are essential for our presentation in the following chapters. Exponential representation (2.5) for an absolutely continuous Cdf *via* the corresponding failure rate plays an important role in understanding, interpreting and applying reliability concepts.

We have considered a number of lifetime distributions which are most popular in applications. Complete information on the subject can be found in Johnson *et al.* (1994, 1995).

The classical Glaser result (Theorem 2.3) helps to analyse the shape of the failure rate, which is important for understanding the ageing properties of distributions. Various generalizations and extensions can be found, *e.g.*, in Lai and Xie (2006). The shape of the failure rate can also be analysed using properties of underlying stochastic processes (Aalen and Gjeissing, 2001). Some examples of this approach are considered in Chapter 10.

In Section 2.4.1, several of the simplest, most popular classes of ageing distributions were defined. It is clear that the IFR ($\lambda(t) \in \mathrm{I}$) property is the simplest and the most natural one for describing deterioration. On the other hand, the decreasing in time mean remaining lifetime also shows a monotone deterioration of an item. Note that Theorem 2.5 states that the decreasing MRL defines a more general type of ageing than the increasing failure rate.

The properties of the reversed failure (hazard) rate have recently attracted considerable interest. Although the corresponding definition seems to be rather artificial, the concept of the waiting time described in Section 2.5.2 makes it relevant for reliability applications. Another possible advantage of the reversed failure rate is that the analysis of parallel systems can usually be simpler using this characteristic than using the 'ordinary' failure rate.

3

More on Exponential Representation

The importance of exponential representation (2.5) was already emphasized in Section 2.1. In this chapter, we will consider two meaningful generalizations: the exponential representation for lifetime distributions with covariates and an analogue of the exponential representation for the multivariate (bivariate) case. The first generalization will be used in Chapter 6 for modelling of mixtures and in the last chapter on applications to demography and biological ageing. Other chapters do not directly rely on this material and can therefore be read independently. The bivariate case will also be considered only in Chapter 7, where the competing risks model of the current chapter will be discussed for the case of correlated covariates.

3.1 Exponential Representation in Random Environment

3.1.1 Conditional Exponential Representation

In statistical reliability analysis, the lifetime Cdf $F(t) = \Pr[T \le t]$ is usually estimated on the basis of the failure times of items. On the other hand, there can be other information available and it is unreasonable not to use it. Possible examples of this additional information are external conditions of operation, observations of internal parameters or expert opinions on the values of parameters, *etc.*

Assume that our item is operating in a random environment defined by some (covariate) stochastic process $Z_t, t \ge 0$ (*e.g.*, an external temperature, an electric or mechanical load or some other stress factor). This is often the case in practice. Similar to Equation (2.4), we can formally define (Kalbfleisch and Prentice, 1980) the following conditional failure rate (given a realization of the process in $[0,t]$ $z(u), 0 \le u \le t$):

$$\lambda(t \mid z(u), 0 \le u \le t) = \lim_{\Delta t \to 0} \frac{\Pr[t < T \le t + \Delta t \mid z(u), 0 \le u \le t; T > t]}{\Delta t}. \tag{3.1}$$

This failure rate is obtained for a realization of the covariate process. Strictly speaking, this is not yet a failure rate as defined by Equation (2.4), but rather a conditional risk or conditional hazard. Whether it will become a 'fully-fledged' failure rate depends on the answer to the following question: does an analogue of exponential representation (2.5) hold for realizations $z(u), 0 \leq u \leq t$?

$$\Pr[T > t \mid z(u), 0 \leq u \leq t] \equiv \bar{F}(t \mid z(u), 0 \leq u \leq t)$$
$$= \exp\left\{-\int_0^t \lambda(u \mid z(s), 0 \leq s \leq u)du\right\}. \quad (3.2)$$

When the answer is positive, Equation (3.2) holds and $\lambda(t \mid z(u), 0 \leq u \leq t)$ becomes the 'real' failure rate. This topic was addressed by Kalbfleisch and Prentice (1980) and has been treated on a technical level using a martingale approach in Yashin and Arjas (1988), Yashin and Manton (1997), Aven an Jensen (1999), Singpurwalla and Wilson (1995, 1999) and Kebir (1991). One can find the necessary mathematical details in these references. We, however, will consider this important issue on a heuristic, descriptive level (Finkelstein, 2004b).

An obvious condition for a positive answer is that $F(t \mid z(u), 0 \leq u \leq t)$ should be an absolutely continuous Cdf. In this case, as follows from Section 2.1, the corresponding conditional failure rate $\lambda(t \mid z(u), 0 \leq u \leq t)$ exists. As this property can depend on the environment, it brings into consideration the issue of external and internal covariates. The notions of external and internal covariates are important for survival analysis and reliability theory. As is traditionally done, define the covariate process $Z_t, t \geq 0$ as external if it may influence but is itself not influenced by the failure process of the item. On the other hand, internal covariates are those that directly convey information about the item's survival (*e.g.*, failed or not). In accordance with this useful interpretation (Fleming and Harrington, 1991), the failure time of our item T is a stopping time for the process $Z_t, t \geq 0$ if the information in the history $z(u), 0 \leq u \leq t$ specifies whether an event described by the lifetime random variable T has happened by time t. Therefore, T is not a stopping time for the external covariate process $Z_t, t \geq 0$ and is usually a stopping time for an internal process. For strict mathematical definitions, the reader is referred to, *e.g.*, Aven and Jensen (1999). Examples of internal covariates are blood pressure or body temperature, which when observed as being below a certain level indicate that the individual is not alive. If we are observing a damage accumulation process and the failure occurs when it reaches some predetermined level, then this process also can be considered as an internal covariate. An example of an external covariate in the context of life sciences is the level of radiation individuals are subjected to (Singpurwalla and Wilson, 1999) or the external temperature and humidity in reliability testing.

Let the time-to-failure Cdf of an item in some baseline, deterministic (and, for simplicity, univariate) environment $z_b(t)$ be absolutely continuous, which means that the corresponding baseline failure rate $\lambda_b(t) \equiv \lambda(t \mid z_b(u), 0 \leq u \leq t)$ exists. Let also the influence of the external stochastic covariate process, which models the real operational environment of the component, be *weak* (smooth) in the sense that the resulting conditional failure rate exists. For instance, if this influence could be modelled *via* realizations $z(t)$ directly, *e.g.*, by the proportional hazards model $z(t)\lambda_b(t)$, the additive hazards model $z(t) + \lambda_b(t)$ or the accelerated life model $\lambda_b(z(t))$, then automatically, as the failure rate exists, the corresponding Cdf

$F(t \mid z(u), 0 \le u \le t)$ is absolutely continuous. Note that these three models are very popular in reliability and survival analysis and have been intensively studied in the literature. We will consider all of them in Chapters 6 and 7. However, if, for instance, a jump in $z(t)$ leads to an item's failure with some non-infinitesimal probability (and it is often the case in practice when, *e.g.*, a jump in a stress occurs), then the corresponding Cdf $F(t \mid z(u), 0 \le u \le t)$ is not absolutely continuous and Equation (3.2) does not hold. A jump of this kind indicates a *strong* influence of the external covariate on the item's failure process.

Remark 3.1 Assume first that $Z_t, t \ge 0$ specifies the complete information about the failure process. Conditioning on the trajectory of the internal covariate of this kind results in a distribution function that is not absolutely continuous. More technically- the stopping time T in this case is a predictable one (Aven and Jensen, 1999) and exponential representation (3.2) does not hold. If, for example, $z(t)$ is increasing and the failure of an item occurs when $z(t)$ reaches a positive threshold, then T in this realization is deterministic and therefore, not absolutely continuous. On the other hand, assume now that observation of $Z_t, t \ge 0$ does not provide a complete description of the item's state. More technically, the stopping time T is totally inaccessible (in other words 'sudden') in this case (Aven and Jensen, 1999). It turns out that exponential representation (3.2) could be valid. The corresponding examples are considered in Finkelstein (2004b). A model of an unobserved overall resource in Section 10.2 also offers a relevant example.

3.1.2 Unconditional Exponential Representation

Let $Z_t, t \ge 0$ be, as in the previous section, an external covariate process and assume that conditional exponential representation (3.2) holds. Now we want to obtain the corresponding unconditional characteristic, which will be called the observed (marginal) representation. As Equation (3.2) holds for realizations $z(t)$ of the covariate process $Z_t, t \ge 0$, the observed survival function is obtained formally as the following expectation with respect to $Z_t, t \ge 0$:

$$\bar{F}(t) = E\left[\exp\left\{-\int_0^t \lambda(u \mid Z_s, 0 \le s \le u)du\right\}\right]. \tag{3.3}$$

Equation (3.3) can be written in compact form as

$$\bar{F}(t) = E\left[\exp\left\{-\int_0^t \lambda_u du\right\}\right], \tag{3.4}$$

where $\lambda_u = \lambda(u \mid Z_s, 0 \le s \le u)$ is usually (Kebir, 1991; Aven and Jensen, 1999) referred to as the *hazard (failure) rate process* (or random failure rate). A similar notion for repairable systems is usually called the *intensity process (stochastic intensity)*. It will be defined in the next chapter for general point processes without multiple occurrences.

There is a slight temptation to obtain the observed failure rate $\lambda(t)$ as $E[\lambda_u]$, but obviously it is not true, as the failure rate itself is a conditional characteristic. Therefore, if we want to write Equation (3.4) in terms of the expectation of the hazard rate process $\lambda_u = \lambda(u \mid Z_s, 0 \le s \le u)$, it should be done conditionally on survival in $[0,t]$, *i.e.*,

$$\bar{F}(t) = \exp\left\{-\int_0^t E[\lambda_u \mid T > u]du\right\}, \tag{3.5}$$

where $\lambda_t \mid T > t, t \ge 0$ denotes the conditional hazard rate process (on condition that the item did not fail in $[0,t)$). Thus, taking into account exponential representation (2.5), the definition of the observed failure rate $\lambda(t)$ *via* the conditional hazard rate process can formally be written as

$$\lambda(t) = E[\lambda_u \mid T > t]. \tag{3.6}$$

We have presented certain heuristic considerations for obtaining this very important result, which will often be used in this book for different settings. The strict mathematical proof can be found in Yashin and Manton (1997). The meaning of the 'compact' Equation (3.6) will become more evident when considering the examples in the next section.

As the exponential function is a convex one, Jensen's inequality can be used for obtaining the lower (conservative) bound for $\bar{F}(t)$ in Equation (3.4), *i.e.*,

$$\bar{F}(t) \ge \exp\left\{-\int_0^t E[\lambda_u]du\right\}. \tag{3.7}$$

Note that the expectation in (3.7) is defined with respect to the process $\lambda_t, t \ge 0$ (see Equation (6.3) and the corresponding discussion). Computations, in accordance with Equations (3.5) and (3.6), are usually cumbersome and can be performed explicitly only in a few special cases. Some meaningful examples are considered in the next section. These examples will be used throughout this book.

3.1.3 Examples

Example 3.1 Consider a special case of Model (3.3)–(3.5) when $Z_t \equiv Z$ is a positive random variable (external covariate) with the pdf $\pi(z)$. It is convenient now to use different notation for the conditional failure rate, *i.e.*,

$$\lambda(t \mid Z = z) \equiv \lambda(t,z),$$

which means that the failure rate is indexed by the parameter z. This example is crucial for the presentation of Chapter 6 and we will often refer to it.

The conditional Cdf $F(t,z)$ can be obtained *via* $\lambda(t,z)$ using the corresponding exponential representation. As usual, $f(t,z) = F_t'(t,z)$. The observed (mixture) $F(t)$ and $f(t)$ are given by the following expectations:

$$\bar{F}(t) = \int_0^t \bar{F}(t,z)\pi(z)dz,$$

$$f(t) = \int_0^t f(t,z)\pi(z)dz,$$

respectively. In accordance with the definition of the failure rate (2.4), the observed (mixture) failure rate can be defined directly as

$$\lambda(t) = \frac{\int_0^\infty f(t,z)\pi(z)dz}{\int_0^\infty \bar{F}(t,z)\pi(z)dz}. \tag{3.8}$$

Using the general relationship $f(t) = \lambda(t)\bar{F}(t)$, it is easy to transform formally the observed failure rate (3.8) into the conditional form (2.11) (Lynn and Singpurwalla, 1997; Finkelstein and Esaulova, 2001):

$$\lambda(t) = \int_0^\infty \lambda(t,z)\pi(z \mid t)dz, \tag{3.9}$$

where $\pi(z \mid t)$ denotes the conditional pdf of Z on condition that $T > t$, *i.e.*,

$$\pi(z \mid t) = \frac{\pi(z)\bar{F}(t,z)}{\int_0^\infty \bar{F}(t,z)\pi(z)dz}. \tag{3.10}$$

Equation (3.9) is an explicit form of Equation (3.6) for the special case under consideration. Thus, $\pi(z \mid t)dz$ is the conditional probability that a realization of the covariate random variable Z belongs to the interval $(z + dz]$ on condition that $T > t$. As Z is an external covariate, this is just the product of $\pi(z)dz$ and of the following probability:

$$\Pr[T > t] = \frac{\bar{F}(t,z)}{\int_0^\infty \bar{F}(t,z)\pi(z)dz}.$$

This useful interpretation explains the simple and self-explanatory form of the observed failure rate given by Equation (3.9).

Example 3.2 In this example, we assume a specific form of $\lambda(t,z)$ and choose the corresponding specific distributions. Let

$$\lambda(t,z) = z\lambda_b(t),$$

where $\lambda_b(t)$ is the failure rate of an item in a baseline environment. Let Z be a gamma-distributed random variable (Equation (2.22)) with shape parameter α and scale parameter β and let $\lambda_b(t) = \gamma t^{\gamma-1}, \gamma > 1$ be the increasing failure rate of the Weibull distribution (in a slightly different notation to that of (2.25)). The observed failure rate $\lambda(t)$ in this case, can be obtained by the direct integration in Equation (3.8), as in Finkelstein and Esaulova (2001) (see also Gupta and Gupta, 1996):

$$\lambda(t) = \frac{\alpha\beta\gamma t^{\gamma-1}}{1+\beta t^{\gamma}}. \tag{3.11}$$

Note that the shape of $\lambda(t)$ in this case differs dramatically from the shape of the increasing baseline failure rate $\lambda_b(t)$. This function is equal to 0 at $t = 0$, increases to a maximum at

$$t_{\max} = \left(\frac{\gamma-1}{\beta}\right)^{\frac{1}{\gamma}}$$

and then decreases to 0 as $t \to \infty$.

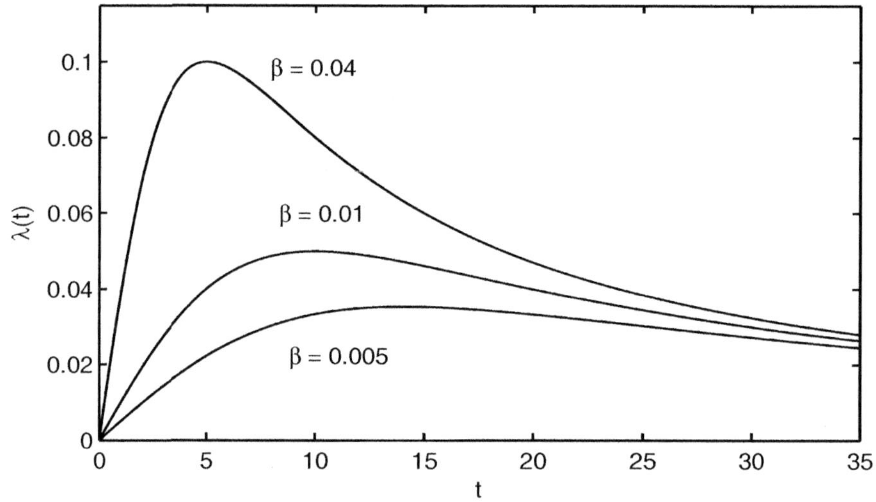

Figure 3.1. The observed failure rate for the Weibull baseline distribution, $\gamma = 2, \alpha = 1$

Example 3.3 Assume that Z is a non-negative discrete random variable with the probability mass $\pi(z_k)$ at $z = z_k$, $k \geq 1$. Then:

$$F(t) = \sum_k F(t, z_k)\pi(z_k),$$

$$f(t) = \sum_k f(t, z_k)\pi(z_k)$$

and Equations (3.8)–(3.9) are transformed into

$$\lambda(t) = \frac{\sum_k f(t, z_k)\pi(z_k)}{\sum_k \overline{F}(t, z_k)\pi(z_k)} = \sum_k \lambda(t, z_k)\pi(z_k \mid t)dz , \tag{3.12}$$

where

$$\pi_k(z \mid t) = \frac{\pi(z_k)\overline{F}(t, z_k)}{\sum_k \overline{F}(t, z_k)\pi(z_k)} \tag{3.13}$$

is the conditional (on condition that $T > t$) probability mass at $z = z_k$.

In Example 10.1 of Chapter 10, devoted to demographic applications, we use Equation (3.12) for obtaining the observed failure (mortality) rate of a parallel system of $Z = N$, $N = 1,2,...$ i.i.d. components with exponentially distributed lifetimes. The distribution of N in this case follows the Poisson law on condition that the system is operating at $t = 0$, which means that $N \neq 0$.

Example 3.4 Assume that the random failure rate $\lambda_t, t \geq 0$ is defined by the Poisson process with rate λ. The definition and simplest properties of the Poisson process are given in Section 4.3.1. Realizations of this process are non-decreasing step functions with unit jumps. They can be caused, *e.g.*, by the corresponding jumps in a stress applied to an item.

The following is obtained by direct computation (Grabski, 2003):

$$\overline{F}(t) = E\left[\exp\left\{-\int_0^t \lambda_u du\right\}\right]$$

$$= \exp\{-\lambda(t - 1 + \exp\{-t\})\} . \tag{3.14}$$

This means that

$$\lambda(t) = \lambda(1 - \exp\{-t\}) \tag{3.15}$$

is the observed failure rate in this case. It follows from Equation (3.15) that

$$\lambda(0) = 0, \quad \lim_{t \to \infty} \lambda(t) = \lambda ,$$

which agrees with the intuitive reasoning for this setting.

3.2 Bivariate Failure Rates and Exponential Representation

This book is mostly devoted to 'univariate reliability'. In this section, however, we will show how the failure rate and the exponential representation can be generalized to multivariate distributions. We will mostly consider the bivariate case and will only remark on the multivariate case where appropriate.

The importance of the failure rate and of the exponential representation for the univariate setting was already discussed in this chapter, as well as in previous chapters. In the multivariate case, however, the corresponding generalizations, although meaningful, usually do not play a similar pivotal role. This is because now there is no unique failure rate and because the probabilistic interpretations of the corresponding notions are often not as simple and appealing as in the univariate case.

3.2.1 Bivariate Failure Rates

The univariate failure rate $\lambda(t)$ of an absolutely continuous Cdf $F(t)$ uniquely defines $F(t)$ *via* exponential representation (2.5). The situation is more complex in the bivariate case. In this section, we will consider an approach to defining multivariate analogues of the univariate failure rate function, which can be used in applications related to analysis of data involving dependent durations. Other relevant approaches and results can be found in Barlow and Proschan (1975), Block and Savits (1980) and Lai and Xie (2006), among others.

Let $T_1 \geq 0, T_2 \geq 0$ be the possibly dependent random variables (describing lifetimes of items) and let

$$F(t_1,t_2) = \Pr[T_1 \leq t_1, T_2 \leq t_2],$$

$$F_i(t_i) = \Pr[T_i \leq t_i],\ i = 1,2$$

be the absolutely continuous bivariate and univariate (marginal) Cdfs, respectively. For convenience and following the conventional notation (Yashin and Iachine, 1999), denote the bivariate (joint) survival function by

$$S(t_1,t_2) \equiv \Pr[T_1 > t_1, T_2 > t_2] = 1 - F_1(t_1) - F_2(t_2) + F(t_1,t_2) \tag{3.16}$$

and the univariate (marginal) survival functions $\bar{F}_i(t), i = 1,2$ with the corresponding failure rates $\lambda_i(t_i), i = 1,2$ by

$$S_1(t_1) \equiv \Pr[T_1 > t_1, T_2 > 0] = \Pr[T_1 > t_1] = S(t_1,0),$$

$$S_2(t_2) \equiv \Pr[T_1 > 0, T_2 > t_2] = \Pr[T_2 > t_2] = S(0,t_2),$$

respectively.

It is natural to define the bivariate failure rate, as in Basu (1971), generalizing the corresponding univariate case:

$$\lambda(t_1,t_2)=\lim_{\Delta t_1,\Delta t_2\to 0}\frac{\Pr(t_1\le T_1<t_1+\Delta t_1, t_2\le T_2<t_2+\Delta t_2\mid T_1>t_1,T_2>t_2)}{\Delta t_1\Delta t_2}$$

$$=\frac{f(t_1,t_2)}{S(t_1,t_2)}. \tag{3.17}$$

Thus, $\lambda(t_1,t_2)dt_1dt_2+o(dt_1dt_2)$ can be interpreted as the probability of the failure of both items in intervals of time $[t_1,t_1+dt_1),[t_2,t_2+dt_2)$, respectively, on the condition that they did not fail before. It is convenient to use reliability terminology in this context, although other interpretations can be employed as well.

Equation (3.17) can be written as

$$f(t_1,t_2)=\lambda(t_1,t_2)S(t_1,t_2),$$

which resembles the univariate case, but the solution to this equation is not defined and therefore cannot be written in a form similar to (2.5). Therefore, a different approach should be developed.

Remark 3.2 Note that, although the failure rate $\lambda(t_1,t_2)$ does not define $F(t_1,t_2)$ in closed form (*e.g.*, in the desired form of some exponential representation), it can be proved that under some additional assumptions (Navarro, 2008) it uniquely defines the bivariate distribution $F(t_1,t_2)$.

Two types of conditional failure rates associated with $F(t_1,t_2)$ play an important role in applications related to analysis of data involving dependent durations (Yashin and Iachine, 1999):

$$\bar{\lambda}_i(t_1,t_2)=\lim_{\Delta t\to 0}\frac{1}{\Delta t}\Pr(t_i\le T_i<t_i+\Delta t\mid T_1>t_1,T_2>t_2)$$

$$=-\frac{\partial}{\partial t_i}\ln S(t_1,t_2);\ i=1,2, \tag{3.18}$$

$$\hat{\lambda}_i(t_1,t_2)=\lim_{\Delta t\to 0}\frac{1}{\Delta t}\Pr(t_i\le T_i<t_i+\Delta t\mid T_i>t_i,T_j=t_j)$$

$$=-\frac{\partial}{\partial t_i}\ln\left(-\frac{\partial}{\partial t_j}S(t_i,t_j)\right);\ i,j=1,2,\ i\ne j. \tag{3.19}$$

These univariate failure rates describe the chance of failure at age t of the i th item given the failure history of the j th item ($i,j=1,2,\ i\ne j$). For instance, $\bar{\lambda}_1(t_1,t_2)dt$ can be interpreted as the probability of failure of the first item in $(t_1,t_1+dt]$ on the condition that it did not fail in $[0,t_1]$ and that the second item also did not fail in $[0,t_2]$. Similarly, $\hat{\lambda}_1(t_1,t_2)dt$ is the probability of failure of the first item in $(t_1,t_1+dt]$ on the condition that it did not fail in $[0,t_1]$ and that the second item had failed in $(t_2,t_2+dt]$. The vector $((\bar{\lambda}_1(t_1,t_2),\bar{\lambda}_2(t_1,t_2))$ sometimes is called the hazard gradient (Johnson and Kotz, 1975) and it has been shown that it uniquely defines the bivariate distribution $F(t_1,t_2)$.

It is clear that if T_1 and T_2 are independent, then $\bar{\lambda}_i(t_1,t_2)=\hat{\lambda}_i(t_1,t_2)$, whereas $\bar{\lambda}_i(t_1,t_2)/\hat{\lambda}_i(t_1,t_2)$ can be considered as a measure of correlation between T_1 and T_2 in the general case.

Failure rates (3.17) and (3.18) are already sufficient for obtaining an analogue of exponential representation (2.5). On the other hand, failure rate (3.19) is important in defining and understanding the dependence structure of bivariate distributions.

Remark 3.3 The bivariate failure rate presented here can easily be generalized to the multivariate case $n>2$ (Johnson and Kotz, 1975).

Remark 3.4 Similar to the hazard gradient vector $(\bar{\lambda}_1(t_1,t_2),\bar{\lambda}_2(t_1,t_2))$ defined by Equation (3.18), the corresponding analogues for the conditional mean remaining lifetime functions exist (compare with Equation (2.7)), *i.e.*,

$$m_i(t_1,t_2)=E[T_i-t_i \mid T_1>t_1,T_2>t_2),\ i=1,2\,.$$

It can be proved that these functions are connected to $\bar{\lambda}_i(t_1,t_2)$ (Arnold and Zahedi, 1988) *via* the following relationships:

$$\bar{\lambda}_1(t_1,t_2)=\frac{1+(\partial/\partial t_1)m_1(t_1,t_2)}{m_1(t_1,t_2)},$$

$$\bar{\lambda}_2(t_1,t_2)=\frac{1+(\partial/\partial t_2)m_2(t_1,t_2)}{m_2(t_1,t_2)}.$$

It has been shown by these authors that the vector $(m_1(t_1,t_2),m_2(t_1,t_2))$ also uniquely defines the bivariate distribution $F(t_1,t_2)$.

3.2.2 Exponential Representation of Bivariate Distributions

Any bivariate survival function can formally be represented by the following simple identity (Yashin and Iachine, 1999):

$$S(t_1,t_2)=S_1(t_1)S_2(t_2)\exp\{A(t_1,t_2)\}, \tag{3.20}$$

where

$$A(t_1,t_2)=\ln\frac{S(t_1,t_2)}{S_1(t_1)S_2(t_2)}.$$

Equation (3.20) can be easily proved taking the logs from both sides. It is clear that the function $A(t_1,t_2)$ can be viewed as a measure of dependence between T_1 and T_2. When these variables are independent, $A(t_1,t_2)=0,\ t_1,t_2\ge 0$. Lehmann (1966) discussed a similar ratio of distribution functions under the title “quadrant dependence”. The following result was proved in Finkelstein (2003d).

Theorem 3.1. Let $F(t_1,t_2) = \Pr[T_1 \le t_1, T_2 \le t_2]$ and $F_i(t_i) = \Pr[T_i \le t_i]$, $i = 1,2$ be absolutely continuous bivariate and univariate (marginal) Cdfs, respectively.

Then the following bivariate exponential representation of the corresponding survival function holds:

$$S(t_1,t_2) = \exp\left\{-\int_0^{t_1} \lambda_1(u)du\right\}\exp\left\{-\int_0^{t_2} \lambda_2(u)du\right\}$$
$$\times \exp\left\{\int_0^{t_1}\int_0^{t_2} (\lambda(u,v) - \overline{\lambda}_1(u,v)\overline{\lambda}_2(u,v))dudv\right\}, \tag{3.21}$$

where $\lambda_i(u)$, $\underline{i} = 1,2$ are the failure rates of marginal distributions and the failure rates $\lambda(u,v)$, $\overline{\lambda}_i(u,v)$ are defined by Equations (3.17) and (3.18), respectively.

Proof. As $F_i(t_i)$, $i = 1,2$ and $A(t_1,t_2)$ are absolutely continuous (Yashin and Iachine, 1999),

$$S_i(t_i) = \exp\left\{-\int_0^{t_i} \lambda_i(u)du\right\},$$
$$A(t_1,t_2) = \int_0^{t_1}\int_0^{t_2} \varphi(u,v)dudv, \tag{3.22}$$

where $\varphi(u,v)$ is some bivariate function.

Rewrite Equation (3.20) in the following way:

$$S(t_1,t_2) = \exp\{-H(t_1,t_2)\}, \tag{3.23}$$

where

$$H(t_1,t_2) \equiv \int_0^{t_1} \lambda_1(u)du + \int_0^{t_2} \lambda_2(u)du - \int_0^{t_1}\int_0^{t_2} \varphi(u,v)dudv.$$

From the definitions of $\overline{\lambda}_i(t_1,t_2)$ and $H(t_1,t_2)$, the following useful relationship can be obtained:

$$\overline{\lambda}_i(t_1,t_2) = \frac{\partial}{\partial t_i} H(t_1,t_2)$$
$$= \lambda_i(t_i) - \frac{\partial}{\partial t_i} A(t_1,t_2),\ i = 1,2. \tag{3.24}$$

Differentiating both sides of this equation and using (3.18) and (3.22) yields

$$\frac{\partial^2}{\partial t_1 \partial t_2} A(t_1,t_2) = \frac{f(t_1,t_2)}{S(t_1,t_2)} - \frac{\partial}{\partial t_1} \ln S(t_1,t_2) \frac{\partial}{\partial t_2} \ln S(t_1,t_2),$$

which, given our notation, can be written as (see also Gupta, 2003)

$$\varphi(u,v) = \lambda(u,v) - \overline{\lambda}_1(u,v)\overline{\lambda}_2(u,v), \tag{3.25}$$

and eventually we arrive at the important exponential representation (3.21) of the bivariate survival function. ■

Before generalizing this result, let us consider several simple and meaningful examples.

Example 3.5 Gumbel Bivariate Distribution
This distribution is widely used in reliability and survival analysis. It defines a simple, self-explanatory correlation between two lifetime random variables. The survival function for this distribution is given by

$$S(t_1,t_2) = \exp\{-t_1 - t_2 - \delta t_1 t_2\}, \tag{3.26}$$

where $0 \le \delta \le 1$. Thus

$$\varphi(u,v) = -\delta, \; A(t_1,t_2) = -\delta t_1 t_2$$

and

$$\overline{\lambda}_i(t_1,t_2) = 1 + \delta t_j;\, i,j = 1,2;\, i \ne j,\, \lambda(t_1,t_2) = -\delta + (1+\delta t_1)(1+\delta t_2),$$

whereas the failure rates of the marginal distributions are $\lambda_i(t) = 1,\, i = 1,2$.

Note that the survival function for this distribution is already given by Equation (3.26) and we are just obtaining the corresponding failure rates. The next example, by contrast, is based on the relationship between the failure rates, which eventually defines the corresponding exponential representation.

Example 3.6 Clayton Bivariate Distribution
Let the dependence structure of the bivariate distribution be given by the following constant ratio:

$$\frac{\lambda(u,v)}{\overline{\lambda}_1(u,v)\overline{\lambda}_2(u,v)} = 1 + \theta, \tag{3.27}$$

where $\theta > -1$. Equation (3.25) for this special case becomes

$$\varphi(u,v) = \theta\overline{\lambda}_1(u,v)\overline{\lambda}_2(u,v)$$

or, equivalently,

$$\varphi(u,v)=\frac{\theta}{1+\theta}\lambda(u,v)\,. \tag{3.28}$$

These equations describe a meaningful proportionality between different bivariate failure rates. For $\theta>0$ (positive correlation), the corresponding bivariate survival function is uniquely defined (up to marginal distributions), and it can be shown that the function $H(t_1,t_2)$ is given by the following expression:

$$H(t_1,t_2)=\theta^{-1}\ln\left(\exp\left\{\theta\int_0^{t_1}\lambda_1(u)du\right\}+\exp\left\{\theta\int_0^{t_2}\lambda_2(u)du\right\}-1\right),$$

which eventually defines the well-known Clayton bivariate survival function (Clayton, 1978; Clayton and Cusick, 1985):

$$S(t_1,t_2)=\left(S(t_1)^{-\theta}+S(t_2)^{-\theta}-1\right)^{-\frac{1}{\theta}}\,. \tag{3.29}$$

This family of distributions was also studied by Cox and Oakes (1984), Cook and Johnson (1986), Oakes (1989) and Hougaard (2000), to name a few. With appropriate marginals, it can define several well-known bivariate distributions (*e.g.*, bivariate logistic distribution of Gumbel (1960), the bivariate Pareto distribution of Mardia (1970)).

Example 3.7 Marshall–Olkin Bivariate Distribution
This distribution is defined by the following survival function:

$$S(t_1,t_2)=\exp\{-\lambda_1 t_1-\lambda_2 t_2-\lambda_{12}\max(t_1,t_2)\}\,, \tag{3.30}$$

where $\lambda_1,\lambda_2,\lambda_{12}$ are positive constants. It cannot be transformed into a form defined by Equation (3.21), as it is not absolutely continuous since $\max(t_i,t_2)$ cannot be written as

$$\int_0^{t_1}\int_0^{t_2}\varphi(u,v)dudv$$

for some bivariate function $\varphi(u,v)$.

A rather general bivariate distribution can be constructed using exponential representation (3.21) and additional 'coefficients of proportionality'. Consider the following bivariate function:

$$S_{\alpha_1\alpha_2\beta_1\beta_2}(t_1,t_2)=S_1^{\alpha_1}(t_1)S_2^{\alpha_2}(t_2)\exp\left\{\int_0^{t_1}\int_0^{t_2}(\beta_1\lambda(u,v)-\beta_2\overline{\lambda}_1(u,v)\overline{\lambda}_2(u,v))dudv\right\},$$

where $\alpha_i>0,\ \beta_i\geq 0;\ i=1,2$.

The following theorem states the sufficient conditions for the function $1-S_{\alpha_1\alpha_2\beta_1\beta_2}(t_1,t_2)$ to be a bivariate Cdf. It is a generalization of Theorem 1 in Yashin and Iachine (1999).

Theorem 3.2. Let $S(t_1,t_2)$ be a bivariate survival function defined by exponential representation (3.21). Let

- $\beta_2 \geq \beta_1$;
- $\alpha_i - \beta_2 \geq 0,\ i=1,2$;
- $\dfrac{\lambda(u,v)}{\overline{\lambda}_1(u,v)\overline{\lambda}_2(u,v)} \geq \dfrac{\beta_2}{\beta_1};\ u,v \geq 0$.

Then $S_{\alpha_1\alpha_2\beta_1\beta_2}(t_1,t_2)$ defines the bivariate survival function for random durations $T_{1\alpha\beta}, T_{2\alpha\beta}$ with marginal survival functions $S_1^{\alpha_1}(t_1)$ and $S_2^{\alpha_2}(t_2)$, respectively.

The proof of this theorem is rather technical and can be found in Finkelstein (2003d).

Remark 3.5 The results of this section can be generalized to the multivariate case when $n>2$ (Finkelstein, 2004d). Similar to Equations (3.20), (3.22) and (3.23),

$$S(t_1,\dots,t_n) = S(t_1)\cdots S(t_2)\exp\{A(t_1,\dots,t_n)\}, \tag{3.31}$$

where

$$A(t_1,\dots,t_n) = \ln\frac{S(t_1,\dots,t_n)}{S(t_1)\cdots S(t_n)},$$

and $S(t_i)=S(0,\dots,0,t_i,0,\dots,0);\ i=1,2,\dots,n$ are the corresponding marginal survival functions. Assume that $S(t_i)$ and $A(t_1,\dots,t_n)$ are absolutely continuous functions. Similar to the bivariate case,

$$S(t_i) = \exp\left\{-\int_0^{t_i}\lambda_i(u)du\right\},$$

$$A(t_1,\dots,t_n) = \int_0^{t_1}\cdots\int_0^{t_n}\varphi(u_1,\dots,u_n)du_1\cdots du_n,$$

where $\varphi(u_1,\dots,u_n)$ is an n-variate function. It is convenient to use the following notation:

$$H(t_1,\dots,t_n) \equiv \int_0^{t_1}\lambda_1(u)du + \cdots + \int_0^{t_n}\lambda_n(u)du - \int_0^{t_1}\cdots\int_0^{t_n}\varphi(u_1,\dots,u_n)du_1\cdots du_n.$$

Therefore, the following exponential representation can be considered the formal generalization of the bivariate case:

$$S(t_1,...,t_n) = \exp\{-H(t_1,...,t_n)\}. \tag{3.32}$$

The analogues of failure rates (3.17)–(3.19) can also be formally defined (Finkelstein, 2004d). For example, the failure rate of Basu (3.17) obviously turns into

$$\lambda(t_1,..,t_n) = \frac{f(t_1,..t_n)}{S(t_1,...,t_n)} = (-1)^n \frac{\partial^n}{\partial t_1 \cdots \partial t_n} \ln S(t_1,...,t_n),$$

where $\lambda(t_1,...,t_n)dt_1 \cdots dt_n + o(dt_1 \cdots dt_n)$ can be interpreted as the probability of failure of all items in the intervals of time $[t_1, t_1 + dt_1),...,[t_n, t_2 + dt_n)$, respectively, on condition that they did not fail before. Using these failure rates, the function $H(t_1,...,t_n)$ can explicitly be obtained, although even for the case of $n = 3$, the corresponding expression is cumbersome and is not as convenient for analysis as Representation (3.21).

3.3 Competing Risks and Bivariate Ageing

3.3.1 Exponential Representation for Competing Risks

In this section, we will use the approach of the previous section for discussing the corresponding bivariate competing risks problem in reliability interpretation: the failure of a series system of possibly dependent components occurs when the first component failure occurs. A detailed treatment of the competing risks theory can be found, *e.g.*, in the books by David and Moeschberger (1978) and by Crowder (2001).

As previously, consider the lifetimes of the components T_1, T_2 with supports in $[0,\infty)$. Assume that they are described by the absolutely continuous univariate $F_i(t_i),\ i = 1,2$ and bivariate $F(t_1,t_2)$ distribution functions. It seems that everything is similar to the usual bivariate case, but there is one important distinction: now we cannot observe T_1 and T_2. What we observe is the following random variable:

$$T = \min\{T_1, T_2\}. \tag{3.33}$$

Therefore, these variables now have the following meaning:

T_i = *the hypothetical time to failure of the* i *th component in the absence of a failure of the* j *th component,* $i, j = 1,2; i \neq j$.

We are interested in the survival of our series system in $[0,t)$. The corresponding survival function is obtained by equating $t_1 = t$ and $t_2 = t$. In this way, it becomes a univariate function. Now we are ready to apply the reasoning of the previous section to the described setting. Adjusting Equations (3.20)–(3.25):

$$\tilde{S}(t) \equiv S(t,t) = S_1(t)S_2(t)\exp\{B(t)\}, \tag{3.34}$$

where

$$B(t) \equiv A(t,t) = \ln \frac{S(t,t)}{S_1(t)S_2(t)} \\ = \int_0^t \int_0^t \varphi(u,v)dudv = \int_0^t \phi(u)du, \tag{3.35}$$

and $\widetilde{S}(t)$ denotes the survival function of our series system. Therefore, (3.21) can be written as the following exponential representation:

$$\widetilde{S}(t) = \exp\left\{-\int_0^t \lambda_1(u)du\right\} \exp\left\{-\int_0^t \lambda_2(u)du\right\} \exp\left\{\int_0^t \phi(u)du\right\}. \tag{3.36}$$

The function $\phi(t)$ formally results after 'transforming' the double integral in (3.35). By differentiating $B(t)$, the following relation between $\phi(u)$ and $\varphi(u,v)$ is obtained:

$$\phi(t) = \int_0^t (\varphi(u,t) + \varphi(t,u))du . \tag{3.37}$$

This means that Equation (3.37) defines the univariate function $\phi(t)$ *via* the bivariate function $\varphi(u,v)$.

Denote the failure rate of our system by $\widetilde{\lambda}(t) = -\ln' \widetilde{S}(t)$. It follows from Equation (3.36) that

$$\widetilde{\lambda}(t) = \lambda_1(t) + \lambda_2(t) - \phi(t) . \tag{3.38}$$

When the components are independent, $\widetilde{\lambda}(t) = \lambda_1(t) + \lambda_2(t)$. Thus, the function $\phi(t)$ can also be viewed as the corresponding measure of dependence.

Remark 3.6 The marginal survival functions $S_i(t), i = 1,2$ are often called the *net survival functions*.

3.3.2 Ageing in Competing Risks Setting

In this section, we will consider a specific approach to describing the bivariate (multivariate) ageing for series systems based on the exponential representations (Finkelstein and Esaulova, 2005). Detailed information on the properties of different univariate and multivariate ageing classes and the related theory can be found, *e.g.*, in Lai and Xie (2006).

In Section 2.4.1, the simplest IFR (DFR) and DMRL (IMRL) classes of distributions were discussed. The formal definitions are as follows.

Definition 3.1. The Cdf $F(x)$ is said to be IFR (DFR) if the survival function of the remaining lifetime T_t defined by Equation (2.3), *i.e.*,

$$\bar{F}_t(x) = \Pr[T_t > x] = \frac{\bar{F}(x+t)}{\bar{F}(t)}$$

is decreasing (increasing) in $t \in [0,\infty)$ for each $x \geq 0$.

Equivalently, it can be seen easily that $F(x) \in$ IFR (DFR) if and only if $-\log \overline{F}(x)$ is convex (concave). When $F(x)$ is absolutely continuous and therefore the failure rate $\lambda(t)$ exists, the increasing (decreasing) property of the failure rate obviously defines the IFR (DFR) classes.

Definition 3.2. The Cdf $F(x)$ is said to be DMRL (IMRL) if the MRL function

$$m(t) = \int_0^\infty \overline{F}_t(u)du$$

is decreasing (increasing) in t.

It was stated in Theorem 2.4 that an increasing (decreasing) failure rate always results in a decreasing (increasing) MRL function (but not *vice versa*). We consider an increasing failure rate and a decreasing MRL function as characteristics of *positive ageing* (or just ageing), whereas a decreasing failure rate and an increasing MRL function describe *negative ageing*. This useful terminology is due to Spizzichino (1992, 2001) (see also Shaked and Spizzichino, 2001 and Basan *et al.*, 2002). It will be shown in Chapter 6 that mixtures of IFR distributions can decrease at least in some intervals of time. For example, it is a well-known fact (Barlow and Proschan, 1975) that mixtures of exponential distributions have a decreasing failure rate and therefore possess the negative ageing property.

Consider a system of two components in series and let the initial age of the i th component be $t_i, i = 1,2$. Therefore, the system starts operating with these initial ages. A natural generalization of Definition 3.1 to this case is the following (Brindley and Thomson, 1972).

Definition 3.3. The Cdf $F(t_1,t_2)$ is a bivariate IFR (DFR) distribution if

$$\frac{S(t_1+x,t_2+x)}{S(t_1,t_2)} \text{ is decreasing (increasing) in } t_1,t_2 \geq 0 \text{ for } \forall x \geq 0. \quad (3.39)$$

Thus, $S(t_1+x,t_2+x)/S(t_1,t_2)$ is the joint probability of surviving an additional x units of time given that the component i survived up to time (age) t_i, $i = 1,2$.

There are several other similar definitions in the literature, but this definition seems to be the most important (Lai and Xie, 2006) owing to its reliability interpretation. Before interpreting (3.39), we must define the following basic stochastic ordering:

Definition 3.4. A random variable X with the Cdf $F_X(x)$ is said to be larger in (usual) stochastic order than a random variable Y with the Cdf $F_X(x)$, $x \geq 0$, if

$$\overline{F}_X(x) \geq \overline{F}_Y(x), \forall x \geq 0. \quad (3.40)$$

The conventional notation for this stochastic order is

$$X \geq_{st} Y .$$

Stochastic ordering plays an important role in reliability, actuarial science and other disciplines. There are numerous types of stochastic ordering (see Shaked and Shanthikumar (2007) for an up-to-date mathematical treatment of the subject). We will use only several relevant stochastic orders to be defined in the appropriate parts of this text. In what follows, when we refer to "stochastic order", it means the order defined by (3.40).

In accordance with this definition and (3.39), the univariate lifetime of the series system under consideration decreases (increases) stochastically as the ages of the components increase.

Similar to (3.39), the following definition generalizes the univariate MRL ageing of Definition 3.2.

Definition 3.5. The Cdf $F(t_1,t_2)$ is a bivariate DMRL (IMRL) distribution if

$$m(t_1,t_2) = \frac{\int_0^\infty S(t_1+u,t_2+u)du}{S(t_1,t_2)} \text{ is decreasing (increasing) in } t_1,t_2 \geq 0 . \qquad (3.41)$$

As in the univariate case (Theorem 2.4), it follows from Definitions 3.3 and 3.5 that

$$\text{Bivariate IFR (DFR)} \Rightarrow \text{Bivariate DMRL (IMRL)}.$$

Let our series system start operating at $t = 0$ when both components are 'new'. The corresponding distribution of the remaining lifetime is

$$\frac{\bar{F}(x+t)}{\bar{F}(t)} = \frac{S(t+x,t+x)}{S(t,t)}, \qquad (3.42)$$

where the left-hand side describes this random variable in the univariate interpretation ($\bar{F}(x)$ is the survival function of the system considered as a 'black box'), whereas the right hand side is written in terms of the corresponding bivariate survival function for $t_1 = t_2 = t$. Therefore, it describes the system's dependence structure in the competing risks setting.

Definition 3.6. (Finkelstein and Esaulova, 2005). A series system of two possibly dependent components is IFR (DFR) if (3.39) holds for equal ages $t_1 = t_2 = t$, *i.e.*,

$$\frac{S(t+x,t+x)}{S(t,t)} \text{ is decreasing (increasing) in } t \text{ for } \forall x \geq 0 . \qquad (3.43)$$

In this case, the corresponding Cdf $F(t_1,t_2)$ is called the bivariate weak IFR (DFR) distribution.

This definition tells us that the remaining lifetime is stochastically decreasing (increasing) in t (in terms of Definition 3.4) and that the univariate failure rate of a system is increasing (decreasing).

Definition 3.7. A series system of two possibly dependent components is DMRL (IMRL) if (3.41) holds for equal ages $t_1 = t_2 = t$, *i.e.*,

$$\frac{\int_0^\infty S(t+u, t+u)du}{S(t,t)} \text{ is decreasing (increasing) in } t. \tag{3.44}$$

In this case, the corresponding Cdf $F(t_1, t_2)$ will be called the bivariate weak DMRL (IMRL) distribution.

In what follows in this section, we will discuss ageing properties of the bivariate Cdf $F(t,t)$. When the components are independent, the ageing properties of a system are defined by the ageing properties of the components, as the system's failure rate is just the sum of the failure rates of the components. For the dependent case, however, the dependence structure can play an important role, and Equations (3.36) and (3.38) should be taken into account. One can assume, *e.g.*, that both marginal distributions are IFR, whereas specific dependence could result in the negative ageing (DFR) of a system.

We are now interested in simple, sufficient conditions for $\tilde{\lambda}(t)$ of our series system to be monotone, which means that the Cdf $F(t_1, t_2)$, in this case, is the bivariate weak IFR (DFR) distribution. The proof of the following theorem is obvious.

Theorem 3.3. Let $F(t_1, t_2)$ be an absolutely continuous bivariate Cdf with exponential marginals and the function $\varphi(u,v)$, defined by Equation (3.25), be decreasing (increasing) in each of its arguments.

Then, as follows from Equations (3.37) and (3.38), the failure rate $\tilde{\lambda}(t)$ is increasing (decreasing), and therefore $F(t_1, t_2)$ is the bivariate weak DFR (IFR) distribution.

It is obvious that the IFR part of Theorem 3.3 holds for IFR marginal distributions as well.

The next result is formulated in terms of copulas. A formal definition and numerous properties of copulas can be found, *e.g.*, in Nelsen (2001). Copulas create a convenient way of representing multivariate distributions. In a way, they 'separate' marginal distributions from the dependence structure. It is more convenient for us to consider the survival copulas based on marginal survival functions. Copulas based on marginal distribution functions are absolutely similar (Nelsen, 2001). As we are dealing with the bivariate competing risks model, we will define the bivariate copula. The case $n > 2$ is similar. Assume that the bivariate survival function can be represented as a function of $S_i(t_i), i = 1,2$ in the following way:

$$S(t_1, t_2) = C_S(S_1(t_1), S_2(t_2)), \tag{3.45}$$

where the survival copula $C_S(u,v)$ is a bivariate function in $[0,1]\times[0,1]$. Note that such a function always exists when the inverse functions for $S_i(t_i), i=1,2$ exist:

$$S(t_1,t_2)=S(S_1^{-1}(t_1),S_1^{-1}(t_2))=C_S(S_1(t_1),S_2(t_2)).$$

It can be shown (Schweizer and Sklar, 1983) that the copula $C_S(u,v)$ is a bivariate distribution with uniform [0,1] marginal distributions. When the lifetimes are independent, the following obvious relationship holds:

$$S(t_1,t_2)=S_1(t_1)S_2(t_2)\Leftrightarrow C_S(u,v)=uv.$$

Substituting different marginal distributions, we obtain different bivariate distributions with the same dependence structure. In many instances, copulas are very helpful in multivariate analysis.

The following specific theorem gives an example of the preservation of the weak IFR (DFR) ageing property (the proof can be found in Finkelstein and Esaulova (2005)).

Theorem 3.4. Let the Cdf $F(t_1,t_2)$ with identical exponential marginal distributions be the weak IFR (DFR) bivariate distribution.

Then the bivariate Cdf with the same copula and with identical IFR (DFR) marginal distributions is also weak IFR (DFR).

Example 3.8 Gumbel Bivariate Distribution
This distribution was defined by Equation (3.26) of Example 3.5. As the marginal distributions are exponential and $\varphi(u,v)=-\delta<0$, it follows from Equations (3.37) and (3.38) that this bivariate distribution is weak IFR and that the corresponding univariate failure rate is a linearly increasing function, *i.e.*,

$$\tilde{\lambda}(t)=2(1+\delta t).$$

Example 3.9 Farlie–Gumbel–Morgenstern Distribution
This distribution is defined as (Johnson and Kotz, 1975)

$$F(t_1,t_2)=F_1(t_1)F_2(t_2)(1+\alpha(1-F_1(t_1))(1-F_2(t_2))),$$

where $-1\le\alpha\le 1$. The corresponding bivariate survival function is

$$S(t_1,t_2)=S_1(t_1)S_2(t_2)(1+\alpha(1-S_1(t_1))(1-S_2(t_2))).$$

In accordance with Equation (3.20),

$$S(t_1,t_2)=S_1(t_1)S_2(t_2)\exp\{\ln(1+\alpha(1-S_1(t_1))(1-S_2(t_2)))\}.$$

When $t_1=t_2=t$ (competing risks) and $S_1(t)=S_2(t)=S(t)$, this equation can be simplified to

$$\tilde{S}(t)\equiv S(t,t)=S^2(t)\exp\{\ln(1+\alpha(1-S(t))^2)\}.$$

Direct calculation (Finkelstein and Esaulova, 2005) gives

$$\tilde{\lambda}'(t) = (-\ln S(t,t))'' = \lambda'(t)(1-\alpha S(t)(1-S(t)) + \alpha(1-S(t))^2(1+\alpha(1-S(t))^2)$$
$$\times 2S^4(t)\tilde{S}^2(t) + \alpha\lambda^2(t)S(t)(1-2S(t)+\alpha(1-S(t))^2)2S^4(t)\tilde{S}^2(t).$$

By analysing this function it can be seen that if $S(t)$ is IFR and $\alpha \geq 0$, the function $\tilde{\lambda}(t)$ ultimately (for sufficiently large t) increases, whereas for the DFR $S(t)$ and $\alpha \leq 0$, the function $\tilde{\lambda}(t)$ ultimately decreases.

Another specific case with exponential $S_1(t)$ and $S_2(t)$ results in the following conclusion: if $\alpha \geq 0$ and $S_1(t) + S_2(t) \leq 1$, then the corresponding bivariate Cdf is weak IFR.

Example 3.10 Durling–Pareto Distribution
This distribution is defined by the following survival function:

$$S(t_1,t_2) = (1+t_1+t_2+kt_1t_2)^{-\alpha}, \alpha > 0, 0 \leq k \leq \alpha+1.$$

For the competing risk setting:

$$\tilde{S}(t) = (1+2t+kt^2)^{-\alpha}.$$

The system's failure rate and its derivative are given by

$$\tilde{\lambda}(t) = 2\alpha\frac{1+kt}{1+2t+kt^2}, \quad \tilde{\lambda}'(t) = 2\alpha\frac{k-2-k^2t^2}{(1+2t+kt^2)^2},$$

respectively. Thus, if $\alpha \leq 1$, this bivariate distribution is weak DFR, and if $\alpha > 1$, it is ultimately weak DFR (increasing for $t \leq \sqrt{k-2/k}$ and decreasing for $t > \sqrt{k-2/k}$).

3.4 Chapter Summary

Exponential representation (2.5) defines the meaningful characterization of a lifetime univariate distribution *via* the corresponding failure rate. It turns out that this representation also holds when the covariates are 'smooth', whereas a strong dependence on covariates can result in non-absolutely continuous distributions. The failure rate does not exist in the latter case, although the corresponding conditional probability (risk) of failure in the infinitesimal interval of time can always be defined.

As the failure rate is a conditional characteristic, the observed (or marginal) failure rate should be obtained as a conditional expectation with respect to the external random covariate on condition that the item survived to time t. Section 3.1.3 gives several meaningful examples of this conditioning. It turns out that the shape of the observed failure rate can differ dramatically from the shape of the baseline failure rate. This topic will be considered in more detail in Chapter 6.

There could be different failure-rate-type functions in the multivariate case. We derive exponential representation (3.21) for a bivariate distribution that involves two types of failure rates. This representation is a convenient tool for analysing data with dependent durations. The corresponding generalization to the multivariate ($n > 1$) case is rather cumbersome and presents mostly a theoretical interest. When $t_1 = t_2 = t$, the bivariate setting can be interpreted in terms of the corresponding competing risks problem. For this case, we defined the notion of bivariate weak IFR (DFR) ageing and considered several examples.

4

Point Processes and Minimal Repair

4.1 Introduction – Imperfect Repair

As minimal repair (see Section 4.4 for a formal definition) is a special case of imperfect repair, this section is, in fact, an introduction to both Chapters 4 and 5, which are devoted to imperfect repair modelling. Whereas the current chapter focuses mostly on some basic properties of the simplest point processes and on a detailed discussion of minimal repair, the next chapter deals with more general models of imperfect repair.

Performance of repairable systems is usually described by renewal processes or alternating renewal processes. This means that a repair action is considered to be perfect, *i.e.*, returning the system to a state that is *as good as new*. In many instances, this assumption is reasonable and it is used in practice as an adequate model for describing the quality of repair. However, in general, perfect repairs do not exist in real life. Even a complete overhaul of a system by means of spare parts is not ideal, as the spare parts can age during storage. We will use the term *imperfect repair* for each repair that is not perfect and the terms *minimal repair* and *general repair* for some specific cases of imperfect repair to be defined later. Note that repair in degrading systems usually decreases the accumulated amount of corresponding wear or degradation.

For the proper modelling of imperfect repair, it is reasonable to assume that the cycles, *i.e.*, the times between successive instantaneous repairs, form a sequence of decreasing (in a suitable probabilistic sense) random variables. Denote by $F_i(t)$ the Cdf of the i th cycle duration, $i = 1,2,...$. All cycles of an ordinary renewal process (see Section 4.3.2 for a formal definition) are i.i.d. random variables with a common Cdf $F(t)$. It is reasonable to assume that a process of imperfect repairs is defined by the durations of the cycles that are stochastically decreasing with i. Therefore, in accordance with Definition 3.4,

$$F_1(t) \le_{st} F_2(t) \le_{st} F_3(t) \le ... \ .$$

Other types of stochastic ordering can also be used for this purpose. For example, one of the weakest stochastic orderings when the corresponding random variables are ordered with respect to their means is definitely suitable for describing deterioration of a system with each repair.

A large number of models have been suggested for modelling imperfect repair processes. Most of the models may be classified into two main groups:

- Models where the repair actions reduce the value of the failure rate prior to a failure;
- Models where the repair actions reduce the age of a system prior to a failure.

An exhaustive survey of available imperfect repair (maintenance) models can be found in Wang and Pham (2006). We will present a detailed bibliography later when describing the corresponding models.

To illustrate these informal definitions, assume that the failure rate of a repairable item $\lambda(t)$ is an increasing function. Therefore, it is suitable for modelling lifetimes of degrading objects. Most of the imperfect repair models assume this simplest class of underlying lifetime distributions. For simplicity, let $\lambda(t) = t$. Consider first the ordinary renewal process (perfect repair). The graph of the corresponding realization of a random failure rate λ_t with renewal times $S_i, i = 1,2,...$ is presented in Figure 4.1.

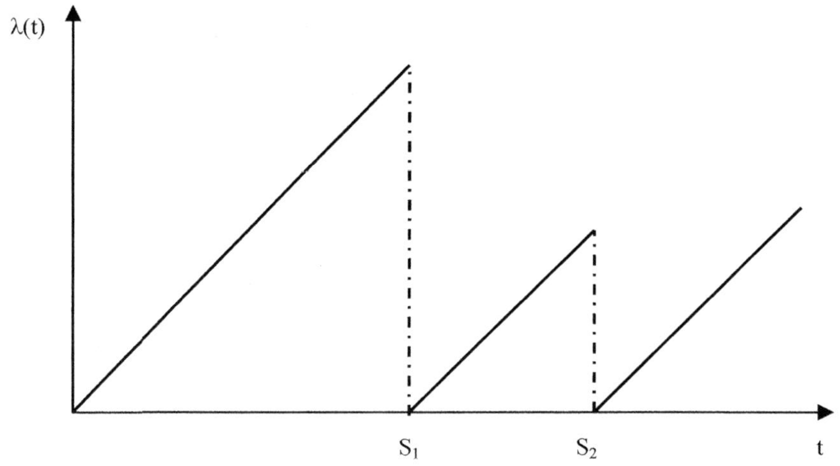

Figure 4.1. Realization of a random failure rate for the renewal process with linear $\lambda(t)$

As the repairable system is 'new' after each repair, its age is just the time elapsed since the last renewal. Assume now that each repair decreases this age by half. This assumption defines a specific case of an age reduction model. We also assume that after the age reduction the failure rate is parallel to the initial $\lambda(t) = t$. Therefore, it is also the failure rate reduction model. This can be illustrated by the following graph:

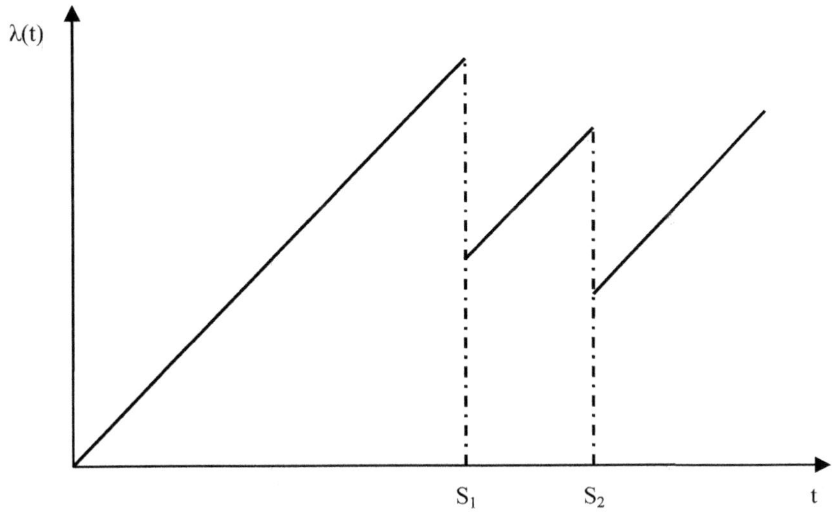

Figure 4.2. Realization of a random failure rate for the imperfect repair process with linear failure rate

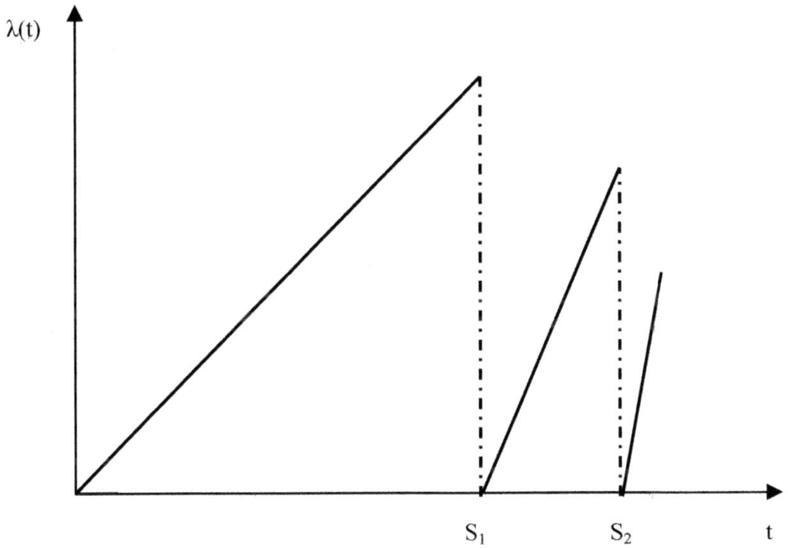

Figure 4.3. Geometric model with linear $\lambda(t)$

On the other hand, let each repair increase the entire failure rate function in the following way: the failure rate that corresponds to the random duration of the second cycle is $2\lambda t$, the third cycle is characterized by $2^2\lambda t$, *etc*. Therefore, at each subsequent cycle, the failure rate is larger than at the previous one. The corresponding graph is given in Figure 4.3.

These graphs give a simple illustration of some of the possible models of imperfect repair. A variety of more general models will be described and analysed in this and the next chapter.

The age reduction and the failure rate reduction define the main approaches to imperfect repair modelling. Note that these are rather formal stochastic models, whereas repair in degrading systems is usually an operation of decreasing the accumulated wear or deterioration of some kind. When, *e.g.*, this wear is decreased to an initial value, the system returns to the *as good as new* state. This means perfect repair; otherwise, imperfect repair is performed. Therefore, stochastic deterioration processes should be used for developing more adequate models of imperfect repair. As far as we know, not much has been done in this prospective direction. In Section 4.6, we consider some initial simplified models of this kind.

Imperfect repair has been studied in numerous publications. In what follows, we will discuss or mention most of the relevant important papers in this field. However, except for the recent monograph by Wang and Pham (2006) devoted to a rather close subject of imperfect maintenance, there is no other reliability-oriented monograph that presents a systematic treatment of this topic. Short sections on imperfect repair can also be found in recent books by Nachlas (2005) and Rausand and Houland (2004). Wang and Pham (2006) consider many useful specific models, whereas we mostly focus on discussing approaches, methods and their interpretation. The forthcoming detailed discussion of the subject intends to fill (to some extent) the gap in the literature devoted to imperfect repair modelling. Note that, in accordance with our methodology, most of the imperfect repair models considered in this book are directly or indirectly exploit the notion of a stochastic failure rate (intensity process).

Instants of repair in technical systems can be considered as points of the corresponding point process. Therefore, before addressing the subject of this chapter, we must briefly describe the main stochastic point processes that are essential for the presentation of this book. Definitions of the compound Poisson process and the gamma process will be given in Section 5.6. These jump (point) processes can also be used for imperfect repair modelling. The rest of this chapter will be devoted to the minimal repair models and some extensions, whereas Chapter 5 will deal with more general imperfect repair models. Note that minimal repair was the first imperfect repair model to be considered in the literature (Barlow and Hunter, 1960).

4.2 Characterization of Point Processes

The randomly occurring time points (instantaneous events) can be described by a stochastic point process $N(t),\ t \ge 0$ with a state space $\{0,1,2,...\}$ as a sequence of increasing random variables. For any $s,t \ge 0$ with $s < t$, the increment

$$N(s,t) \equiv N(t) - N(s)$$

is equal to the number of points that occur in $[s,t)$ and $N(s) \le N(t)$ for $s \le t$.

Assume that our process is orderly (or simple), which means that there are no multiple occurrences, *i.e.*, the probability of the occurrence of more than one event in a small interval of length Δt is $o(\Delta t)$. Assuming the limits exist, the rate of this process $\lambda_r(t)$ is defined as

$$\lambda_r(t) = \lim_{\Delta t \to 0} \frac{\Pr[N(t,t+\Delta t)=1]}{\Delta t}$$
$$= \lim_{\Delta t \to 0} \frac{E[N(t,t+\Delta t)]}{\Delta t}. \tag{4.1}$$

We use a subscript r, which stands for "rate", to avoid confusion with the notation for the 'ordinary' failure rate of an item $\lambda(t)$. Thus, $\lambda_r(t)dt$ can be interpreted as an approximate probability of an event occurrence in $[t+dt)$. The mean number of events in $[0,t)$ is given by the cumulative rate

$$E[N(0,t)] \equiv \Lambda_r(t) = \int_0^t \lambda_r(u)du .$$

The rate $\lambda_r(t)$ does not completely define the point process, and therefore a more detailed description should be used for this type of characterization. The heuristic definition of this stochastic process that is sufficient for our presentation (see Aven and Jensen, 1999; Anderson *et al.*, 1993 for mathematical details) is as follows.

Definition 4.1. An intensity process (stochastic intensity) $\lambda_t, t \ge 0$ of an orderly point process $N(t), t \ge 0$ is defined as the following limit:

$$\lambda_t = \lim_{\Delta t \to 0} \frac{\Pr[N(t,t+\Delta t)=1 \mid \mathrm{H}_t]}{\Delta t}$$
$$= \lim_{\Delta t \to 0} \frac{E[N(t,t+\Delta t) \mid H_t]}{\Delta t}, \tag{4.2}$$

where $\mathrm{H}_t = \{N(s): 0 \le s < t\}$ is an internal filtration (history) of the point process in $[0,t)$, *i.e.*, the set of all point events in $[0,t)$.

This definition can be written in a compact form *via* the following conditional expectation:

$$\lambda_t dt = E[dN(t) \mid \mathrm{H}_t]. \tag{4.3}$$

Note that, as the end point of the interval $[0,t)$ is not included in the history, the notation H_{t-} is also often used in the literature. Intensity process (stochastic intensity) completely defines (characterizes) the corresponding point process. We will consider several meaningful examples of $\lambda_t, t \ge 0$ in Section 4.3, whereas some informal illustrations were already given in the previous section. We will mostly use the term *intensity process* in what follows.

It is often more convenient in practical applications to interpret Definition 4.1 in terms of realizations of history. To distinguish it from the intensity process, we will call the corresponding notion a conditional intensity function (CIF).

Definition 4.2. Similar to (4.2), a CIF of an orderly point process $N(t), t \ge 0$ is defined for each fixed t as

$$\lambda(t \mid H(t)) = \lim_{\Delta t \to 0} \frac{\Pr[N(t, t+\Delta t) = 1 \mid \mathrm{H}(t)]}{\Delta t}$$
$$= \lim_{\Delta t \to 0} \frac{E[N(t, t+\Delta t) \mid \mathrm{H}(t)]}{\Delta t}, \tag{4.4}$$

where $\mathrm{H}(t)$ is a realization of H_t : the observed (known) history of a point process in $[0, t)$, *i.e.*, the set of all events that occurred before t.

Note that the terms “intensity process” and “CIF” are often interchangeable in the literature (Cox and Isham, 1980; Pulchini, 2003).

It follows from the foregoing considerations that the rate of the orderly point process $\lambda_r(t)$ can be viewed as the expectation of the intensity process $\lambda_t, t \geq 0$ over the entire space of possible histories, *i.e.*,

$$\lambda_r(t) = E[\lambda_t].$$

In the next section, we will consider several meaningful examples of point processes.

4.3 Point Processes for Repairable Systems

4.3.1 Poisson Process

The simplest point process is one where points occur ‘totally randomly’. The following definition is formulated in terms of conditional characteristics and is equivalent to the standard definitions of the Poisson process (Ross, 1996).

Definition 4.3. The non-homogeneous Poisson process (NHPP) is an orderly point process such that its CIF and intensity process are equal to the rate, *i.e.*,

$$\lambda_t = \lambda(t \mid \mathrm{H}(t)) = \lambda_r(t). \tag{4.5}$$

The corresponding probabilities in general Definitions 4.1 and 4.2 do not depend on the history, and therefore the property of independent increments holds automatically for this process. When $\lambda_r(t) \equiv \lambda_r$, the process is called the homogeneous Poisson process, or just the Poisson process. The number of events in any interval of length d is given by

$$\Pr[N(d) = n] = \exp\{-\Lambda_r(d)\} \frac{(\Lambda_r(d))^n}{n!}, \tag{4.6}$$

where $\Lambda_r(t)$ is the cumulative rate defined in the previous section. The distribution of time since $t = x$ up to the next event, in accordance with Equation (2.2), is

$$F(t \mid x) = 1 - \exp\left\{ - \int_{x}^{x+t} \lambda_r(u) du \right\}. \tag{4.7}$$

Therefore, the time to the first event for a Poisson process that starts at $t = 0$ is described by the Cdf with the failure rate $\lambda_r(t)$. Note that, although the NHPP $N(t),\ t \geq 0$ has independent increments, the times between successive events, as follows from (4.6), are not independent.

Assume, *e.g.*, that $\lambda_r(t)$ is an increasing function. In accordance with Definition 3.4 and Equation (4.7), the time to the next failure is stochastically decreasing in x, *i.e.*,

$$F(t \mid x_1) \geq F(t \mid x_2), \quad 0 \leq x_1 \leq x_2 .$$

This property, similar to that in Section 4.1, can already be used for defining the simplest model of imperfect repair.

Let the arrival times in the NHPP with rate $\lambda_r(t)$ be denoted by $S_i,\ i = 1,2,...,$ $S_0 = 0$. The following property will be used in Section 4.3.5. Consider the time-transformed process with arrival times

$$\tilde{S}_0 = 0, \quad \tilde{S}_i = \Lambda_r(S_i) \equiv \int_0^{S_i} \lambda_r(u) du .$$

It can be shown (Ross, 1996) that the process defined by $\tilde{S}_i$ is a homogeneous Poisson process with the rate equal to 1, *i.e.*, $\tilde{\lambda}_r(t) = 1$.

4.3.2 Renewal Process

As the generalization of a renewal process is the main goal of these two chapters, we will consider this process in detail. In addition, we will often use most of the results of this section in what follows.

Let $\{X_i\}_{i \geq 1}$ denote a sequence of i.i.d. lifetime random variables with common Cdf $F(t)$. Therefore, $X_i,\ i \geq 1$ are the copies of some generic X. Let the waiting (arrival) times be defined as

$$S_0 = 0, \quad S_n = \sum_1^n X_i ,$$

where X_i can also be interpreted as the interarrival times or cycles, *i.e.*, times between successive renewals. Obviously, this setting corresponds to perfect, instantaneous repair. Define the corresponding point process as

$$N(t) = \sup\{n : S_n \leq t\} = \sum_1^{\infty} I(S_n \leq t) ,$$

where, as usual, the indicator is equal to 1 if $S_n \leq t$ and is equal to 0 otherwise.

Definition 4.4. The described counting process $N(t), t \geq 0$ and the point process $S_n, n = 0,1,2,...$ are both called *renewal processes*.

The rate of the process defined by Equation (4.1) is called the *renewal density function* in this specific case. Denote this function by $h(t)$. Similar to the general setting, the corresponding cumulative function defines the mean number of events (renewals) in $[0,t)$, *i.e.*,

$$H(t) = E[N(t)] = \int_0^t h(u)du .$$

The function $H(t)$ is called the *renewal function* and is the main object of study in renewal theory. This function also plays an important role in different applications, as, *e.g.*, it defines the mean number of repairs or overhauls of equipment in $[0,t)$. Applying the operation of expectation to $N(t)$ results in the following relationship for $H(t)$:

$$H(t) = \sum_1^\infty F^{(n)}(t), \tag{4.8}$$

where $F^{(n)}(t)$ denotes the n-fold convolution of $F(t)$ with itself. Assume that $F(t)$ is absolutely continuous so that the density $f(t)$ exists. Denote by

$$H^*(s) = \int_0^\infty \exp\{-st)H(t)dt \quad \text{and} \quad f^*(s) = \int_0^\infty \exp\{-st)f(t)dt$$

the Laplace transforms of $H(t)$ and $f(t)$, respectively.

Applying the Laplace transform to both sides of (4.8) and using the fact that the Laplace transform of a convolution of two functions is the product of the Laplace transforms of these functions, we arrive at the following equation:

$$H^*(s) = \frac{1}{s}\sum_{k=1}^\infty (f^*(s))^k = \frac{f^*(s)}{s(1 - f^*(s))}. \tag{4.9}$$

As the Laplace transform uniquely defines the corresponding distribution, (4.9) implies that the renewal function is uniquely defined by the underlying distribution $F(t)$ *via* the Laplace transform of its density.

The functions $H(t)$ and $h(t)$ satisfy the following integral equations:

$$H(t) = F(t) + \int_0^t H(t - x)f(x)dx, \tag{4.10}$$

$$h(t) = f(t) + \int_0^t h(t - x)f(x)dx. \tag{4.11}$$

These renewal equations can be formally proved using Equation (4.8) (Ross, 1996), but here we are more interested in the meaningful probabilistic reasoning that also leads to these equations. Let us prove Equation (4.10) by conditioning on the time of the first renewal, *i.e.*,

$$\begin{aligned} H(t) &= \int_0^t E[N(t) \mid X_1 = x] f(x)dx \\ &= \int_0^t [1 + H(t-x)] f(x)dx \\ &= F(t) + \int_0^t H(t-x) f(x)dx . \end{aligned} \tag{4.12}$$

If the first renewal occurs at time $x \le t$, then the process simply restarts and the expected number of renewals after the first one in the interval $(x,t]$ is $H(t-x)$. Note that Equation (4.9) can also be obtained by applying the Laplace transform to both parts of Equation (4.10). In a similar way, the equation

$$h(t) = \int_0^t \frac{d}{dt}(E[N(t) \mid X_1 = x]) f(x)dx$$

eventually results in (4.11).

Denote, as usual, the failure rate of the underlying distribution $F(t)$ by $\lambda(t)$. The intensity process, which corresponds to the renewal process, is

$$\lambda_t = \sum_{n \ge 0} \lambda(t - S_n) I(S_n \le t < S_{n+1}),\ t \ge 0, \tag{4.13}$$

and the CIF for this case is defined by

$$\lambda(t \mid \mathrm{H}(t)) = \sum_{s_i < t} \lambda(t - s_i) I(s_i \le t < s_{i+1}),\ t \ge 0, \tag{4.14}$$

where $\mathrm{H}(t) = 0 \le s_1 < s_2 < \ldots < s_{n(t)}$ is the observed history of the renewal process in $[0,t)$ and s_i is the realization of the arrival time S_i, $i \ge 1$. Thus, at each fixed t the CIF of the renewal process is defined by $\lambda(t - s_{n(t)})$, where $s_{n(t)}$ denotes the observed time of the last (before t) $n(t)$th renewal. On the other hand, the intensity process at each fixed t can also be compactly written as $\lambda(t - S_{N(t)})$, where $S_{N(t)}$ is the random time of the last renewal. Therefore, for brevity, where reasonable, we will use the following representations for the intensity process and the CIF:

$$\lambda_t = \lambda(t - S_{N(t)}), \tag{4.15}$$

$$\lambda(t \mid \mathrm{H}(t)) = \lambda(t - s_{n(t)}) .$$

Note that the graph of the CIF for the linear underlying failure rate is presented in Figure 4.1.

In contrast to the Poisson process, when the underlying Cdf $F(t)$ is non-exponential, a renewal process does not possess the Markov property and therefore its increments are not independent. However, the Markov property is preserved only at renewal times, as the process restarts after each renewal.

The asymptotic behaviour of renewal processes is also usually of interest in different applications. A well-known result (Ross, 1996) states the intuitively expected asymptotic properties for the renewal function and the renewal density function as $t \to \infty$, *i.e.*,

$$H(t) = \frac{t}{m}[1 + o(1)], \quad h(t) = \frac{1}{m}[1 + o(1)], \tag{4.16}$$

where we assume that $E[X] = m < \infty$ exists. Thus, in contrast to the Poisson process with the rate defined by an 'arbitrary' function $\lambda_r(t)$, the rate of the renewal process tends to a constant as $t \to \infty$.

The following relationship defines the asymptotic behaviour of the standard deviation of $N(t)$ as $t \to \infty$ (Ross, 1996):

$$\sigma_{N(t)} = \frac{\sigma\sqrt{t}}{\sqrt{m^3}}[1 + o(1)],$$

where σ is the standard deviation that corresponds to the Cdf $F(t)$.

Combining the asymptotic expressions for $H(t) = E[N(t)]$ and $\sigma_{N(t)}$ results in

$$\frac{\sigma_{N(t)}}{H(t)} = \frac{\sigma}{\sqrt{mt}}[1 + o(1)] \to 0$$

as $t \to \infty$. This means that the random variable $N(t)$ becomes asymptotically 'relatively less dispersed' and therefore tends (in some sense) to a linear deterministic function.

4.3.3 Geometric Process

The geometric process is a meaningful generalization of a renewal process. In contrast to a renewal process, which models a perfect repair, the geometric process can already be useful for modelling an imperfect repair as its cycles are not identically distributed. However, the cycle's durations are 'governed' by the same generic distribution in the following way.

Definition 4.5. Let $\{X_n\}_{n\geq 1}$ be a sequence of independent lifetime random variables with the corresponding distributions $F_n(t)$ defined by the underlying distribution $F(t)$ as

$$F_n(t) = F(a^{n-1}t), \quad n = 1,2,..., \tag{4.17}$$

where a is a positive constant. Then the sequence $\{X_n\}_{n\geq 1}$ is called a geometric process.

Geometric processes in a reliability context were introduced by Lam (1988a) (see also Lam, 1988b, 1996, 1997; and Lam *et al.*, 2002). Finkelstein (1993) considered some generalizations of (4.17) to non-linear scale transformations. Wang and Pham (2006) call a similar process a *quasi-renewal process*. When $a = 1$, a geometric process reduces to a renewal process.

An important feature of this model is that, as in the case of a renewal process, it is also governed by one underlying distribution $F(t)$. It is clear that, *e.g.*, for $a > 1$, in accordance with Definition 3.4, the cycles of this process are stochastically decreasing in n, *i.e.*,

$$F(a^n t) > F(a^{n-1} t) \Rightarrow X_{n+1} <_{st} X_n, \quad t > 0, n = 1,2,\ldots .$$

Therefore, this process can already model an imperfect repair action when after each repair a system's 'quality' is worse than at the previous cycle. When $a < 1$, a system is improving with each repair, which is not often seen in practice.

Let $E[X_1] = m, Var(X_1) = \sigma^2$. It follows from (4.17) that

$$E[X_n] = \frac{m}{a^{n-1}}, \quad Var(X_n) = \frac{\sigma^2}{a^{2(n-1)}} .$$

The density function and the failure rate are

$$f_n(t) = a^{n-1} f(a^{n-1} t), \quad \lambda_n(t) = a^{n-1} \lambda(a^{n-1} t), \quad n = 1,2\ldots, \tag{4.18}$$

where $f(t)$ and $\lambda(t)$ denote the density and the failure rate of the underlying distribution $F(t)$, respectively. Therefore, for $a > 1$, in contrast to a renewal process and to the case $a < 1$, the sum of expectations is converging, *i.e.*,

$$\sum_1^\infty E[X_n] = \frac{am}{(a-1)} < \infty . \tag{4.19}$$

The counting process $N(t)$ and the renewal function $H(t) = E[N(t)]$ are defined similarly to the renewal case. However, the corresponding convolutions in (4.8) should be substituted (Lam, 1988a) by

$$F^{(n)}(t) = \int_0^t F^{(n-1)}(t-x) dF(a^{(n-1)} x)$$

$$= \int_0^t F^{(n-1)}(a(t-x)) f(x) dx .$$

Using this property and a similar argument to that used to obtain Equations (4.10) and (4.11), the following renewal-type equations with a convolution in the right-hand side are derived:

$$H(t) = F(t) + \int_0^t H(a(t-x)) f(x) dx , \tag{4.20}$$

$$h(t) = f(t) + a\int_0^t h(a(t-x))f(x)dx. \tag{4.21}$$

Although it seems that the difference between, *e.g.*, Equations (4.20) and (4.10) is not so important, it prevents us from obtaining the solution in terms of the Laplace transform in a simple form, similar to Equation (4.9). However, formally, the Laplace transform $H^*(s)$ (and therefore $h^*(s)$ as well) can be obtained as an infinite series. It can be seen from (4.17) that

$$F_n^*(s) = F^*(s/a^{n-1}),\ f_n^*(s) = sF_n^*(s);\quad n = 1,2,\ldots\ .$$

Therefore, after applying the Laplace transform to both sides of equation, which is similar to Equation (4.8), we obtain the Laplace transform of the renewal function of a geometric process as (compare with (4.9))

$$H^*(s) = \frac{1}{s}\sum_{n=1}^{\infty}\prod_{j=1}^{n} f^*(s/a^{j-1}). \tag{4.22}$$

Equation (4.22) can be inverted numerically (Nachlas, 2005). Note that the renewal function $H(t)$ for $a > 1$ and for sufficiently large t can be non-finite (Lam, 1988a). However, it is always finite for $0 < a \le 1$ and the series (4.22) is always converging in this case.

Taking Equation (4.18) into account, it is easy to modify the intensity process (4.13) for the case of a geometric process, *i.e.*,

$$\lambda_t = \sum_{n\ge 0} a^n \lambda(a^n(t - S_n))I(S_n \le t < S_{n+1}),\ t \ge 0. \tag{4.23}$$

The CIF (4.14) becomes

$$\lambda(t \mid \mathrm{H}(t)) = \sum_{s_i < t} a^n \lambda(a^n(t - s_i))I(s_i \le t < s_{i+1}),\ t \ge 0. \tag{4.24}$$

In accordance with (4.15), the intensity process at each fixed t can be compactly written as

$$\lambda_t = a^{N(t)}\lambda(a^{N(t)}(t - S_{N(t)})),$$

where $S_{N(t)}$ is the random time of the last renewal and $N(t)$ is the random number of this renewal. Similarly,

$$\lambda(t \mid \mathrm{H}(t)) = a^{n(t)}\lambda(a^{n(t)}(t - s_{n(t)})),$$

where $s_{n(t)}$ denotes the observed time of the last renewal.

The graph of the CIF for the linear underlying failure rate is similar to the one in Figure 4.3. For $\lambda(t) = t, a = 2$, the CIF is

$$\lambda(t \mid \mathrm{H}(t)) = \sum_{s_i<t} 2^{2n}(t - s_i))I(s_i \le t < s_{i+1}),\ t \ge 0 .$$

Therefore, the failure rate in this special case is linear at each cycle and the slope is increasing in accordance with the factor 2^{2n}, $n = 0,1,2...$.

A decreasing geometric process ($a > 1$) can be used for modelling an imperfect repair when each subsequent cycle is stochastically smaller than the current one (Finkelstein, 1993). If repair is not instantaneous, an increasing geometric process ($0 < a < 1$) can also be used for modelling a stochastically increasing sequence of repair times. Various optimal maintenance problems for this setting were considered in Lam (1997, 1988a,b).

Note that, the history of a renewal process is just the time since the last renewal, whereas the history of a geometric process is defined by the time since the last renewal and additionally, by the number of the last renewal.

4.3.4 Modulated Renewal-type Processes

In accordance with his idea of the proportional hazards model, Cox (1972) suggested the following generalization of the renewal intensity process (4.15), which in our notation defines a modulated renewal process, *i.e.*,

$$\lambda_t = z(t)\lambda(t - S_{N(t)}), \tag{4.25}$$

where $z(t)$ is a deterministic function of a calendar time t (age of a repairable system since inception into operation) that usually captures the impact of external factors (*e.g.*, temperature, stress, humidity) on the failure rate. The proportional hazards model is often used for statistical inference in regression analysis, and therefore the function $z(t)$ is usually defined by a vector of factors $\{z_i(t)\}$, $i = 1,2,...,n$, and by a vector of unknown regression coefficients $\{\beta_i\}$, $i = 1,2,...,n$, in the following way:

$$z(t) = \exp\left\{-\sum_1^n \beta_i z_i(t)\right\}.$$

There is no need for this 'structure' of the function $z(t)$ for our purpose of general modelling, therefore, we continue with (4.25). Observe that the history H_t in this model is defined by the time since the last renewal and by the calendar time t .

If the function $z(t)$ is increasing with time, then, similar to a geometric process with $a > 1$, the cycles of the modulated renewal process are stochastically decreasing. To show this simple fact, assume that a cycle had start at time t_1 . This means that in s units of time the corresponding failure rate will be $z(t_1 + s)\lambda(s)$. For another cycle with a starting calendar time $t_2, t_2 > t_1$, the failure rate is $z(t_2 + s)\lambda(s)$. As the function $z(t)$ is increasing,

$$\exp\left\{-\int_0^t z(t_1 + s)\lambda(s)\right\} \ge \exp\left\{-\int_0^t z(t_2 + s)\lambda(s)\right\},\quad t \ge 0,$$

which, in accordance with Definition 3.4, states that the second cycle is stochastically smaller than the first one. Therefore, as the cycles are stochastically decreasing, similar to the previous case of a geometric process, the modulated renewal process can also be used for modelling imperfect repair.

Remark 4.1 As $z(t)$ often models the external factors that, in the first place, influence not a repair mechanism as such, but the failure mechanism of items, the usage of this model for imperfect repair modelling is usually formal. This criticism can probably be applied to some extent to a geometric process as well.

Another type of modulation for renewal processes can be defined *via* a trend-renewal process (TRP). It was suggested by Lindqvist (1999) and extensively studied in Lindqvist *et al.* (2003) and Lindqvist (2006). This process generalizes a well-known property of the NHPP, which was formulated in Section 4.3.1, *i.e.*, the specific time transformation of the NHPP results in the homogeneous Poisson process. The formal definition is as follows.

Definition 4.6. Let $z(t)$ be a non-negative function defined for $t \geq 0$ and let $Z(t)$ be an integral of this function:

$$Z(t) = \int_0^t z(u)du \, .$$

A point process $N(t),\ t \geq 0$ with arrival times $S_i, i = 1, 2, \ldots, S_0 = 0$ is called a TRP ($F(t),\ z(t)$) if the arrival times of the transformed process $Z(S_i),\ i = 1, 2, \ldots,$ $Z(S_0) = 0$ form a renewal process with an underlying distribution $F(t)$.

The function $z(t)$ is called a trend function and it can be interpreted as the rate of some baseline NHPP, whereas $F(t)$ is called a renewal distribution. When $F(t) = 1 - \exp\{-\lambda t\}$, the TRP reduces to the NHPP. On the other hand, when $z(t) = const$, the TRP reduces to a renewal process. Therefore, it contains both the NHPP and the renewal processes as special cases.

Similar to Equation (4.15), the intensity process can be defined in this case as

$$\lambda_t = z(t)\lambda(Z(t) - Z(S_{N(t)})) \, . \tag{4.26}$$

Remark 4.2 The modulating structures in Equations (4.25) and (4.26) look rather similar, but the time transformation in the latter equation creates a certain difference. It measures the time elapsed from the last arrival not in chronological time, as in (4.25), but in the transformed time. If, *e.g.*, $z(t) > 1$, then we observe an 'acceleration of the internal time in the renewal process' in the following sense:

$$Z(t) - Z(S_{N(t)}) = \int_{S_{N(t)}}^t z(u)du > t - S_{N(t)} \, .$$

Therefore, Equation (4.26) can loosely be interpreted as a renewal process analogue of the conventional accelerated life model for the scale-transformed (in accordance with $F(Z(t))$) lifetimes. The failure rate that corresponds to this distribution function is $z(t)\lambda(Z(t))$, where $\lambda(t)$ is the failure rate of the baseline Cdf $F(t)$.

Definition 4.6 states that the point process $\tilde{N}(u) = N(Z^{-1}(u))$ is a renewal process with an underlying Cdf $F(t)$ (Lindqvist *et al.*, 2003). Then, *e.g.*, the second equation in (4.16) can be written as $E[N(Z^{-1}(u))/u] \to 1/m$. Substituting $t = Z^{-1}(u)$ in Equations (4.16) results in the following asymptotic (as $t \to \infty$) results for the TRP:

$$E[N(t)] = \frac{Z(t)}{m}[1 + o(1)], \quad \frac{d}{dt}E[N(t)] = \frac{z(t)}{m}[1 + o(1)].$$

These equations show that the TRP can be asymptotically approximated by the NHPP with the rate $z(t)/m$.

With an obvious exception of a renewal process, the point processes considered in this chapter can be used for imperfect repair modelling. Some criticism in this respect was already discussed in Remark 4.1. We now start describing the approaches that were developed specifically for imperfect repair modelling.

4.4 Minimal Repair

The concept of minimal repair is crucial for analysing the performance and maintenance policies of repairable systems. It is the simplest and best understood type of imperfect repair in applications. Minimal repair was introduced by Barlow and Hunter (1960) and was later studied and applied in numerous publications devoted to modelling of repair and maintenance of various systems. It was also independently used in bio-demographic studies (Yashin and Vaupel, 1987). After discussing the definition and interpretations of minimal repair, we consider several important specific models.

4.4.1 Definition and Interpretation

The term *minimal repair* is meaningful. In contrast to an overhaul, it usually describes a minor maintenance or repair operation. The mathematical definition is as follows.

Definition 4.7. The survival function of an item (with the Cdf $F(t)$ and the failure rate $\lambda(t)$) that had failed and was instantaneously minimally repaired at age x is

$$\frac{\bar{F}(x+t)}{\bar{F}(x)} = \exp\left\{-\int_x^{x+t}\lambda(u)du\right\}. \tag{4.27}$$

In accordance with Equation (2.2), this is exactly the survival function of the remaining lifetime of an item of age x. Therefore, the failure rate just after the minimal repair is $\lambda(x)$, *i.e.*, the same as it was just before the repair. This means that minimal repair does not change anything in the future stochastic behaviour of an item, as if a failure did not occur. It is often described as the repair that returns an item to the state it had been in prior to the failure. Sometimes this state is called *as bad as old*. The term *state* should be clarified. In fact, the state in this case depends only on the time of failure and does not contain any additional information. Therefore, this type of repair is usually referred to as *statistical* or *black box* minimal repair (Bergman, 1985; Finkelstein, 1992). To avoid confusion and to comply with tradition, we will use the term *minimal repair* (without adding "statistical") for the operation described by Definition 4.7.

Comparison of (4.27) with (4.6) results in the important conclusion that the process of minimal repair is a non-homogeneous Poisson process with rate $\lambda_r(t) = \lambda(t)$. Therefore, in accordance with Equation (4.5), the intensity process $\lambda_t, t \geq 0$ that describes the process of minimal repairs is also deterministic, *i.e.*,

$$\lambda_t = \lambda(t). \tag{4.28}$$

There are two popular interpretations of minimal repair. The first one was introduced to mimic the behaviour of a large system of many components when one of the components is perfectly repaired (replacement). It is clear that in this case the performed repair operation can be approximately qualified as a minimal repair. We must assume additionally that the input of the failure rate of this component in the failure rate of the system is sufficiently small. The second interpretation describes the situation where a failed system is replaced by a statistically identical one, which was operating in the same environment but did not fail. The following example interprets in terms of minimal repairs the notion of a deprivation of life that is used in demographic literature.

Example 4.1 Let us think of any death in $[t, t+dt)$, whether from accident , heart disease or cancer, as an 'accident' that deprives the person involved of the remainder of his expectation of life (Keyfitz, 1985), which in our terms is the MRL function $m(t)$, defined by Equation (2.7). Suppose that everyone is saved from death once but thereafter is unprotected and is subject to the usual mortality in the population. Then the average deprivation can be calculated as

$$D = \int_0^\infty f(u) m(u) du,$$

where $f(t)$ is the density which corresponds to the Cdf $F(t)$. In our terms, D is the mean duration of the second cycle in the process of minimal repair with rate $\lambda(t)$. Note that the mean duration of the first cycle is $m(0) = m$. The case of several additional life chances or, equivalently, subsequent minimal repairs is considered in Vaupel and Yashin (1987). These authors show that the mortality (failure) rate with a possibility of n minimal repairs is

$$\lambda_n(t) = \lambda(t) \frac{\Lambda^n(t)}{n! \sum_{r=0}^{n} \frac{\Lambda^r(t)}{r!}},$$

where $\lambda(t)$ is the mortality rate without possibility of minimal repairs. Note that, when $\lambda(t) = \lambda$, the right-hand side of this equation becomes the failure rate that corresponds to the Erlangian distribution (2.21).

4.4.2 Information-based Minimal Repair

It is clear that the observed information in the process of operation of repairable systems is an important source for adequate stochastic modelling. This topic was addressed by Aven and Jensen (1999) on a general mathematical level. We will use minimal repair as an example of this reasoning.

It follows from Definition 4.7 that the only available information in the minimal repair model is operational time at failure. On the other hand, other information can also be available. If, *e.g.*, a failure of a multi-component system is caused by a failure of one component and we observe the states (operating or failed) of all components, it is reasonable to repair only this failed component. In accordance with Arjas and Norros (1989), Finkelstein (1992) and Boland and El-Newihi (1998), we define the *information-based minimal* repair for a system as the minimal repair of the failed component. It is interesting to compare the Cdfs of the remaining lifetimes and the failure rates of the system after the minimal and the information-based minimal repairs, respectively. The following examples (Finkelstein, 1992) consider this comparison for the simplest redundant systems.

Example 4.2 Consider a standby system of two components with i.i.d. exponential lifetimes, $F(t) = 1 - \exp\{-\lambda t\}$. Then the Cdf of the system is

$$F_s(t) = 1 - (\exp\{-\lambda t\})(1 + \lambda t).$$

The information-based minimal repair of the system restores it to the state (the number of operational components) it had just before the failure, *i.e.*, one operating component. Therefore, the failure rate $\lambda_{si}(t)$ after the information-based minimal repair is λ, whereas the failure rate of the system after the minimal repair at time t is $\lambda_s(t) = \lambda^2 t/(1 + \lambda t)$. Finally,

$$\lambda_s(t) < \lambda_{si}(t)$$

for this specific case, and therefore the corresponding remaining lifetimes are ordered in the sense of the failure rate ordering that implies the (usual) stochastic ordering (3.40). This means that the remaining lifetime after the minimal repair of the considered standby system is stochastically larger than the remaining lifetime after the described information-based minimal repair. Generalization to the system of one operating component and $n > 1$ standby components is straightforward.

Example 4.3 Consider a parallel system of independent components with exponential lifetimes: $F_i(t) = 1 - \exp\{-\lambda_i t\},\ i = 1,2$, and let $\lambda_1 > \lambda_2$. Denote by $P_i(t),\ i = 1,2$ the probabilities that the described system after the minimal repair at time t is in a state where the i th component is operating (the other has failed) and by $P_{1+2}(t)$ the probability that it is in a state with both operating components. Conditioning on the event that the system is operating at t gives

$$P_i(t) = \frac{\exp\{-\lambda_i t\}(1-\exp\{-\lambda_j t\})}{\exp\{-\lambda_i t\}+\exp\{-\lambda_j t\}-\exp\{-(\lambda_1+\lambda_2)t\}}, \quad i,j=1,2; i \neq j,$$

$$P_{1+2}(t) = \frac{\exp\{-(\lambda_1+\lambda_2)t\}}{\exp\{-\lambda_i t\}+\exp\{-\lambda_j t\}-\exp\{-(\lambda_1+\lambda_2)t\}}, \quad i,j=1,2; i \neq j.$$

After the statistical minimal repair, by definition, our system can obviously be in only one of two states with probabilities denoted by $P_i^{in}(t),\ i=1,2$:

$$P_i^{in}(t) = \frac{\lambda_i \exp\{-\lambda_i t\}(1-\exp\{-\lambda_j t\})}{\lambda_1 \exp\{-\lambda_1 t\}(1-\exp\{-\lambda_2 t\})+\lambda_2 \exp\{-\lambda_2 t\}(1-\exp\{-\lambda_1 t\})}, \quad i,j=1,2,$$

where $i \neq j$. Using the assumption $\lambda_1 > \lambda_2$, it can be seen that $P_1^{in}(t) > P_1(t)$. This means that the information-based minimal repair brings the system to a state where the worst component is functioning with a larger probability than in the case of the minimal repair. Combining this inequality with the following identities:

$$P_1^{in}(t) + P_2^{in}(t) = 1,$$

$$P_1(t) + P_2(t) + P_{1+2}(t) = 1$$

results in the fact that, similar to the previous example, the remaining lifetime after the minimal repair is stochastically larger than that after the information-based minimal repair. This, of course, does not mean that minimal repair is better, as more resources are usually required to perform this operation.

4.5 Brown–Proschan Model

When the rate $\lambda_r(t)$ of the Poisson process is an increasing function, the corresponding interarrival times form a stochastically decreasing sequence (Section 4.3.1), and therefore the minimal repair process can be used for imperfect repair modelling.

Real-life repair is neither perfect nor minimal. It is usually intermediate in some suitable sense. Note that it can even be worse than a minimal repair (*e.g.*, correction of a software bug can result in new bugs).

One of the first imperfect repair models was suggested by Beichelt and Fischer (1980) (see also Brown and Proschan, 1983). This model combines minimal and perfect repairs in the following way. An item is put into operation at $t=0$. Each time it fails, a repair is performed, which is perfect with probability p and is minimal with probability $1-p$. Thus, there can be $k=0,1,2,...$ imperfect repairs between two successive perfect repairs. The sequence of i.i.d. times between consecutive perfect repairs $X_i, i=1,2,...$, as usual, forms a renewal process.

The Brown–Proschan model was extended by Block *et al.* (1985) to an age-dependent probability $p(t)$, where t is the time since the last perfect repair. Therefore, each repair is perfect with probability $p(t)$ and is minimal with probability $1-p(t)$. Denote by $F_p(t)$ the Cdf of the time between two consecutive perfect repairs. Assume that

$$\int_0^\infty p(u)\lambda(u)du = \infty \ , \tag{4.29}$$

where $\lambda(t)$ is the failure rate of our item. Then

$$F_p(t) = 1 - \exp\left\{ -\int_0^t p(u)\lambda(u)du \right\} . \tag{4.30}$$

Note that Condition (4.29) ensures that $F_p(t)$ is a proper distribution ($F_p(\infty)=1$). Thus, the failure rate $\lambda_p(t)$ that corresponds to $F_p(t)$ is given by the following meaningful, simple relationship:

$$\lambda_p(t) = p(t)\lambda(t) .$$

The formal proof of (4.30) can be found in Beichelt and Fischer (1980) and Block *et al.* (1985). On the other hand, the following simple general reasoning leads to the same result. Let an item start operating at $t=0$ and let T_p denote the time to the first perfect repair. We will now 'construct' the failure rate $\lambda_p(t)$ in a direct way. Owing to the properties of the process of minimal repairs, we can reformulate the described model in a more convenient way. Assume that events are arriving in accordance with the NHPP with rate $\lambda(t)$. Each event independently from the history 'stays in the process' with probability $1-p(t)$ and terminates the process with probability $p(t)$. Therefore, the random variable T_p can now be interpreted as the time to termination of our point process. The intensity process that corresponds to the NHPP is equal to its rate and does not depend on the history H_t of the point process of minimal repairs. Moreover, owing to our assumption, the probability of termination also does not depend on this history. Therefore,

$$\lambda_p(t)dt = \Pr[T_p \in [t, t+dt) \mid \mathrm{H}_t, T_p \geq t] = p(t)\lambda(t)dt . \tag{4.31}$$

In Section 8.1, we present a more detailed proof of Equation (4.31) for a slightly different (but mathematically equivalent) setting.

4.6 Performance Quality of Repairable Systems

In this section, we will generalize the Brown–Proschan model to the case where the quality of performance of a repairable system is characterized by some decreasing function or by a monotone stochastic process that describes degradation of this system. Along with the minimal (probability $1-p(t)$) or perfect (probability $p(t)$) repair considered earlier, the perfect or imperfect 'restoration' of a degradation function will be added to the model. In order to proceed with this imperfect repair model, the case of a perfect repair for repairable systems characterized by a performance quality function should be described first.

4.6.1 Perfect Restoration of Quality

Consider first a non-repairable system, which starts operating at $t = 0$. Assume that the quality of its performance is characterized by some function of performance $Q(t)$ to be called the quality function. It is often a decreasing function of time, and this assumption is quite natural for describing the degrading system. In applications, the function $Q(t)$ can describe some key parameter of a system, *e.g.*, the decreasing in time accuracy of the information measuring system or effectiveness (productivity) of some production process. Assume, for simplicity, that $Q(t)$ is a deterministic function. Let the system's time-to-failure distribution be $F(t)$ and assume that the quality function is equal to 0 for the failed system. Then the expected quality of the system at time t is

$$Q_E(t) = E[Q(t)I(t)],$$

where $I(t) = 1$ if the system is operable at t and $I(t) = 0$ when it fails.

Now, let the described system be instantly and perfectly repaired at each moment of failure. This means that the quality function is also restored to its initial value $Q(0)$. Therefore, failures occur in accordance with a renewal process defined by i.i.d. cycles with the Cdf $F(t)$. Denote by $\tilde{Q}(t) \equiv Q(Y)$ a random value of the quality function at time t, where Y is the random time since the last renewal. Using similar arguments as when deriving Equations (4.10) and (4.11), the following equation for the expected value of $\tilde{Q}(t)$ can be derived:

$$Q_E(t) \equiv E[\tilde{Q}(t)] = \bar{F}(t)Q(t) + \int_0^t h(x)\bar{F}(t-x)]Q(t-x)dx\,. \tag{4.32}$$

The first term on the right-hand side of Equation (4.32) is the probability that there were no failures in $[0,t)$, whereas $h(x)\bar{F}(t-x)dx$ defines the probability that the last failure before t had occurred in $[x, x+dx)$. Therefore, the quality function at t is equal to $Q(t-x)$. The expected quality $Q_E(t)$ is an important performance characteristic. Obviously, when $Q(t) \equiv 1$, it reduces to the 'classical' availability function.

In practice, as in the case of a time-dependent availability, the corresponding numerical methods should be used for obtaining $Q_E(t)$ defined by Equation (4.32). On the other hand, there exists a simple stationary solution. After applying the key renewal theorem (Ross, 1996), the following stationary value ($t \to \infty$) of the expected quality Q_{ES} can be derived:

$$Q_{ES} = \frac{1}{m}\int_0^\infty \bar{F}(x)Q(x)dx\,, \tag{4.33}$$

where m is the mean that corresponds to the Cdf $F(t)$.

Another important performance characteristic is the probability that $\tilde{Q}(t)$ exceeds some acceptable level of performance Q_0. Assume that $Q(t)$ is strictly decreasing and that $Q(\infty) < Q_0 < Q(0)$. Similar to Equation (4.33), the stationary probability of exceeding level Q_0 is

$$P_S(Q_0) = \frac{1}{m}\int_0^{t_0} \overline{F}(x))dx\,, \tag{4.34}$$

where t_0 is uniquely determined from the equation $Q(t_0) = Q_0$.

Example 4.4 Let $F(t) = 1 - \exp\{-\lambda t\}; Q(t) = \exp\{-\alpha t\}, \alpha > 0$. Then

$$Q_{ES} = \frac{\lambda}{\lambda + \alpha}, \tag{4.35}$$

$$P_S(Q_0) = \lambda \int_0^{-\frac{\ln Q_0}{\alpha}} \exp\{-\lambda x\}dx = 1 - Q_0^{\frac{\lambda}{\alpha}}\,. \tag{4.36}$$

Let $Q_t, t \geq 0$ be a stochastic process with decreasing continuous realizations and let it be independent from the considered renewal process of system failures (repairs). Equations (4.33) and (4.34) are generalized in this case to

$$Q_{ES} = \frac{1}{m}\int_0^{\infty} \overline{F}(x) E[Q_x]dx \tag{4.37}$$

and

$$P_S(Q_0) = \frac{1}{m}\int_0^{\infty} \overline{F}(x)\Pr[Q_x \geq Q_0]dx\,, \tag{4.38}$$

respectively. For obtaining $P_S(Q_0)$, we need the distribution of the first passage time $S(x, Q_0)$ *i.e.*, the distribution function of time to the first crossing of level Q_0. Therefore,

$$P_S(Q_0) = \frac{1}{m}\int_0^{\infty} \overline{F}(x)(1 - S(x, Q_0))dx\,.$$

Example 4.5 Let $F(t) = 1 - \exp\{-\lambda t\}$, $Q(t, Z) = 1 - \exp\{-Zt\}$, where the random variable Z is uniformly distributed in $[0, a]$, $a > 0$. Then

$$S(t, Q_0) = \begin{cases} 0, & t \leq d \\ 1 + \dfrac{\ln Q_0}{at}, & t > d \end{cases},$$

where $d = -\ln Q_0 / a$. Finally,

$$P_S(Q_0) = 1 - (Q_0)^{\frac{\lambda}{a}} + \lambda d \int_d^{\infty} \frac{\exp\{-\lambda x\}}{x} dx\,.$$

Remark 4.3 The discussion in this section can be considered a special case of the renewal reward processes (Ross, 1996).

4.6.2 Imperfect Restoration of Quality

The results of the previous section were obtained under the assumption that the repair action is perfect. Therefore, after the perfect repair of the described type, the system is in an *as good as new* state: the Cdf of the current cycle duration is the same as for the previous cycle and the quality of the performance function is also the same at each cycle.

Following Finkelstein (1999), consider now a generalization of the Brown–Proschan model of Section 4.5. As in this model, the perfect repair performs the renewal in a statistical sense and restores the quality function to its initial level $Q(0)$, whereas the minimal repair, defined in statistical terms by Definition 4.7, performs this restoration to a lower (intermediate) level to be specified later. We will call this type of repair the *minimal-imperfect repair*: it is minimal with respect to the cycle distribution function and is imperfect with respect to the quality function. As a special case, the quality function could be restored to the level it was at just prior to the failure (*minimal-minimal repair*), but a more general situation is of interest.

We will combine the results of Sections 4.5 and 4.6.1. Equation (4.30) defines the Cdf of the time between consecutive perfect repairs. Therefore, the renewal process of instants of perfect repairs is defined by the interarrival times with the Cdf $F_p(t)$. We will consider only the stationary value of the quality function in this case, but an analogue of Equation (4.32) can also be derived easily.

It follows from Equations (4.30) and (4.33) that the stationary value of the quality function is

$$Q_{ES} = \frac{1}{m_p} \int_0^{\infty} \exp\left\{ - \int_0^x p(u)\lambda(u)du \right\} E[\hat{Q}(x)]dx , \tag{4.39}$$

where m_p is the mean defined by the Cdf $F_p(t)$ and $\hat{Q}(x)$ is the value of the performance function in x units of time after the last perfect repair. This function is now random, as a random number of minimal-imperfect repairs was performed since the last perfect repair. Different reasonable models for $\hat{Q}(x)$ can be suggested (Finkelstein, 1999). The following model is already defined in terms of the corresponding expectation and is probably the simplest:

$$E[\hat{Q}(x)] = \exp\left\{ - \int_0^x \lambda(u)du \right\} Q(x) + \int_0^x \lambda(y) \exp\left\{ - \int_y^x \lambda(u)du \right\} Q(x,y)dy . \tag{4.40}$$

The first term on the right-hand side of Equation (4.40) corresponds to the event when there are no minimal repairs in $[0, x)$. The integrand of the second term defines the probability that the last minimal-imperfect repair occurred in $[y, y + dy)$, multiplied by a quality function $Q(x, y)$, which depends now on the time since the last perfect repair x and on the time of the last minimal-imperfect repair y. The simplest model for $Q(x, y)$ is

$$Q(x,y)=\frac{C(y)}{Q(0)}Q(x-y), \tag{4.41}$$

where $C(y)$ is the level of the minimal-imperfect repair performed at time y after the last perfect repair. We also assume that the function $C(y)$ is monotonically decreasing and $C(y)>Q(y);\ y>0;\ C(0)=Q(0)$.

Example 4.6 Let $Q(x)=\exp\{-\alpha_1 x\};\quad C(y)=\exp\{-\alpha_2\},\quad \alpha_1>\alpha_2$. Then

$$Q(x,y)=\exp\{-\alpha_1 x\}\exp\{-(\alpha_1-\alpha_2)y\}.$$

Let $\lambda(x)\equiv\lambda$ and $p(x)\equiv p$. Performing simple calculations in accordance with Equations (4.39)–(4.41) results in

$$Q_{ES}=\frac{\lambda p}{\alpha_1-\alpha_2-\lambda}\left[\frac{\alpha_1-\alpha_2}{\lambda+\lambda p+\alpha_1}-\frac{\lambda}{2\alpha_1-\alpha_2+\lambda p}\right]. \tag{4.42}$$

If $\alpha_1=\alpha_2=\alpha$ and $p=1$, Equation (3.42) reduces to $Q_{ES}=\lambda/(\lambda+\alpha)$, which coincides with Equation (4.35).

Similar to Equation (4.38), the stationary probability of exceeding the fixed level Q_0 can also be derived (Finkelstein, 1999).

4.7 Minimal Repair in Heterogeneous Populations

Chapters 6 and 7 of this book are entirely devoted to mixture failure rate modelling in heterogeneous populations. The discussion of minimal repair in this section is based on definitions and results for mixture failure rates of Chapter 6, which are essential for the presentation in this section. Therefore, it is reasonable to read Chapter 6 first. Some of the relevant equations were also given in the introductory Example 3.1. Note that generalization of the notion of minimal repair to the heterogeneous setting is not straightforward, and we present here only some initial findings (Finkelstein, 2004c).

For explanatory purposes, we start with the following reasoning. Consider a stock of n substocks of 'identical' items, which are manufactured by n different manufacturers, and therefore their failure rates $\lambda_i, i=1,2,...,n$ differ. Assume that at $t=0$ one item is picked up from a randomly chosen (in accordance with some discrete distribution) substock. It is put into operation, whereas all other items are kept in a 'hot' standby. It is clear that the lifetime Cdf of the chosen item can be defined by the corresponding discrete mixture. The following scenarios for repair (replacement) actions are of interest:

- We do not (or cannot) observe the choice (the manufacturer, or equivalently, the value of λ_i). An operating item is replaced on failure by the standby one, which is chosen in accordance with the same random procedure (as at $t=0$);

- The same as in the first scenario, but the failed item is replaced with one of the same make;
- The initial choice is observed as we 'observe' i, and therefore we 'know' λ_i and use items from this stock for replacements.

Thus, we have described three types of minimal repair for heterogeneous population to be described mathematically in what follows.

Consider an item with the Cdf $F_m(t)$ defined by Equation (6.4) that describes a lifetime in a heterogeneous population. Let $S_1 = t_1$ be the realization of the time to the first failure (repair). Then the (usual) minimal repair is obviously defined by Equation (4.27), where $F(t)$ is substituted by $F_m(t)$ and x by t_1, whereas the process of minimal repairs of this kind is a NHPP with rate $\lambda_m(t)$. This is a continuous version of the first scenario of the above reasoning.

It is much more interesting to define the *information-based minimal* repair for the heterogeneous setting. In accordance with the general definition of the information-based minimal repair, an object is restored to the 'defined' state it had been in just prior to the failure. It is reasonable to assume in this case that the state is defined by the value of the frailty parameter Z. As we observe only the failures at arrival times $S_i, i = 1,2,...$, the intensity process in $[0,t_1)$ is deterministic and is equal to the mixture failure rate $\lambda_m(t)$ defined by Equation (6.5). Denote this function in $[0,t_1)$ by $\lambda_m(t) \equiv \lambda_m^1(t)$. As the unobserved $Z = z$ 'was chosen' at $t = 0$, the *information-based minimal* repair restores it to the state defined by $Z = z$. This means that the intensity process in $[t_1,t_2)$ is

$$\lambda_m^2(t, t-t_1) = \int_a^b \lambda(t,z)\pi(z \mid t-t_1)dz, \tag{4.43}$$

where the mixing density $\pi(z \mid t-t_1)$ is given by the adjusted Equation (3.10) in the following way:

$$\pi(z \mid t-t_1) = \pi(z)\frac{\bar{F}(t-t_1,z)}{\int_a^b \bar{F}(t-t_1,z)\pi(z)dz}. \tag{4.44}$$

The fact that Z is unobserved does not prevent us from performing and interpreting the information-based minimal repair of the described type. Similar to the (usual) minimal repair case, we can substitute the failed object by the statistically identical one which had also started operating at $t = 0$ and did not fail in $[0,t_1)$. The term "statistically identical" means the same Cdf $F(t,z)$ in this case.

In accordance with Equations (4.43) and (4.44), the corresponding intensity process is

$$\lambda_t = \sum_{n=1}^{\infty} \lambda_m^n(t, t-S_{n-1}) I(S_{n-1} \le t < S_n), \quad S_0 = 0, \tag{4.45}$$

where

$$\lambda_m^n(t, t - S_{n-1}) = \int_a^b \lambda(t, z)\pi(z \mid t - S_{n-1})dz\,. \tag{4.46}$$

Note that, as $\pi(z \mid 0) \equiv \pi(z)$, the intensity process (4.45) is equal at failure (renewal) points to the 'unconditional mean' of $\lambda(t, Z)$, *e.g.*,

$$\lambda_m^n(S_n, 0) = \int_a^b \lambda(S_n, z)\pi(z)dz\,.$$

Therefore, the function

$$\lambda_p(t) = \int_a^b \lambda(t, z)\pi(z)dz\,,$$

which defines some 'unconditional mixture failure rate', is important for describing the model under investigation. The subscript " P ", as in Chapter 6, here stands for "Poisson", as this equation defines the mean intensity function for the doubly stochastic Poisson process (Cox and Isham, 1980).

The model defined by the mixture failure rate $\lambda_P(t)$ is relevant when Z is observed, and this corresponds to the last scenario in our introductory reasoning. The following examples (Finkelstein, 2004) deal with comparison of $\lambda_m(t)$, $\lambda_P(t)$ and λ_t.

Example 4.7 Let $F(t, z)$ be an exponential distribution with the failure rate $\lambda(t, z) = z\lambda$ and let $\pi(z)$ be an exponential density in $[0, \infty)$ with parameter ϑ. Therefore, $\lambda_m(t) = \lambda/(\lambda t + \vartheta)$, which is a special case of Equation (3.11). It can easily be seen that $\lambda_P(t) = \lambda/\vartheta$. The corresponding intensity process is

$$\lambda_t = \sum_{n=1}^{\infty} \frac{\lambda}{\lambda(t - S_{n-1}) + \vartheta} I(S_{n-1} \le t < S_n), \quad S_0 = 0.$$

Thus,

$$\lambda_m(t) \le \lambda_t \le \lambda_P(t), t > 0 \tag{4.47}$$

and $\lambda_t = \lambda_P(t)$ only at failure points $S_n, n \ge 1$, whereas $\lambda_t = \lambda_m(t)$ in $[0, S_1)$.

The failure rates $\lambda(t, z)$ in the previous example were ordered in z, *i.e.*, the larger value of z corresponds to the larger value of $\lambda(t, z)$ for all $t \ge 0$. The following example shows that Relationship (4.47) does not hold when the failure rates are not ordered in the described sense.

Example 4.8 Consider a simple case of a discrete mixture of two distributions with periodic failure rates:

$$\lambda_1(t) = \begin{cases} \lambda, & 0 \le t < a \\ 2\lambda, & a \le t < 2a \\ \ldots \end{cases}, \qquad \lambda_2(t) = \begin{cases} 2\lambda, & 0 \le t < a \\ \lambda, & a \le t < 2a \\ \ldots \end{cases},$$

where $2a > 0$ is a period. Therefore, these failure rates are not ordered. Assume that the discrete mixing distribution is defined by the probabilities $P(Z = z_1) = P(Z = z_2) = 0.5$. Thus, the function $\lambda_P(t)$ is a constant: $\lambda_P(t) = 1.5\lambda$. The corresponding mixture failure rate $\lambda_m(t)$ is also a periodic function with the period $2a$ and is defined in $[0,2a)$ as

$$\lambda_m(t) = \begin{cases} \dfrac{\lambda + 2\lambda\exp\{-\lambda t\}}{1+\exp\{-\lambda t\}}, & 0 \le t < a, \\ \dfrac{2\lambda + \lambda\exp\{-2\lambda a\}\exp\{\lambda t\}}{1+\exp\{-2\lambda a\}\exp\{\lambda t\}}, & a \le t < 2a. \end{cases}$$

It can be shown that the inequality $\lambda_m(t) < \lambda_P(t),\ t > 0$ ($\lambda_m(0) = 1.5\lambda$) does not hold in this case.

4.8 Chapter Summary

Performance of repairable systems is usually described by renewal processes or alternating renewal processes. Therefore, a repair action in these models is considered to be perfect, *i.e.*, returning a system to an *as good as new* state. This assumption is not always true, as repair in real life is usually imperfect. The minimal repair is the simplest case of imperfect repair and we consider this topic in detail. It restores a failed system to the state it was in just prior to a failure. We discuss several types of minimal repair that are defined by a different meaning of "the state just prior to repair". An information-based minimal repair, for example, takes into account the real (not statistical) state of a system on failure, and this creates a basis for more adequate modelling. In the last section, we consider the minimal repair in heterogeneous populations when there are different possibilities for defining this repair action.

Instants of repair in technical systems can be considered as points of the corresponding point process. Therefore, the first part of this chapter is devoted to a brief, necessary introduction to the theory of point processes. We focus on a description of the renewal-type processes keeping in mind that the recurring theme in this book is the importance of the complete intensity function (4.4) or, equivalently, of the intensity process (4.2).

5

Virtual Age and Imperfect Repair

5.1 Introduction – Virtual Age

In accordance with Equation (2.7), the MRL function of a non-repairable object $m(t)$ is defined by the Cdf $F(x)$ and the current time t. Therefore, the 'statistical' state of an operating item with a given Cdf is defined by t. What happens for a repairable item? Sections 5.2–5.6 of this chapter answer this question. We will show that the notion of *virtual age*, to be defined later, will be a substitute for t in this case. Note that our discussion of this notion will combine 'physical' reasoning (sometimes heuristic) with the corresponding probabilistic modelling.

Let a repairable item start operating at time $t = 0$. As usual, we assume (for simplicity) that repair is instantaneous. Generalization to the non-instantaneous case is straightforward. The time t since an item started operating will be called the *calendar* (chronological) age of the repairable item. We will assume usually that an item is deteriorating in some suitable stochastic sense, which is often manifested by an increasing failure rate $\lambda(t)$ or by a decreasing MRL function at each cycle. As in the previous chapter, by cycle we mean the time between successive repairs. In contrast to the calendar age t, it is reasonable to consider an age that describes in probabilistic terms the state of a repairable item at each calendar instant of time. It is clear that this age should depend at least on the moments and quality of previous repairs. It is also obvious that both ages coincide for non-repairable items.

If the repair is perfect, this 'new' age is just the time elapsed since the last repair, as in the case of renewal processes defined by stochastic intensity (4.15). Minimal repair does not change the statistical state of an item, and therefore, as in the non-repairable case, this age is equal to the calendar age t. As follows from Section 4.3.1, the instants of minimal repair follow the NHPP defined by deterministic stochastic intensity (4.5).

Various models can be suggested for defining the corresponding 'equivalent' age of a repairable item when a repair is imperfect in a more general sense. In accordance with the established terminology, we will call it the *virtual* age. A more suitable term would probably be the *real* age, as it is defined by the *real* state of an item (*e.g.*, by a level of deterioration). The term *virtual age* was suggested by Kijima (1989) (see also Kijima *et al.*, 1988) for a meaningful, specific model of imperfect repair, but we will use it in a broader sense. An important feature of this model is the assumption that the repair action does not change the baseline Cdf $F(x)$ (or the baseline failure rate $\lambda(x)$) and only the 'initial time' changes after each repair. Therefore, the Cdf of a lifetime after repair in Kijima's model is defined as a remaining lifetime distribution $F(x \mid t)$. Note that there is no change in the initial age after minimal repair and that it is 0 after each perfect repair. A similar model was independently developed by Finkelstein (1989).

The virtual age concept can be relevant for stochastic modelling of non-repairable items as well, but in this case we must compare the states of identical items operating in different environments. Assume, for example, that the first item is operating in a baseline (reference) environment and the second (identical) item is operating in a more severe environment. It seems natural to define the virtual age of the second item *via* the comparison of its level of deterioration with the deterioration level of the first item. If the baseline environment is 'equipped' with the calendar age, then it is reasonable to assume that the virtual age of an item in the second environment, which was operating for the same amount of time as the first one, is larger than the corresponding calendar age. In Section 5.1, we develop formal models for the described age correspondence. Some results of this section will be used in other sections devoted to repairable items modelling. However, it should be noted that the repairable item is operating in one fixed environment and its virtual age depends on the quality of repair actions.

Remark 5.1 Several qualitative approaches to understanding and describing the notion of biological age, which is, in fact, a synonym to virtual age, have been developed in the life sciences (see, *e.g.*, Klemera and Daubal, 2006 and references therein). These authors write: "The concept of biological age can be found in the literature throughout the last 30 years. Unfortunately, the concept lacks a precise and generally accepted definition. The meaning of biological age is often explained as a quantity expressing the 'true global state' of an ageing organism better than the corresponding chronological age." If, for example, someone 50 years old looks like and has vital characteristics (blood pressure, level of cholesterol *etc.*) of a 'standard' 35-year-old individual, we can say that this observation indicates that his virtual (biological) age can be estimated 35. His lifestyle (environment, diet) is probably very healthy. These are, of course, rather vague statements, which will be made more precise in mathematical terms for some simple settings to be considered in this chapter and in Chapter 10.

Kijima's virtual age concept is not the only one used for describing imperfect repair modelling. For example, several failure rate reduction models are developed in the literature. In Section 5.5, we present a brief overview of these models and also perform a comparison with the age reduction (virtual age) models.

Most of the imperfect repair models can be used for modelling the corresponding imperfect maintenance actions. Note that repair is often called *corrective* or *unplanned* maintenance, whereas the scheduled actions are called *preventive* maintenance. Different combinations of imperfect (perfect) repair with imperfect (perfect) maintenance and various optimal maintenance policies have been considered in the literature. The interested reader is referred to a recent book by Wang and Pham (2006), where a detailed analysis of this topic with numerous references is given.

Remark 5.2 In this chapter, we do not consider statistical inference for imperfect repair modelling. The corresponding results can be found in Guo and Love (1992), Kaminskij and Krivtsov (1998, 2006), Dorado *et al.* (1997), Hollander and Sethuraman (2002), Kahle and Love (2003) and Kahle (2006), among others.

5.2 Virtual Age for Non-repairable Objects

Two main approaches to defining virtual age will be considered. The first one is based on an assumption that lifetimes in different environments are ordered in the sense of the (usual) stochastic ordering of Definition 3.4, which will also be interpreted *via* the accelerated life model. This reasoning helps in recalculating age when one regime (stress) is switched to another. In the second approach, an observed value of some overall parameter of degradation is compared with the expected value, and the information-based virtual age is defined on the basis of this comparison.

5.2.1 Statistical Virtual Age

Consider a degrading item that operates in a baseline environment and denote the corresponding Cdf of time to failure by $F_b(t)$. We will use the terms *environment, regime* and *stress* interchangeably. By "degrading" we mean that that the quality of performance of an item is decreasing in some suitable sense, *e.g.*, the corresponding wear is increasing or some damage is accumulating. We will implicitly assume that degradation or wear is additive, but formally the virtual age can be defined without this assumption.

Let another statistically identical item be operating in a more severe environment with the Cdf of time to failure denoted by $F_s(t)$. Assume for simplicity that environments are not varying with time and that distributions are absolutely continuous. Denote by $\lambda_b(t)$ and $\lambda_s(t)$ the failure rates in two environments, respectively. The time-dependent stresses can also be considered (Finkelstein, 1999a). We want to establish an age correspondence between the systems in two regimes by considering the baseline as a reference. It is reasonable to assume that degradation in the second regime is more intensive, and therefore the time for accumulating the same amount of degradation or wear is smaller than in the baseline regime. Therefore, in accordance with Definition 3.4, assume that the lifetimes in two environments are ordered in terms of (usual) stochastic ordering as

$$\bar{F}_s(t) < \bar{F}_b(t),\ \ t \in (0, \infty). \tag{5.1}$$

Note that this is our assumption. Although Inequality (5.1) naturally models the impact of a more severe environment, other weaker orderings can, in principle, describe probabilistic relationships between the corresponding lifetimes in two regimes (*e.g.*, ordering of the mean values, which, in fact, does not lead to the forthcoming results).

Inequality (5.1) implies the following equation:

$$F_s(t) = F_b(W(t)), \quad W(0) = 0, t \in (0, \infty), \tag{5.2}$$

where the function $W(t) > t$ is strictly increasing. The latter property obviously follows after applying the inverse function to both sides of (5.2), *i.e.*,

$$W(t) = F_b^{-1}(F_s(t))$$

and noting that the superposition of two increasing functions is also increasing. Equation (5.2) can be interpreted as a general Accelerated Life Model (ALM) (Cox and Oakes, 1984; Meeker and Escobar, 1998; Finkelstein, 1999, to name a few) with a time-dependent scale-transformation function $W(t)$. As this function is differentiable, it can be interpreted as an additive cumulative degradation function:

$$W(t) = \int_0^t w(u)du, \tag{5.3}$$

where $w(t)$ has the same meaning as that of a degradation rate. Without losing generality, we assume for convenience that the degradation rate in the baseline environment is equal to 1. In fact, by doing this we define $W(t)$ and $w(t)$ as the *relative* cumulative degradation and the *relative* rate of degradation, respectively.

Definition 5.1. Let t be the calendar age of a degrading item operating in a baseline environment. Assume that ALM (5.2) describes the lifetime of another statistically identical item, which operates in a more severe environment for the same duration t.

Then the function $W(t)$ defines the *statistical* virtual age of the second item, or, equivalently, the inverse function $W^{-1}(t)$ defines the *statistical* virtual age of the first item when a more severe environment is set as the baseline environment.

This definition means that an item that was operating in a more severe environment for the time t 'acquires' the statistical virtual age $W(t) > t$. On the other hand, if we define a more severe regime as the baseline regime, the corresponding acquired statistical virtual age in a lighter regime would be $W^{-1}(t) < t$. This can easily be seen after substituting into Equation (5.2) the inverse function $W^{-1}(t)$ instead of t.

Definition 5.1 is, in fact, about the *age correspondence* of statistically identical items operating in different environments. When the failure rates or the corresponding Cdfs are given (or estimated from data), the ALM defined by (5.2) can be viewed as an equation for obtaining $W(t)$, *i.e.*,

$$\exp\left\{-\int_0^t \lambda_s(u)du\right\} = \exp\left\{-\int_0^{W(t)} \lambda_b(u)du\right\}$$

$$\Rightarrow \int_0^t \lambda_s(u)du = \int_0^{W(t)} \lambda_b(u)du. \tag{5.4}$$

Hence, the statistical virtual age $W(t)$ is uniquely defined by Equation (5.4). Similar to (5.4), the 'symmetrical' statistical virtual age $W^{-1}(t)$ is obtained from the following equation:

$$\int_0^t \lambda_b(u)du = \int_0^{W^{-1}(t)} \lambda_s(u)du .$$

Remark 5.3 Equation (5.4) can be interpreted in terms of the cumulative exposure model (Nelson, 1990), *i.e.*, the virtual age $W(t)$ 'produces' the same population cumulative fraction of units failing in a more severe environment as the age t does in the baseline environment (see also the next section). This age (time) correspondence concept was widely used in the literature on accelerated life testing. However, it does not necessarily lead to our degradation-based virtual age, but just defines the time (age) correspondence in different regimes based on equal probabilities of failure.

The problem of age correspondence for different populations is very important in demographic applications, especially for modelling possible changes in the retirement age. Populations in developed countries are ageing, which means that the proportion of old people is increasing. Therefore, the increase in the retirement age from 65 to 65+ has already been considered as an option in some of the European countries. Equation (5.4) can be used for the corresponding modelling of two populations: one with the 'old' mortality rate $\lambda_s(t)$ and the other the contemporary mortality rate $\lambda_b(t)$. As $\lambda_b(t) < \lambda_s(t),\ t > 0$, the value $W(65) > 65$ obtained from Equation (5.4) defines the new retirement age. Other approaches to the age correspondence problem in demography are considered, for example, in Denton and Spencer (1999).

Example 5.1 Let the failure rates in both regimes be increasing, positive power functions (the Weibull distributions), which are often used for lifetime modelling of degrading objects, *i.e.*,

$$\lambda_b(t) = \alpha t^{\beta} \ , \ \lambda_s(t) = \mu t^{\eta} \, , \ \alpha, \beta, \mu, \eta > 0 \, .$$

The statistical virtual age $W(t)$ is defined by Equation (5.4) as

$$W(t) = \left(\frac{\mu(\beta+1)}{\alpha(\eta+1)} \right)^{\frac{1}{\beta+1}} t^{\frac{\eta+1}{\beta+1}} .$$

In order for the inequality $W(t) > t$ to hold, the following restrictions on the parameters are sufficient: $\eta \geq \beta,\ \mu(\beta+1) > \alpha(\eta+1)$.

As follows from Equation (5.2), the failure rate that corresponds to the Cdf $F_s(t)$ is

$$\lambda_s(t) = \frac{dF_b(W(t))}{dt\bar{F}_b(W(t))} = w(t)\lambda_b(W(t)) . \tag{5.5}$$

If, for example, the failure rate in a baseline regime is constant, then $\lambda_s(t)$ is proportional to the rate of degradation $w(t)$.

Remark 5.4 The assumption of degradation is important for our model. The statistical virtual age is defined in (5.4) by equating the same amount of degradation in different environments. We implicitly assume that the accumulated failure rate is a measure of this degradation, which often (but not always) can be considered as a reasonably appropriate model.

5.2.2 Recalculated Virtual Age

The previous section was devoted to age correspondence in different environments. It is more convenient now to use the term *regime* instead of *environment*. What happens when the baseline regime is switched to a more severe one? The answer to this question is considered in this section.

Let an item start operating in a baseline regime at $t = 0$, which is switched at $t = x$ to a more severe regime. In accordance with Definition 5.1, the statistical virtual age immediately after the switching is $V_x = W^{-1}(x)$, where the new notation V_x is used for convenience.

Assume now that the governing Cdf after the switching is $F_s(t)$ and that the Cdf of the remaining lifetime is $F_s(t \mid V_x)$, *i.e.*,

$$F_s(t \mid V_x) = 1 - \frac{\bar{F}_s(t + V_x)}{\bar{F}_s(V_x)}, \tag{5.6}$$

as defined by Equation (2.7). Thus, an item starts operating in the second regime with a starting age V_x defined with respect to the Cdf $F_s(t)$. Note that the form of the lifetime Cdf after the switching given by Equation (5.6) is our assumption and that it does not follow directly from ALM (5.2). In general, the starting age could differ from V_x, or (and) the governing distribution could differ from $F_s(t)$. Alternatively, we can proceed starting with ALM (5.2) and obtain the Cdf of an item's lifetime for the whole interval $[0, \infty)$, and this will be performed in what follows.

According to our interpretation of the previous section, the rate of degradation is 1 in $t \in [0, x)$. Assume that the switching at $t = x$ results in the rate $w(t) > 1$ in $[x, \infty)$, where $w(t)$ is defined by ALM (5.2) and (5.3). Note that this is an important assumption on the nature of the impact of regime switching in the context of the ALM.

Remark 5.5 An alternative option, which is not discussed here, is the jump from the curve $\lambda_b(t)$ to the curve $\lambda_s(t)$ at $t = x$. This option can be interpreted in terms of the proportional hazards model, which is usually not suitable for lifetime modelling of degrading objects (Bagdonavicius and Nikulin, 2002).

Under the stated assumptions, the item's lifetime Cdf in $[0, \infty)$, to be denoted by $F_{bs}(t)$, can be written as (Finkelstein, 1999)

$$F_{bs}(t) = \begin{cases} F_b(t), & 0 \le t < x, \\ F_b\left(x + \int_x^t w(u))du\right), & x \le t < \infty. \end{cases} \tag{5.7}$$

Transformation of the second row on the right-hand side of this equation results in

$$\begin{aligned} F_b\left(x + \int_x^t w(u)du\right) &= F_b\left(\int_{\tau(x)}^t w(u))du\right) \\ &= F_b\big(W(t) - W(\tau(x))\big), \end{aligned} \tag{5.8}$$

where $\tau(x) < x$ is uniquely defined from the equation

$$x = \int_{\tau(x)}^x w(u)du = W(x) - W(\tau(x)). \tag{5.9}$$

It follows from Equation (5.9) that the cumulative degradation in $[\tau(x), x)$ in the second regime is equal to the cumulative degradation in the baseline regime in $[0, x)$, which is x. Therefore, the age of an item just after switching to a more severe regime can be defined as $\tilde{V}_x = x - \tau(x)$. Let us call it the *recalculated virtual age.*

Definition 5.2. Let a degrading item start operating at $t = 0$ in the baseline regime and be switched to a more severe regime at $t = x$. Assume that the corresponding Cdf in $[0, \infty)$ is given by Equation (5.7), which follows from ALM (5.2) and (5.3).

Then the recalculated virtual age $\tilde{V}_x$ after switching at $t = x$ is defined as $x - \tau(x)$, where $\tau(x)$ is the unique solution to Equation (5.9).

Remark 5.6 It can be shown that $\tilde{V}_x$ uniquely defines the state of an item in the described model only for linear $W(t)$. For a general case, the vector $(\tilde{V}_x, \tau(x))$ should be considered.

We are now interested in comparing the statistical virtual age V_x with the recalculated virtual age $\tilde{V}_x$ and will show that under certain assumptions these quantities are equal. Equation (5.9) has the following solution:

$$\tau(x) = W^{-1}(W(x) - x).$$

As $V_x = W^{-1}(x)$, the equation $V_x = \tilde{V}_x$ can be written in the form of the following functional equation:

$$x - W^{-1}(x) = W^{-1}(W(x) - x).$$

Applying operation $W(\cdot)$ to both parts of this equation gives

$$W(x - W^{-1}(x)) = W(x) - x .$$

It is easy to show (see also Example 5.2) that the linear function $W(t) = wt$ is a solution to this equation. It is also clear that it is the unique solution, as the functional equation $f(x+y) = f(x) + f(y)$ has only a linear solution. Therefore, the recalculated virtual age in this case is equal to the statistical virtual age. The following example shows that the function defined by the second row in the right-hand side of Equation (5.7) is a *segment* of the Cdf $F_s(t)$ for $t \geq x$ only for this specific linear case.

Example 5.2 In accordance with Equations (5.2) and (5.8),

$$F_b(w \cdot (t - \tau(x))) = F_s(t - \tau(x)) ,$$

where $\tau(x)$ is obtained from a simplified version of Equation (5.8), *i.e.*,

$$x = \int_{\tau(x)}^{x} w du \Rightarrow \tau(x) = \frac{x(w-1)}{w}$$

and

$$\tilde{V}_x = x - \tau(x) = x / w ,$$

$$V_x = W^{-1}(x) = x / w .$$

Note that the virtual age in this case does not depend on the distribution functions. It also follows from this example that the Cdf $F_{bs}(t)$ for the linear $W(t)$ can be defined in the way most commonly found in the literature on accelerated life testing (*e.g.*, Nelson, 1990; Meeker and Escobar, 1998), *i.e.*,

$$F_{bs}(t) = \begin{cases} F_b(t), & 0 \leq t < x, \\ F_s(t - \tau(x)), & x \leq t < \infty. \end{cases}$$

This Cdf can be equivalently written as

$$F_{bs}(t) = \begin{cases} F_b(t), & 0 \leq t < x, \\ F_s(t - x + \tilde{V}_x), & x \leq t < \infty. \end{cases}$$

The Cdf of the remaining time at $t = x$, in accordance with this equation, is

$$\frac{F_s(t - x + \tilde{V}_x) - F_b(t)}{\bar{F}_b(t)} = F_s(t' \mid V_x) ,$$

where the notation $t-x \equiv t' \geq 0$ and equations $F_b(x) = F_s(V_x)$, $V_x = \widetilde{V}_x$ were used. Therefore, the remaining lifetimes obtained *via* the rate-of-degradation concept and *via* Equation (5.6) are equal for the linear scale function $W(t) = wt$. Moreover, the Cdf after switching is just the shifted $F_s(t)$ in this particular case. The failure rate that corresponds to the Cdf $F_{bs}(t)$ is

$$\lambda_{bs}(x) = \begin{cases} \lambda_b(t), & 0 \leq t < x, \\ \lambda_s(t-\tau(x)) = \lambda_s(t-x+V_x), & x \leq t < \infty. \end{cases}$$

This form of the failure rate often defines the 'Sedjakin Principle' (Bagdonavicius and Nikulin, 2002; Finkelstein, 1999a). In his original seminal work, Sedjakin (1966) defines the notion of a *resource* in the form of a cumulative failure rate. He assumes that after switching, the operation of the item depends on the history only *via* this resource and does not depend on how it was accumulated. This assumption, in fact, leads to Equation (5.4), which describes the equality of resources for different regimes, and eventually to the definition of the virtual age in our sense of the term. This paper played an important role in the development of accelerated life testing as a field. For example, the cumulative exposure model of Nelson (1990) is a reformulation of the Sedjakin Principle.

When $W(t)$ is a non-linear function, the statistical virtual age $V_x = W^{-1}(x)$ is not equal to the recalculated virtual age $\widetilde{V}_x = x - \tau(x)$, and the second row in the right-hand side of Equation (5.7) cannot be transformed into a segment of the Cdf $F_s(t)$. Therefore, the appealing virtual age interpretation of the age recalculation model with a governing Cdf $F_s(t)$ no longer exists in the described simple form. Note that we can still formally define a different Cdf after switching and the corresponding virtual age as a starting age for this distribution, but this approach needs more clarification and additional assumptions (Finkelstein, 1997).

The considered virtual age concept makes sense only for degrading items. Assume now that an item is not degrading and is described by exponential distributions in both regimes, *i.e.*,

$$\overline{F}_b(t) = \exp\{-\lambda_b t\},\ \overline{F}_s(t) = \exp\{-\lambda_s t\},\quad \lambda_b < \lambda_s.$$

Equation (5.1) holds for this setting, and therefore, taking into account (5.4), the scale transformation is also linear, *i.e.*, $W(t) = wt$, where $w = \lambda_s / \lambda_b$. We can formally define V_x and $\widetilde{V}_x$, but these quantities now have nothing to do with the virtual age concept, as they describe only the correspondence between the times of exposure in the two regimes (Nelson, 1990). Therefore, the increasing with time cumulative failure rate is not a good choice for 'resource function' in this case. A possible alternative approach dealing with this problem is based on considering the decreasing MRL function as a measure of degradation. The corresponding recalculated virtual age can also be defined for this setting (Finkelstein, 2007a).

Remark 5.7 The virtual age concept of this section can also be applied to repairable systems. Keeping the notation but not the literal meaning, assume that initially the lifetime of a repairable item is characterized by the Cdf $F_b(t)$ and the imperfect repair changes it to $F_s(t \mid V_x)$, where V_x is the virtual age just after repair at $t = x$.

The special case $F_s(t) = F_b(t)$ will be the basis for age reduction models of imperfect repair to be considered later in this chapter. Thus, we have two factors that define a distribution after repair. First, the imperfect repair changes the Cdf from $F_b(t)$ to $F_s(t)$, and it is reasonable to assume that the corresponding lifetimes are ordered as in (5.1). As an option, parameters of the Cdf $F_b(t)$ can be changed by the repair action. If, *e.g.*, $F_b(t) = 1 - \exp\{-\lambda t^{\alpha}\}; \lambda, \alpha > 0$ is a Weibull distribution, then a smaller value of parameter λ will result in (5.1). Secondly, the model includes the virtual age V_x as the starting (initial) age for an item described by the Cdf $F_s(t)$, which was called in Finkelstein (1997) "the hidden age of the Cdf after the change of parameters". This model describes the dependence between lifetimes before and after repair that usually exists for degrading repairable objects. If $V_x = 0$, the lifetimes are independent, but the model still can describe an imperfect repair action, as Ordering (5.1) holds. Specifically, the consecutive cycles of the geometric process of Section 4.3.3 present a relevant example.

5.2.3 Information-based Virtual Age

An item in the previous section was considered as a 'black box' and no additional information was available. However, deterioration is a stochastic process, and therefore individual items age differently. Observation of the state of an item at a calendar time t can give an indication of its virtual age defined by the level of deterioration. This reasoning is somehow similar to the approach used in Chapter 2 for describing the information-based MRL (Example 2.1) and in Chapter 4 for the information-based minimal repair (Section 4.4.2). Note that we discuss this topic here mostly on a heuristic level that can be made mathematically strict using an advanced theory of stochastic processes (Aven and Jensen, 1999).

We start with a meaningful reliability example that will help us to understand the notion of the information-based virtual age. The number of operating components in a system k at the time of observation t defines the corresponding level of deterioration in this example. We want to compare k with the expected number of operating components $D(t)$. Therefore, $D(t)$ is just a scale transformation of the calendar age t, whereas k is defined as the same scale transformation of the corresponding information-based virtual age.

Example 5.3 Consider a system of $n+1$ i.i.d. components (one operating at $t = 0$ and n standby components) with constant failure rates λ. Denote the system's lifetime random variable by T_{n+1}. The system lifetime Cdf is defined by the Erlangian distribution as

$$F_{n+1}(t) \equiv \Pr[T_{n+1} \le t] = 1 - \exp\{-\lambda t\} \sum_0^n \frac{(\lambda t)^i}{i!}$$

with the increasing failure rate

$$\lambda_{n+1}(t) = \frac{\lambda \exp\{\lambda t\} (\lambda t)^n / n!}{\exp\{-\lambda t\} \sum_0^n \frac{(\lambda t)^i}{i!}}.$$

For this system, the number of failed components observed at time t is a natural measure of accumulated degradation in $[0,t]$. In order to define the corresponding information-based virtual age to be compared with the calendar age t, consider, firstly, the following conditional expectation:

$$D(t) \equiv E[N(t) \mid N(t) \leq n] = \frac{\exp\{-\lambda t\}\sum_0^n i\frac{(\lambda t)^i}{i!}}{\exp\{-\lambda t\}\sum_0^n \frac{(\lambda t)^i}{i!}}, \tag{5.10}$$

where $N(t)$ is the number of events in $[0,t]$ for the Poisson process with rate λ. The function $D(t)$ is monotonically increasing, $D(0)=0$ and $\lim_{t\to\infty} D(t)=n$. The unconditional expectation $E[N(t)]=\lambda t$ is a linear function and exhibits a shape that is different from $D(t)$. The function $D(t)$ defines an average degradation curve for the system under consideration. If our observation $0 \leq k \leq n$, *i.e.*, the number of failed components at time t 'lies' on this curve, then the information-based virtual age is equal to the calendar age t.

Denote the information-based virtual age by $V(t)$ and define it (for the considered specific model) as the following inverse function:

$$V(t) = D^{-1}(k). \tag{5.11}$$

If $k = D(t)$, then $V(t) = D^{-1}(D(t)) = t$. Similarly,

$$k < D(t) \Rightarrow V(t) < t, \quad k > D(t) \Rightarrow V(t) > t,$$

which is illustrated by Figure 5.1.

The approach to defining the virtual age considered in Example 5.3 can be generalized to a monotone, smoothly varying stochastic process of degradation (wear). We also assume for simplicity that this is a process with independent increments, and therefore it possesses the Markov property.

Definition 5.3. Let $D_t, t \geq 0$ be a monotone, predictable, smoothly varying stochastic process of degradation with independent increments and a strictly monotone mean $D(t)$, and let d_t be its realization (observation) at calendar time t. Then the information-based virtual age at t is defined by the following function:

$$V(t) = D^{-1}(d_t). \tag{5.12}$$

Note that, in accordance with the corresponding definition (Aven and Jensen, 1999), the failure time of the system in Example 5.3 is a stopping time for the degradation process, as observation of this process indicates whether a failure had occurred or not. Definition 5.3 refers to the case of a stochastic process without a stopping time. However, if this is the case and the failure time T is a stopping time, this definition should be modified by using $E[D_t \mid T > t]$ instead of $D(t)$.

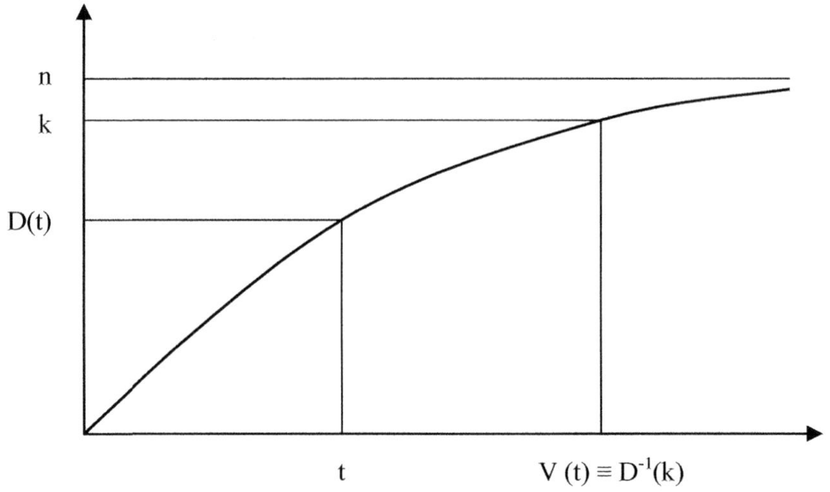

Figure 5.1. Degradation curve for the system with standby components

Remark 5.8 $V(t)$ is a realization of the corresponding information-based virtual age process $V_t, t \geq 0$ that can be defined as

$$V_t = D^{-1}(D_t).$$

The process $V_t - t$ shows the deviation of the information-based virtual age from the calendar age t.

An alternative way of defining the information-based virtual age $V(t)$ is *via* the information-based remaining lifetime (Example 2.1). The conventional mean remaining lifetime (MRL) $m(t)$ of an item with the Cdf $F(x)$ is defined by Equation (2.7). We will compare $m(t)$ with the information-based MRL denoted by $m_I(t)$. In this case, the observed level of degradation d_t is considered a new initial value for a corresponding degradation process. Therefore, $m_I(t)$ defines the mean time to failure for this setting. If $d_t = k$ is the number of failed components, as in Example 5.3, then $m_I(t) = (n+1-k)/\lambda$.

Definition 5.4. The information-based virtual age of a degrading system is given by the following equation:

$$V(t) = t + (m(t) - m_I(t)). \tag{5.13}$$

Thus, the information-based virtual age in this case is the chronological age plus the difference between the conventional and the information-based MRLs. It is clear that $V(t)$ can be positive or negative. If, *e.g.*, $m(t) = t_1 < t_2 = m_I(t)$, then $V(t) = t - (t_2 - t_1) < t$ and we have an additional $t_2 - t_1$ expected years of life of our system, as compared with the 'no information' version. It follows from Equation (2.9) that $dm(t)/dt > -1$, and therefore, under some reasonable assumptions, $m_I(t) - m(t) < t$ (Finkelstein, 2007). This ensures that $V(t)$ is positive.

Note that the meaning of Definition 5.4 is in adding (subtracting) to the chronological age t the gain (loss) in the remaining lifetime owing to additional information on the state of a degradation process at time t. The next example illustrates this definition.

Example 5.4 Consider a system of two i.i.d. components in parallel with exponential Cdfs. Then $\bar{F}(t) = \exp\{-2\lambda t\} - 2\exp\{-\lambda t\}$ and

$$\frac{1}{\lambda} < m(t) = \int_0^\infty \frac{2\exp\{-\lambda t\} - \exp\{-2\lambda t\}\exp\{-\lambda x\}}{2 - \exp\{-\lambda x\}} dx < \frac{1.5}{\lambda}.$$

If we observe at time t two operating components, then $m_I(t) > m(t)$, and the information-based virtual age in this case is smaller than the calendar age t. If we observe only one operating component, then $V(t) > t$.

We have discussed several different definitions of virtual age. The approach to be used usually depends on information at hand and the assumptions of the model. If there is no additional information and our main goal is to consider age correspondence for different regimes, then the choice is $W(t)$ of Definition 5.1. When there is a switching of regimes for degrading items, then a possible option is the recalculated virtual age of Definition 5.2. If the degradation curve can be modelled by an observed, monotone stochastic process and the criterion of failure is not well defined, then the first choice is Definition 5.3. Finally, if the failure time distribution of an item is based on a stochastic process with different initial values, and therefore the corresponding mean remaining lifetime can be obtained, then the information-based Definition 5.4 is preferable. These are just general recommendations. The actual choice depends on the specific settings.

5.2.4 Virtual Age in a Series System

In this section, possible approaches to defining the virtual age of a series system with different virtual ages of components will be briefly considered. In a conventional setting, all components have the same calendar age t, and therefore a similar problem does not exist, as the calendar age of a system is also t.

When components of a system can be characterized by virtual ages, it is really challenging in different applications (especially biological) to define the corresponding virtual age of a series system. For example, assume that there are two components in series. If the first one has a much higher relative level of degradation than the second component, the corresponding virtual ages are also different. Therefore, the virtual age of this system should be defined in some way. As usual, when we want to aggregate several measures into one overall measure, some kind of weighting of individual quantities should be used.

We start by considering the statistical virtual age discussed in Section 5.2.1. The survival functions of a series system of n statistically independent components in the baseline environment and in a more severe environment are

$$\bar{F}_b(t) = \prod_1^n \bar{F}_{bi}(t), \quad \bar{F}_s(t) = \prod_1^n \bar{F}_{bi}(W_i(t)),$$

respectively, where $W_i(t)$ is a scale transformation function for the i th component. We assume that Model (5.2) holds for every component. Thus, each component has its own statistical virtual age $W_i(t)$, whereas the virtual age for the system $W(t)$ is obtained from the following equation:

$$F_b(W(t)) = \prod_1^n F_{bi}(W_i(t))$$

or, equivalently, using Equation (5.4),

$$\int_0^{W(t)} \sum_1^n \lambda_{bi}(u)du = \sum_1^n \int_0^{W_i(t)} \lambda_{bi}(u)du. \qquad (5.14)$$

Example 5.5 Let $n = 2$. Assume for simplicity that $W_1(t) = t$ (which means, *e.g.*, that the first component is protected from the environment) and that the virtual age of the second component is $W_2(t) = 2t$. Therefore, the second component has a higher level of degradation. Equation (5.14) turns into

$$\int_0^{W(t)} (\lambda_{b1}(u) + \lambda_{b2}(u))du = \int_0^t \lambda_{bi}(u)du + \int_0^{2t} \lambda_{b2}(u)u.$$

Let the failure rates be linear, *i.e.*, $\lambda_{b1}(t) = \lambda_1 t$, $\lambda_{b2}(t) = \lambda_2 t$, $\lambda_1, \lambda_2 > 0$. Integrating and solving the simple algebraic equation gives

$$W(t) = \left(\sqrt{\frac{\lambda_1 + 4\lambda_2}{\lambda_1 + \lambda_2}} \right) t.$$

If the components are statistically identical in the baseline environment ($\lambda_1 = \lambda_2$), then

$$W(t) = \sqrt{5/2}\, t \approx 1.6t,$$

which means that the statistical virtual age of a system with chronological age t is approximately $1.6t$. The 'weight' of each component is eventually defined by the relationship between λ_1 and λ_2. When, *e.g.*, λ_1/λ_2 tends to 0, the statistical virtual age of a system tends to $2t$, *i.e.*, the statistical virtual age of the second component.

In order to define the information-based virtual age of a series system, we will weight the virtual ages of n degrading components in accordance with the reliability importance (Barlow and Proschan, 1975) of the components with respect to the failure of the system. Let $V_i(t), i = 1,2,...,n$ denote the information-based virtual age of the i th component with the failure rate $\lambda_i(t)$ in a series system of n statistically independent components. The virtual age of a system at time t can be defined as the expected value of the virtual age of the failed in $[t,t+dt)$ component, *i.e.*,

$$V(t)=\sum_{1}^{n}\frac{\lambda_i(t)}{\lambda_s(t)}V_i(t), \tag{5.15}$$

where $\lambda_s(t)=\sum_{1}^{n}\lambda_i(t)$ is the failure rate of the series system.

Similar to the previous section, the second approach is also based on the notion of the MRL function (Finkelstein, 2007).

5.3 Age Reduction Models for Repairable Systems

Our discussion of the virtual age concept in Section 5.2 was mostly based on the age recalculation technique for non-repairable items with a single regime change point. Remark 5.7 already presented some initial reasoning concerning the application of the virtual age concept to repairable objects. We now start with a description of several imperfect repair models, where each repair decreases the age of the operating item to a value always to be called the virtual age. When a repair is perfect, the virtual age is 0; when it is minimal, the virtual age is equal to the calendar age. Our interest is in intermediate cases. We study properties of the corresponding renewal-type processes and other relevant characteristics.

5.3.1 G-renewal Process

This model was probably the first mathematically justified virtual age model of imperfect repair, although the authors (Kijima and Sumita, 1986) considered it as a useful generalization of the renewal process not linking it directly with a process of imperfect repair. However, this link definitely exists and can be seen from the following example.

Example 5.6 Suppose that a component with an absolutely continuous Cdf $F(t)$ is supplied with an infinite number of 'warm standby' components with Cdfs $F(qt)$, where $0<q\leq 1$ is a constant. This system starts operating at $t=0$. The first component operates in a baseline regime, whereas the standby components operate in a less severe regime. Upon each failure in the baseline regime, the component is instantaneously replaced by a standby one, which is switched into operation in the baseline regime. Therefore, the calendar age of the standby component should be recalculated. This is exactly the setting considered in Example 5.2 with an obvious change of w to $1/q$, as the baseline regime is now more severe. Thus, the virtual age (which was called the recalculated virtual age in Section 5.2.2) V_x of a standby component that had replaced the operating one at $t=x$ is qx. The corresponding remaining lifetime Cdf, in accordance with Equation (2.7), is

$$F(t\mid V_x)=F(t\mid qx)=\frac{F(t+qx)-F(qx)}{\bar{F}(qx)}. \tag{5.16}$$

Note that Equation (5.16) is obtained using the age recalculation approach of Section 5.2.1, which is based on the specific linear case of Equation (5.2). When $q = 1$, (5.16) defines minimal repair; when $q = 0$, the components are in cold standby (perfect repair).

The age recalculation in this model is performed upon each failure. The corresponding sequence of interarrival times $\{X_i\}_{i\geq 1}$ forms a generalized renewal process. Recall that the cycles of the ordinary renewal process are i.i.d. random variables. In the g-renewal process, the duration of the $(n+1)$th cycle, which starts at $t = s_n \equiv x_1 + x_2 + ... + x_n$, $n = 0,1,2...$, $s_0 = 0$, is defined by the following conditional distribution:

$$\Pr[X_{n+1} \leq t] = F(t \mid qs_n),$$

where, as usual, s_n is a realization of the arrival time S_n.

An obvious and practically important interpretation of the model considered in Example 5.6 is when the standby components are interpreted as the spares for the initial component. The imperfect repair in this case is just an imperfect overhaul, as the spare parts are also ageing. Statistical estimation of q in this specific model was studied by Kaminskij and Krivtsov (1998, 2006).

We will now generalize Example 5.6 to the case of non-linear ALM (5.2). Let a failure, not necessarily the first one, occur at $t = x$. It is instantaneously imperfectly repaired. In accordance with Equation (5.6), the virtual age after the repair is $V_x = W^{-1}(x) \equiv q(x)$, where $q(x)$ is a continuous increasing function, $0 \leq q(x) \leq x$. As in Equation (5.16), the Cdf of the time to the next failure is $F(t \mid V_x)$. The most important feature of the model is that $F(t \mid V_x)$ depends only on the time x and not on the other elements of the history of the corresponding point process. This property makes it possible to generalize Equations (4.10) and (4.11) to the case under consideration. The point process of imperfect repairs $N(t), t \geq 0$, as in the case of an ordinary renewal process, is characterized by the corresponding renewal function $H(t) = E[N(t)]$ and the renewal density function $h(t) = H'(t)$.

The following generalizations of the ordinary renewal equations (4.10) and (4.11) can be derived:

$$H(t) = F(t) + \int_0^t h(x) F(t - x \mid q(x)) dx, \tag{5.17}$$

$$h(t) = f(t) + \int_0^t h(x) f(t - x \mid q(x)) dx, \tag{5.18}$$

where $f(t - x \mid q(x))$ is the density that corresponds to the Cdf $F(t - x \mid q(x))$.

The strict proof of these equations and the sufficient conditions for the corresponding unique solutions can be found in Kijima and Sumita (1986). This paper is written as an extension of the traditional renewal theory. On the other hand, Equation (5.18) has an appealing probabilistic interpretation, which can be considered a heuristic proof: as usual, $h(t)dt$ defines the probability of repair in $[t, t + dt)$. Using the law of total probability, we split this probability into the probability $f(t)dt$ that the first repair had occurred in $[t, t + dt)$ and the probability $h(x)dx$ that the last before t repair had occurred in $[x, x = dx)$ multiplied by the probability

$f(t-x \mid q(x))dt$ that the last repair had occurred in $[t, t+dt)$. Obviously, this product should be integrated from 0 to t. This brings us to Equation (5.18). Note that the ordinary renewal equation (4.11) also has the same interpretation. This can be seen after the corresponding change of the variable of integration, *i.e.*,

$$\int_0^t h(t-x)f(x)dx = \int_0^t h(x)f(t-x)dx. \tag{5.19}$$

Example 5.7 Let $q(x)=0$. Then $f(t-x \mid q(x)) = f(t-x)$. Taking into account (5.19), it is easy to see that Equation (5.18) becomes Equation (4.11). The same is true for Equation (5.18), which can be seen after changing the variable of integration on the right-hand side of Equation (4.10) and integrating by parts, *i.e.*,

$$\int_0^t H(t-x)f(x)dx = \int_0^t h(x)F(t-x)dx. \tag{5.20}$$

Example 5.8 Let $q(x)=x$ (the minimal repair). Equations (5.17) and (5.18) can be explicitly solved in this case. However, we will only show that the rate of the non-homogeneous Poisson process $\lambda_r(t)$, which is equal to the failure rate $\lambda(t)$ of the governing Cdf $F(t)$ (Section 4.3.1), is a solution to Equation (5.18). Taking into account that $h(t)=\lambda(t)$ and that

$$f(t-x \mid x)) = f(t)/\bar{F}(x),$$

$$(1/\bar{F}(x))' = \lambda(x)/\bar{F}(x),$$

the right-hand side of Equation (5.18) is equal to $\lambda(t)$, *i.e.*,

$$f(t) + \int_0^t h(x)f(t-x \mid q(x))dx = f(t) + f(t)\int_0^t \frac{\lambda(x)}{\bar{F}(x)}dx = \lambda(t),$$

as the process of minimal repairs is the NHPP.

A crucial feature of the g-renewal model is a specific simple dependence of the virtual age V_x after the repair on the chronological time $t=x$ only of this repair. This allows us to derive the renewal equations in the form given by Equations (5.17) and (5.18). Although these equations cannot be solved explicitly in terms of Laplace transforms, they are integral equations of the Volterra type and can be solved numerically.

In what follows we will consider models with a more complex dependence on the past.

5.3.2 'Sliding' Along the Failure Rate Curve

The g-renewal process of the previous section possesses another important feature. Each cycle of this renewal-type process is defined by the same governing Cdf

$F(t)$ with the failure rate $\lambda(t)$ and only the starting age for this distribution is given by the virtual age $V_x = q(x)$. Therefore, the cycle duration after the repair at $t = x$ is described by the Cdf $F(t \mid V_x)$. The formal definition of the g-renewal process can now be given *via* the corresponding intensity process.

Definition 5.5. The g-renewal process is defined by the following intensity process:

$$\lambda_t = \lambda(t - S_{N(t)} + q(S_{N(t)})), \tag{5.21}$$

where, as usual, $S_{N(t)}$ denotes the random time of the last renewal.

In the imperfect repair setting, $q(x)$ is usually a continuous, increasing function and $0 \le q(x) \le x$. When $q(x) = 0$, Equation (5.21) reduces to renewal intensity process (4.15), and when $q(x) = x$, we arrive at the rate of the NHPP. In the spare parts example, the function V_x is linearly increasing in x.

Thus, as in the case of an ordinary renewal process, the intensity process is defined by the same failure rate $\lambda(t)$, only the cycles now start with the initial failure rate $\lambda(q(S_{n(t)}),\ n(t) = 1,2,\ldots$.

One of the important restrictions of this model is the assumption of the 'fixed' shape of the failure rate. However, this assumption is well motivated, *e.g.*, for the spare-parts setting. Another strong assumption states that the future performance of an item repaired at $t = x$ depends on the history of a point process only *via* x. Therefore, we will keep the 'sliding along the $\lambda(t)$ curve' reasoning and will generalize it to a more complex case than the g-renewal case dependence on a history of the point process of repairs.

Assume that each imperfect repair reduces the virtual age of an item in accordance with some recalculation rule to be defined for specific models. As the shape of the failure rate is fixed, the virtual age at the start of a cycle is uniquely defined by the 'position' of the corresponding point on the failure rate curve after the repair. Therefore, Equation (5.21) for the intensity process can be generalized to

$$\lambda_t = \lambda(t - S_{N(t)} + V_{S_{N(t)}}), \tag{5.22}$$

where $V_{S_{N(t)}}$ is the virtual age of an item immediately after the last repair before t. From now on, for convenience, the capital letter V will denote a random virtual age, whereas v will denote its realization. Equation (5.22) gives a general definition for the models with a fixed failure rate shape. It should be specified by the corresponding virtual age, *e.g.*, as in Equation (5.21). In a rather general model considered by Uematsu and Nishida (1987), the virtual age in (5.22) was defined as an arbitrary positive and continuous function of all previous cycle durations and of the corresponding repair factors. These authors assumed that the function $q(x)$ is linear, *i.e.*, $q(x) = qx$ and that the repair factor q is different for different cycles. It is clear that one cannot derive useful properties from a general setting like this. The relevant special cases will be considered later in this section. It follows from Equation (5.22) that the intensity process between consecutive repairs can be 'graphically' described as *horizontally parallel* to the initial failure rate $\lambda(t)$ as all corresponding shifts are in the argument of the function $\lambda(t)$ (Doyen and Gaudoin, 2004, 2006).

Before considering specific models, we define a simple but important notion of a virtual age process, which will be used for discussing the ageing properties of the renewal-type processes.

Definition 5.6. Let the intensity process of the imperfect repair model be given by Equation (5.22). Then the corresponding virtual age process is defined by the following equation:

$$A_t = t - S_{N(t)} + V_{S_{N(t)}}. \tag{5.23}$$

It follows immediately from this definition and Equations (4.5) and (4.15) that the virtual age processes for the minimal repair and the ordinary renewal processes are

$$A_t = t, \tag{5.24}$$

$$A_t = t - S_{N(t)}, \tag{5.25}$$

respectively. Thus, as the shape of the failure rate is fixed, A_t is just a random argument for intensity process (5.22), *i.e.*, $\lambda_t = \lambda(A_t)$. Obviously, this process reduces to the virtual age $V_{S_{N(t)}}$ at the moments of repair $t = S_{N(t)}$. We now start describing some important specific models for $V_{S_{N(t)}}$. The following model (and its generalizations) is the main topic of the rest of this chapter.

Let an item start operating at $t = 0$. Therefore, the first cycle duration is described by the Cdf $F(t)$ with the corresponding failure rate $\lambda(t)$. Let the first failure (and the instantaneous imperfect repair) occur at $X_1 = x_1$. Assume that the imperfect repair decreases the age of an item to $q(x_1)$, where $q(x)$ is an increasing continuous function and $0 \le q(x) \le x$. Values exceeding x can also be considered, but for definiteness we deal with a model that decreases the age of a failed item. Thus the second cycle of the point process starts with the virtual age $v_1 = q(x_1)$ and the cycle duration X_2 is distributed as $F(t \mid v_1)$ with the failure rate $\lambda(t + v_1)$, $t \ge 0$. Therefore, the virtual age of an item just before the second repair is $v_1 + x_2$ and it is $q(v_1 + x_2)$ just after the second repair, where we assume for simplicity that the function $q(x)$ is the same at each cycle. The sequence of virtual ages after the i th repair $\{v_i\}_{i \ge 0}$ at the start of the $(i+1)$ th cycle in this model is defined for realizations x_i as

$$v_0 = 0,\ v_1 = q(x_1),\ v_2 = q(v_1 + x_2), \ldots,$$
$$v_i = q(v_{i-1} + x_i), \tag{5.26}$$

or, equivalently,

$$V_n = q(V_{n-1} + X_n),\ n \ge 1,$$

where the distributions of the corresponding interarrival times X_i are given by

$$F_i(t) \equiv F(t \mid v_{i-1}) = \frac{F(v_{i-1} + t) - F(v_{i-1})}{\bar{F}(v_{i-1})},\ i \ge 1. \tag{5.27}$$

For the specific linear case, $q(x) = qx,\ 0 < q < 1$, this model was considered on a descriptive level in Brown *et al.* (1983) and Bai and Jun (1986). Following the publication of the paper by Kijima (1989) it usually has been referred to as the Kijima II model, whereas the Kijima I model describes a somewhat simpler version of age reduction when only the duration of the last cycle is reduced by the corresponding imperfect repair (Baxter *et al.*, 1996; Stadje and Zuckerman, 1991). The latter model was first described by Malik (1979). The Kijima II model and its probabilistic analysis was also independently suggested in Finkelstein (1989) and later considered in numerous subsequent publications. We will give relevant references in what follows. The term 'virtual age' in connection with imperfect repair models was probably used for the first time in Kijima *et al.* (1988), but the corresponding meaning was already used in a number of publications previously.

When $q(x) = qx$, the intensity process λ_t can be defined in the explicit form. After the first repair the virtual age v_1 is qx_1, after the second repair $v_2 = q(qx_1 + x_2) = q^2 x_1 + qx_2, \ldots$, and after the nth repair the virtual age is

$$v_n = q^n x_1 + q^{n-1} x_2 + \ldots + qx_n = \sum_{i=0}^{n-1} q^{n-i} x_{i+1}, \tag{5.28}$$

where $x_i, i \geq 1$ are realizations of interarrival times X_i in the point process of imperfect repairs. Therefore, in accordance with the general Equation (5.22), the intensity process for this specific model with a linear $q(x) = qx$ is

$$\lambda_t = \lambda\left(t - S_{N(t)} + \sum_{i=0}^{N(t)-1} q^{n-i} X_{i+1}\right). \tag{5.29}$$

A similar equation in a slightly different form was obtained by Doyen and Gaudoin (2004). Note that the 'structure' of the right-hand side of Equation (5.29) in our notation explicitly defines the corresponding virtual age.

Example 5.9 Whereas the repair action in the Kijima II model depends on the whole history of the corresponding stochastic process, the dependence in the Kijima I model is simpler and takes into account the reduction of the last cycle increment only. Similar to (5.26),

$$v_0 = 0,\ \ v_1 = qx_1,\ \ v_2 = v_1 + qx_2, \ldots,$$
$$v_n = v_{n-1} + qx_n. \tag{5.30}$$

Therefore,

$$v_n = q(x_1 + x_2 + \ldots + x_n),\ V_n = q(X_1 + X_2 + \ldots + X_n),$$

and we arrive at the important conclusion that this is exactly the same model as the one defined by the g-renewal process of the previous section (Kijima *et al.*, 1988). These considerations give another motivation for using the Kijima I model for obtaining the required number of ageing spare parts. Moreover, Shin *et al.* (1996) had developed an optimal preventive maintenance policy in this case.

In accordance with Equations (5.22) and (5.30), the intensity process for this model is

$$\lambda_t = \lambda(t - S_{N(t)} + V_{S_{N(T)}}) = \lambda(t - S_{N(t)} + qS_{N(t)})$$
$$= \lambda(t - (1-q)S_{N(t)}) .$$

The obtained form of the intensity process suggests that the calendar age t is decreased in this model by an increment proportional to the calendar time of the last imperfect repair. Therefore, Doyen and Gaudoin (2004) call it the "arithmetic age reduction model".

The two types of the considered models represent two marginal cases of history for the corresponding stochastic repair processes, *i.e.*, the history that 'remembers' all previous repair times and the history that 'remembers' only the last repair time, respectively. Intermediate cases are analysed in Doyen and Gaudoin (2004). Note that, as q is a constant, the repair quality does not depend on calendar time, or on the repair number.

The original models in Kijima (1989) were, in fact, defined for a more general setting when the reduction factors $q_i, i \geq 1$ are different for each cycle (the case of independent random variables $Q_i, i \geq 1$ was also considered). The quality of repair that is deteriorating with i can be defined as $0 < q_1 < q_2 < q_3,...,$ which is a natural ordering in this case. Equation (5.28) then becomes

$$v_n = x_1 \prod_{i=1}^{n} q_i + x_2 \prod_{i=2}^{n} q_i + ... + q_n x_n = \sum_{i=1}^{n} x_i \prod_{k=i}^{n} q_k , \tag{5.31}$$

and the corresponding intensity process is similar to (5.29), *i.e.*,

$$\lambda_t = \lambda\left(t - S_{N(t)} + \sum_{i=1}^{N(t)} X_i \prod_{k=i}^{N(t)} q_k \right). \tag{5.32}$$

The virtual age in the Kijima I model is

$$v_0 = 0,\ \ v_1 = q_1 x_1,\ \ v_2 = v_1 + q_2 x_2, ...,$$
$$v_n = v_{n-1} + q_n x_n = \sum_{1}^{n} q_i x_i ,$$

and the corresponding intensity process is defined by

$$\lambda_t = \lambda\left(t - S_{N(t)} + \sum_{i=1}^{N(t)} q_i X_i \right). \tag{5.33}$$

The practical interpretation of (5.31) is quite natural, as the degree of repair at each cycle can be different and usually deteriorates with time. The practical application of Model (5.33) is not so evident. Substitution of a random Q_i instead of a deterministic q_i in (5.32) and (5.33) results in general relationships for the intensity processes in this case.

Note that, when $Q_i \equiv Q, i = 1,2,...$ are i.i.d. Bernoulli random variables, the Kijima II model can be interpreted *via* the Brown–Proschan model of Section 4.5. In this model the repair is perfect with probability p and is minimal with probability $1-p$.

Example 5.10 We will now derive Equation (4.30) for the Brown–Proschan model ($p(t) \equiv p$) in a direct way. Denote by $S_n^P(x)$ the Cdf of the arrival time S_n in the Poisson process with rate $\lambda(t)$. Therefore, in accordance with (4.6),

$$\bar{S}_n^P(x) = \sum_0^n \exp\{-\Lambda(t)\}\frac{(\Lambda(t))^n}{n!}.$$

Thus, the survival function of the time between perfect repairs $\bar{F}_P(t)$ is

$$\begin{aligned}\bar{F}_p(t) &= \sum_0^\infty \exp\{-\Lambda(t)\}\frac{(\Lambda(t))^n(1-p)^i}{n!} \\ &= \exp\{-\Lambda(t)\}\exp\{(1-p)\Lambda(t)\} \\ &= \exp\left\{-\int_0^t p\lambda(u)du\right\},\end{aligned}$$

where the term $(1-p)^i$ defines the probability that all $i, i = 1,2,...$ repairs in $[0,t)$ are minimal.

Consider now briefly the comparisons of the relevant characteristics of the described models with respect to the different values of the reduction factor q. With this in mind, denote the virtual age just after the i th repair by V_i^q. Kijima (1989) proved an intuitively expected result stating that in both models, virtual ages for different values of the age reduction factor q are ordered in the sense of the usual stochastic ordering (Definition 3.4), *i.e.*,

$$V_i^{q_1} <_{st} V_i^{q_2}, \quad q_2 > q_1, i \geq 1. \tag{5.34}$$

This means that the larger the value of q, the larger (in the sense of usual stochastic ordering) the random virtual age after each repair. This inequality can be loosely interpreted by noting that larger values of the reduction factor 'push' the process to the right along the time axis.

Denote by $X_i^{q_j}(t)$ the Cdf of $X_i^{q_j}$, $j = 1,2$.

Theorem 5.1. Let $0 < q_1 < q_2 \leq 1$ and the governing $F(t)$ be IFR.

Then the following inequality holds for imperfect repair models (5.26) and (5.30):

$$X_i^{q_1} >_{st} X_i^{q_2}, \ i \geq 1,$$

which means that larger values of q result in stochastically smaller interarrival times.

Proof. Integrating by parts

$$X_i^{q_j}(t) = \int_0^{\infty} d[V_{i-1}^{q_j}(y)]\left(1 - \exp\int_y^{y+t}\lambda(u)du\right)$$

$$= \lim_{y\to\infty}\left(1 - \exp\int_y^{y+t}\lambda(u)du\right) - \int_0^{\infty} V_{i-1}^{q_j}(y)d_y\left(1 - \exp\int_y^{y+t}\lambda(u)du\right),$$

where $V_i^{q_j}(t)$ denotes the Cdf of the virtual age $V_i^{q_j}$. As the governing failure rate $\lambda(t)$ is increasing, the differential d_y in the last integrand is positive. Therefore, comparing $X_i^{q_j}(t)$ for $j = 1$ and $j = 2$ and taking into account Inequality (5.34) proves the theorem. ■

Interpretation of this theorem is also rather straightforward. The larger the (initial) virtual age at the beginning of a cycle, the larger the initial value 'on the failure rate curve' $\lambda(t)$. As $\lambda(t)$ is increasing, this leads to the smaller (in the defined sense) cycle duration. Other more advanced inequalities of a similar type can be found in Kijima (1989) and Finkelstein (1999).

5.4 Ageing and Monotonicity Properties

The content of this section is rather technical and the corresponding proofs of the main results can be omitted at first reading. The presentation mostly follows our recent paper (Finkelstein, 2007). We start by defining some ageing properties of the renewal-type point processes.

Definition 5.7. A stochastic point process is *stochastically ageing* if its interarrival times $\{X_i\}$, $i \ge 1$ are stochastically decreasing, *i.e.*,

$$X_{i+1} \le_{st} X_i,\ i \ge 1. \tag{5.35}$$

Obviously, the renewal process, in accordance with this definition, is not stochastically ageing, whereas the non-homogeneous Poisson process is ageing if its rate is an increasing function. We have chosen the simplest and the most natural type of ordering, but other types of ordering can also be used.

The following definition deals with the ageing properties of the sequence of virtual ages at the start (end) of cycles for the point processes of imperfect repair.

Definition 5.8. The virtual age process $A_t, t \ge 0$ defined by Equation (5.23) is *stochastically increasing* if the (embedded) sequence of virtual ages at the start (end) of cycles is stochastically increasing.

If, *e.g.*, a governing $F(t)$ is IFR, then the stochastically increasing $A_t, t \ge 0$ describes the overall deterioration of our repairable item with time, which is the case in practice for various systems that are wearing out. However, if the failure rate $\lambda(t)$ is decreasing, the stochastically increasing $A_t, t \ge 0$ leads to an 'improvement' of a repairable item. This is similar to the obvious fact that the MRL of an item with a decreasing $\lambda(t)$ is an increasing function. Note that Definition 5.8 is formulated under the assumption of the 'sliding along the failure rate curve' model. Although our interest is mainly in the models with increasing $\lambda(t)$, some results will be given for a more general case as well.

Now we turn to a more detailed study of the generalized Kijima II model with a non-linear quality of repair function $q(t)$ (Finkelstein, 2007). Assume that this is an increasing, concave function that is continuous in $[0,\infty)$ and $q(0)=0$. The assumption of concavity is probably not so natural, but at that time, however, not so restrictive, and we will need it for proving the results to follow. Thus,

$$q(t_1+t_2) \le q(t_1)+q(t_2),\ t_1,t_2 \in [0,\infty). \tag{5.36a}$$

Also, let

$$q(t) < q_0 t\,, \tag{5.36b}$$

where $q_0 < 1$, which shows that repair rejuvenates the failed item, at least to some extent, and that $q(t)$ cannot be arbitrarily close to $q(t)=t$ (minimal repair).

Let a cycle start with a virtual age v. Denote by $X(v)$ the cycle duration with the corresponding survival function given by the right-hand side of Equation (5.27) for $v_{i-1}=v$. The next cycle will start at a random virtual age $q(v+X(v))$. We will be interested in some equilibrium age v^*. Define this virtual age as the solution to the following equation:

$$E[q(v+X(v))]=v\,. \tag{5.37}$$

Thus, if some cycle of a general (imperfect) repair process starts at virtual age v^*, then the next cycle will start with a random virtual age with the expected value v^*, which is obviously a martingale property.

Theorem 5.2. Let $\{X_n\}, n \ge 1$ be a process of imperfect repair, defined by Equations (5.26), where an increasing, continuous quality of repair function $q(t)$ satisfies Equations (5.36a) and (5.36b).

Assume that the governing distribution $F(t)$ has a finite first moment and that the corresponding failure rate is either bounded from below for sufficiently large t by $c>0$ or is converging to 0 as $t \to \infty$ such that

$$\lim_{t\to\infty} t\lambda(t)=\infty\,. \tag{5.38}$$

Then there exists at least one solution to Equation (5.37), and if there is more than one, the set of these solutions is bounded in $[0,\infty)$.

Proof. As $E[X(0)]<\infty$, it is evident that $E[T(v)]<\infty, v>0$. If $\lambda(t)$ is bounded from below by $c > 0$, then

$$E[X(v)] \le \frac{1}{c}.$$

Applying (5.36a), we obtain

$$E[q(v + X(v)] \le q(v) + E[X(v)]. \tag{5.39}$$

It follows from Equations (5.36b) and (5.39) that

$$E[q(v + X(v))] < v$$

for sufficiently large v. On the other hand, $E[q(X(0))] > 0$, which proves the first part of the theorem, as the function $E[q(v + X(v))] - v$ is continuous in v, positive at $v = 0$, and negative for sufficiently large v.

Now, let $\lambda(t) \to 0$ as $t \to \infty$. Consider the following quotient:

$$\frac{E[X(v)]}{v} = \frac{\int_v^{\infty} \exp\left\{-\int_0^x \lambda(u)du\right\}dx}{v \exp\left\{-\int_0^v \lambda(u)du\right\}}.$$

Applying L'Hopital's rule and using Assumption (5.38), we obtain

$$\lim_{v\to\infty} \frac{E[X(v)]}{v} = \lim_{t\to\infty} \frac{1}{\lambda(v)v - 1} = 0. \tag{5.40}$$

Therefore, applying Inequality (5.39) and taking into account (5.36a) and (5.40), we obtain

$$\frac{E[q(v + X(v))]}{v} \le \frac{q(v)}{v} + \frac{E[X(v)]}{v} < 1.$$

The last inequality holds for sufficiently large v. Using the same argument as in the first part of the proof completes our reasoning. ■

Corollary 5.1. If $F(t)$ is IFR, then the conditions of Theorem 5.2 hold and there is at least one solution to Equation (5.37).

Remark 5.9 The sufficient condition (5.38) is a rather weak one stating, in fact, that $t\lambda(t)$ must just have a limit as $t \to \infty$, which should not be finite. It is clear that, for example, for the Weibull distribution with decreasing failure rate, (5.38) holds.

Theorem 5.3. Let $F(t)$ be IFR. Assume that a current cycle starts at virtual age $v^* + \Delta v$, where v^* is an equilibrium solution to Equation (5.37) and $\Delta v > 0$.

Then the expectation of the virtual age at the start of the next cycle will 'be closer' to v^*, *i.e.*,

$$v^* < E[q(v^* + \Delta v + X(v^* + \Delta v))] < v^* + \Delta v. \tag{5.41}$$

Proof. As stated in Corollary 5.1, at least one solution to Equation (5.37) exists in this case. Let us first prove the second inequality in (5.41). Taking into account that $q(t)$ is an increasing function and that the random variables $X(v)$ are stochastically decreasing in v (for increasing $\lambda(t)$), we have

$$E[q(v^* + \Delta v + X(v^* + \Delta v))] < E[q(v^* + \Delta v + X(v^*))].$$

When obtaining this inequality the following simple fact was used. If two distributions are ordered as $\bar{F}_1(t) > \bar{F}_2(t), t \in (0,\infty)$ and $g(t)$ is an increasing function, then by integrating by parts it is easy to see that

$$\int_0^\infty g(t)dF_2(t) < \int_0^\infty g(t)dF_1(t).$$

Finally,

$$\begin{aligned} E[q(v^* + \Delta v + X(v^*))] &\le E[q(v^* + X(v^*))] + q(\Delta v) \\ &= v^* + q(\Delta v) < v^* + \Delta v. \end{aligned}$$

The first inequality in (5.41) is proved using similar arguments. ■

The following corollary is important and will be used for obtaining further results.

Corollary 5.2. If $F(x)$ is IFR, then Equation (5.37) has a unique solution.

Proof. Assume that there are two solutions to Equation (5.37), *i.e.*,

$$E[q(v^* + X(v^*))] = v^*, \tag{5.42}$$

$$E[q(\tilde{v} + X(\tilde{v}))] = \tilde{v}. \tag{5.43}$$

Let $\tilde{v} = v^* + \Delta v, \Delta v > 0$. Then, in accordance with (5.41), we obtain

$$\begin{aligned} E[q(\tilde{v} + X(\tilde{v}))] &= E[q(v^* + \Delta v + X(v^* + \Delta v))] \\ &< v^* + \Delta v = \tilde{v}, \end{aligned}$$

which contradicts (5.43). ■

It can be shown that the results of this section hold when the repair action is stochastic. That is, $\{Q_i\}, i \ge 1$ is a sequence of i.i.d. random variables (independent of other stochastic components of the model) with support in $[0,1]$ and $E[Q_i] < 1$.

We believe that under certain reasonable ordering assumptions these results under reasonable assumptions can also be generalized to a sequence of non-identically distributed random variables.

The described properties show that there is a shift in the direction of the equilibrium point v^* of the starting virtual age of the next cycle compared to the starting virtual age of the current cycle. Note that, for the minimal repair process, the corresponding shift is always in the direction of infinity.

In what follows in this section, we will study the properties of the virtual age process $A_t, t \geq 0$ explicitly defined for the model under consideration by Relationships (5.26). It will be shown under rather weak assumptions that this process is stochastically increasing in terms of Definition 5.2 and that it is becoming stable in distribution (*i.e.*, converges to a limiting distribution as $t \to \infty$). These issues for the linear $q(t)$ were first addressed in Finkelstein (1992b). The rigorous and detailed treatment of monotonicity and stability for rather general age processes driven by the governing $F(t)$ was given by Last and Szekli (1998). The approach of Last and Szekli was based on applying some fundamental probabilistic results: a Lyones-type scheme and Harris-recurrent Markov chains were used. Our approach for a more specific model (but with weaker assumptions on $F(t)$ and with a time dependent $q(t)$) is based on direct probabilistic reasoning and on the appealing 'geometrical' notion of an equilibrium virtual age v^*.

Apart from obvious engineering applications, these results may have some important biological interpretations. Most biological theories of ageing agree that the process of ageing can be considered as process of "wear and tear" (see, *e.g.*, Yashin *et al.*, 2000). The existence of repair mechanisms in organisms decreasing the accumulated damage on various levels is also a well-established fact. As in the case of DNA mutations in the process of cell replication, this repair is not perfect. Asymptotic stability of the repair process means that an organism, as a repairable system, is practically not ageing in the defined sense for sufficiently large t. Therefore, the deceleration of the human mortality rate at advanced ages (see, *e.g.*, Thatcher, 1999) and even the approaching of this rate to the mortality plateau can be explained in this way. This conclusion relies on the important assumption that a repair action decreases the overall accumulated damage and not only its last increment. Another possible source of this deceleration is in the heterogeneity of human populations. This topic is discussed in the next chapter, whereas some biological considerations are analysed in Chapter 10.

Denote the virtual age distribution at the start of the $(i+1)$th cycle by $\theta^S_{i+1}(v)$, $i = 1,2,\ldots$, and denote the corresponding virtual age distribution at the end of the previous, ith cycle by $\theta^E_i(v), i = 1,2,\ldots$. It is clear that, in accordance with (5.26), we have

$$\theta^S_{i+1}(v) = \theta^E_i(q^{-1}(v)),\ i = 1,2,\ldots, \tag{5.44}$$

where the inverse function $q^{-1}(v)$ is also increasing. This can easily be seen, since

$$\theta^S_{i+1}(v) = \Pr[V^S_{i+1} \leq v] = \Pr[q(V^E_i) \leq v] = \Pr[V^E_i \leq q^{-1}(v)],$$

where V_{i+1}^S and V_i^E are virtual ages at the start of the $(i+1)$ th cycle and at the end of the previous cycle, respectively The following theorem states that the age processes under consideration are stochastically increasing.

Theorem 5.4. Virtual ages at the end (start) of each cycle in imperfect repair model (5.26), (5.36a)–(5.36b) form the following stochastically increasing sequences:

$$V_{i+1}^E >_{st} V_i^E, \; V_{i+1}^S >_{st} V_i^S, \; i = 1,2,\ldots .$$

Proof. In accordance with Definition 3.4, we must prove that

$$\bar{\theta}_{i+1}^E(v) > \bar{\theta}_i^E(v), \; \bar{\theta}_{i+2}^S(v) > \bar{\theta}_{i+1}^S(v); \; v > 0, i = 1,2,\ldots . \tag{5.45}$$

We shall prove the first inequality; the second one follows trivially from (5.44). Consider the first two cycles. Let v_1^E be the realization of V_1^E, where V_1^E is the virtual age at the end of the first cycle and at the same time the duration of this cycle. Then (for this realization) the age at the end of the second cycle is

$$q(v_1^E) + X_{(q(v_1^E))},$$

where, as usual, the notation X_v means that this random variable has the Cdf $F(t \mid v)$. It is clear that it is stochastically larger than V_1^E, and, as this property holds for each realization, (5.45) holds for $i = 1$.

Assume that (5.45) holds for $i = n-1, n \geq 3$. Due to the definition of virtual age at the start and the end of a cycle, integrating by parts and using (5.44), we obtain

$$\begin{aligned} \theta_n^E(v) &= \int_0^v \left(1 - \exp\left\{-\int_x^v \lambda(u)du\right\}\right) d\left[\theta_n^S(x)\right], \\ &= \int_0^v \theta_{n-1}^E(q^{-1}(x)) d_x\left(\exp\left\{-\int_x^v \lambda(u)du\right\}\right), \end{aligned} \tag{5.46}$$

$$\begin{aligned} \theta_{n+1}^E(v) &= \int_0^v \left(1 - \exp\left\{-\int_x^v \lambda(u)du\right\}\right) d\left[\theta_{n+1}^S(x)\right] \\ &= \int_0^v \theta_n^E(q^{-1}(x)) d_x\left(\exp\left\{-\int_x^v \lambda(u)du\right\}\right), \end{aligned} \tag{5.47}$$

where we use the fact that

$$\exp\left\{-\int_x^{x+(v-x)} \lambda(u)du\right\} = \exp\left\{-\int_x^v \lambda(u)du\right\}$$

is the probability of survival from initial virtual age x to $v > x$. Taking into account the induction assumption and comparing (5.46) and (5.47), using similar reasoning to that used when obtaining (5.42), we have

$$\theta_n^E(v) < \theta_{n-1}^E(v) \Rightarrow \theta_n^E(q^{-1}(v)) < \theta_{n-1}^E(q^{-1}(v)) \Rightarrow \theta_{n+1}^E(v) < \theta_n^E(v),$$

which completes the proof. ■

The next theorem states that the increasing sequences of distribution functions $\bar{\theta}_i^E(v)$, $\bar{\theta}_i^S(v)$ converge to a limiting distribution function as $i \to \infty$. Thus, the imperfect repair process considered is stable in the defined sense.

Theorem 5.5. Taking into account the conditions of Theorem 5.4, assume additionally that the governing distribution $F(t)$ is IFR.

Then there exist the following limiting distributions for virtual ages at the start and end of cycles:

$$\lim_{i\to\infty} \theta_i^E(v) = \theta_L^E(v) \text{ and } \lim_{i\to\infty} \theta_i^S(v) = \theta_L^S(v). \tag{5.48}$$

Proof. The proof is based on Theorems 5.3 and 5.4. As Sequences (5.45) increase at each $v > 0$, there can be only two possibilities. Either there are limiting distributions (5.48) with uniform convergence in $[0, \infty)$ or the virtual ages grow infinitely, as for the case of minimal repair $(q = 1)$. The latter means that, for each fixed $v > 0$,

$$\lim_{i\to\infty} \theta_i^E(v) = 0 \text{ and } \lim_{i\to\infty} \theta_i^S(v) = 0. \tag{5.49}$$

Assume that (5.49) holds and consider the sequence of virtual ages at the start of a cycle. Then, for an arbitrary small $\varsigma > 0$, we can find n such that

$$\Pr[V_i^S \le v^*] \le \varsigma, \; i \ge n,$$

where v^* is an equilibrium point, which is unique and finite according to Corollary 5.2. It follows from (5.41) that for each realization $v_i^S > v^*$ the expectation of the starting age at the next cycle is smaller than v_i^S. On the other hand, the 'contribution' of ages in $[0, v^*)$ can be made arbitrarily small, if (5.49) holds. Therefore, it can easily be seen that for the sufficiently large i

$$E[V_{i+1}^S] < E[V_i^S].$$

This inequality contradicts Theorem 5.4, according to which expectations of virtual ages form an increasing sequence. Therefore, Assumption (5.49) is wrong and (5.48) holds. As previously, the result for the second limit in (5.48) follows trivially from (5.44). ■

Corollary 5.3. If $F(t)$ is IFR, then the sequence of interarrival lifetimes $\{X_n\}, n \geq 1$ is stochastically decreasing to a random variable with a limiting distribution, *i.e.*,

$$\lim_{i\to\infty} F_i(t) = F_L(t) = \int_0^{\infty}\left(1-\exp\left\{-\int_v^{v+t}\lambda(u)du\right\}\right)d(\theta_L^S(v)). \tag{5.50}$$

Proof. Equation (5.50) follows immediately after taking into account that convergence in (5.48) is uniform. On the other hand, comparing

$$F_i(t) = \int_0^{\infty}\left(1-\exp\left\{-\int_v^{v+t}\lambda(u)du\right\}\right)d(\theta_i^S(v))$$

with

$$F_{i+1}(t) = \int_0^{\infty}\left(1-\exp\left\{-\int_v^{v+t}\lambda(u)du\right\}\right)d(\theta_{i+1}^S(v))$$

it is easy to see, using the same argument as in the proof of Theorem 5.3, that $F_{i+1}(t) > F_i(t), t > 0; i = 1,2,...$ (*i.e.*, a stochastically decreasing sequence of interarrival times), as $\theta_{i+1}^s(v) < \theta_i^s(v)$, and the integrand function is increasing in v for the IFR case. ■

Example 5.11 We will now obtain a stability property for the simplified imperfect maintenance model in a direct way. Note that practically all imperfect repair models can be used for describing imperfect maintenance. Consider the imperfect maintenance actions for a repairable item with an arbitrary lifetime distribution $F(t)$ that are performed at calendar instants of time $nT, n = 1,2,...$ (Kahle, 2007). Assume that all occurring failures are minimally repaired and that at each maintenance the corresponding virtual age is decreased in accordance with the Kijima II model with a constant q, $0 < q < 1$. Therefore, taking into account Equation (5.28), the virtual age after the nth maintenance is

$$v_n = T\sum_{i=0}^{n-1} q^{n-i} = T\sum_1^n q^i = T\frac{1}{1-q}(1-q^n). \tag{5.51}$$

Thus, the virtual age v_n is deterministic and

$$\lim_{n\to\infty} v_n = \frac{1}{1-q},$$

which illustrates the stability property of Theorems 5.4 and 5.5 for this special case.

5.5 Renewal Equations

Renewal equations for g-renewal processes (5.17) and (5.18), or, equivalently, for the age reduction model (5.30), were discussed in Section 5.2.1. We mentioned that although the form of these equations differs from the ordinary renewal equations (4.10) and (4.11), the well-developed numerical methods can be used for obtaining the corresponding solutions. It turns out that renewal equations for the age reduction model (5.26) and (5.27) (the Kijima II model) are more complex.

In order to derive these equations we must assume that a repairable item, in accordance with Model (5.26) and (5.27), starts operating at age (virtual age) x. Let $N(t,x)$ be the number of imperfect repairs in $[0,t)$ for this initial condition. Denote the corresponding renewal function and the renewal density function by $H(t,x)$ and $h(x,t)$, respectively, *i.e.*,

$$H(t,x) = E[N(x,t)],\ h(t,x) = \frac{\partial}{\partial t}H(x,t).$$

Conditioning on the first repair at $t = y$, similarly to Equation (4.12),

$$\begin{aligned} H(t,x) &= \int_0^t E[N(t,x) \mid X_1 = y]\frac{f(y+x)}{\bar{F}(x)}dy \\ &= \int_0^t [1 + H(t-y, q(x+t))]\frac{f(y+x)}{\bar{F}(x)}dy \\ &= F(t \mid x) + \int_0^t H(t-y, q(x+t))f(y \mid x)dy. \end{aligned} \tag{5.52}$$

In a similar way:

$$\begin{aligned} h(t,x) &= \int_0^t \frac{\partial}{\partial t}(E[N(t,x) \mid X_1 = y])f(y \mid x)dx \\ &= f(t \mid x) + \int_0^t h(t-y, q(x+t))f(y \mid x)dy. \end{aligned} \tag{5.53}$$

These equations were first derived in Finkelstein (1992b) and independently by Dagpunar (1997). It can easily be checked that $h(t,x) = \lambda(x+t)$ is the solution to Equation (5.53) for the case of minimal repair when $q(x) = x$. For the case of perfect repair, when $q(x) = 0$, these equations reduce to ordinary renewal equations. Because of the extra dependence on x in the functions $H(x,t)$ and $h(x,t)$, Equations (5.52) and (5.53) are more complex than the corresponding 'univariate' versions (4.10) and (4.11), respectively.

When the function $q(x)$ is linear, Equation (5.53) can be solved numerically for $t \in [0,D],\ D > 0$. Assume that $h(t,x)$ is differentiable with respect to x. Integration by parts (Dagpunar, 1997) yields

$$h(t,x) = f(t \mid x) + h(t,qx) - \lambda(q(t+x))F(t \mid x)$$
$$+ \int_0^t F(t \mid x) dh(t-y, q(x+y)) .$$

Following the approach used by Xie (1991a), the integral in this equation can be approximated by the discrete sum, dividing $[0,D]$ into n subintervals each of length Δ, where $n\Delta = D$.

In Dagpunar (1997), a numerical solution is obtained for $h(t,0)$ for the case of the Weibull $F(x)$. It was shown that $h(t,0)$ rather quickly converges to a constant. In view of our results of the previous section on the stability of the process of imperfect repair, this is not surprising. Corollary 5.3 states that this process converges as $t \to \infty$ to an ordinary renewal process with the Cdf defined by Equation (5.50). Therefore, similar to the asymptotic result (4.16), we have

$$H(t,0) = \frac{t}{m_L}[1+o(1)], \quad h(t,0) = \frac{1}{m_L}[1+o(1)], \tag{5.54}$$

where m_L is the mean defined by the limiting distribution $F_L(t)$ in (5.50). Note that the same results hold for $H(x,t)$ and $h(x,t)$, respectively.

Example 5.12 Consider a system of two identical components with failure rates $\lambda(t)$. The second component is in a state of (cold) standby. After a failure of the main component, the second component is switched into operation, while the failed one is instantaneously minimally repaired. Then the process continues in the same pattern. Let us call the corresponding point process of failures (repairs) the *generalized process of minimal repairs*. Denote by $h(t,x,y)$ the renewal density function for this process, where x is the initial age of the main component and y is the initial age of the standby component at $t=0$. Similar to Equation (5.53),

$$h(t,x,y) = f(t \mid x) + \int_0^t h(t-u, y, x+u) f(u \mid x) du .$$

This integral equation can also be solved using numerical methods. On the other hand, when $x=0, y=0$, a simple approximate solution exists if additional switching (maintenance actions) is allowed. Assume that the main component is operating in the interval of time $[0,\Delta t)$, then it is switched to standby and the former standby component operates in $[\Delta t, 2\Delta t)$, *etc.* When $\lambda(t)$ is increasing, these switching actions increase the reliability of our system. Denote by $\lambda_{\Delta t}(t)$ the resulting failure rate of the system. It can be shown that the following asymptotic relation holds:

$$\lim_{\Delta t \to 0} \mid \lambda_{\Delta t} - \lambda(t/2) \mid = 0 ,$$

which means that asymptotically, as $\Delta t \to 0$, the failure rate of the system can be approximated by the function $\lambda(t/2)$. This operation can be interpreted as the corresponding scale transformation. The failures of the main component are instantaneously repaired by switching to a standby component, which is approximately (for $\Delta t \to 0$) equivalent to minimal repair. Therefore,

$$h(t,0,0) \approx \lambda(t/2)$$

for the sufficiently small Δt.

5.6 Failure Rate Reduction Models

A crucial feature of the age reduction models of the previous sections is the fixed shape of the failure rate $\lambda(t)$ defined by the governing Cdf $F(t)$. The starting point of each cycle 'lies' on the failure rate curve and its position is uniquely defined by the corresponding virtual age v, whereas the duration of the cycle follows the Cdf $F(t \mid v)$. Therefore, imperfect repair rejuvenates an item to some intermediate level between perfect and minimal repair. This approach can be justified in many engineering and biological applications. Another positive feature for modelling is that the corresponding probabilistic model is formalized in terms of the generalized renewal processes. On the other hand, the assumption of the fixed shape of the failure rate is not always convincing and other approaches should be investigated. Before describing the pure failure rate reduction approach, we briefly discuss the model that contains most of the models considered so far as various special cases.

The Dorado–Hollander–Sethuraman (DHS) model (Dorado *et al.*, 1997) is a general model, which describes a departure from the pure age reduction approach. This model assumes that there exist two sequences a_i and $v_i, i = 1,2,...$ such that $a_1 = 1, v_1 = 0$ and the conditional distributions of the cycle durations for the point process of imperfect repairs are given by

$$\Pr[X_i > t \mid a_1,...,a_i, v_1,...,v_i, X_1,...,X_{i-1}] = \frac{\bar{F}(a_i t + v_i)}{\bar{F}(v_i)}, \tag{5.55}$$

where $\bar{F}(t)$ is the survival function for X_1. We see that (5.55) extends (5.27) to additional scale transformations. Therefore, this model generalizes some of the imperfect repair models considered in this and the previous sections. When $v_i = 0$ and $a_i = a^{i-1}, i \geq 1$, we arrive at the geometric process of Section 4.3.3. When $a_i \equiv 1$ and $v_i = q(x_1 + x_2 + ... + x_n)$, we obtain the Kijima I model (5.30) and the Relationship (5.28) results in the Kijima II model (5.26). The minimal repair case also follows trivially from (5.55). Note that Model (5.55) is in turn a specific case of the hidden age model of Finkelstein (1997) discussed by Remark 5.7. The main focus of Dorado *et al.* (1997) was on a nonparametric statistical estimation of a_i and $v_i, i = 1,2,...$. As $F(t)$ in this model can still be considered a governing distribution, the integral equations generalizing Equations (5.52) and (5.53) can also be derived in a formal way.

The intensity process that corresponds to (5.55) is

$$\lambda_t = a_{N(t)+1} \lambda(v_{N(t)+1} + a_{N(t)+1}(t - S_{N(t)})), \tag{5.56}$$

where, as usual, $S_{N(t)}$ denotes the time of the last imperfect repair before t.

Failure rate reduction models differ significantly from age reduction models. Although some of these models can still be governed by an initial (baseline) Cdf and statistical inference of parameters involved can be well defined, a corresponding renewal-type theory cannot be developed. Furthermore, the motivation of the failure rate reduction is usually more formal than that of the age reduction.

Consider, for example, the simplest geometric failure rate reduction model. Assume, as usual, that the first cycle of the process of imperfect repair is described by the Cdf $F(t)$ and the failure rate $\lambda(t)$. Let the failure rate for the second cycle be $a\lambda(t)$, where $0 < a < 1$ with the corresponding survival function $(\bar{F}(t))^a$. The third cycle is described by the failure rate $a^2\lambda(t)$ and the survival function $(\bar{F}(t))^{2a}$, *etc*. The corresponding intensity process is defined as (compare with the intensity process for geometric process (4.23))

$$\lambda_t = a^{N(t)}\lambda(t - S_{N(t)}). \tag{5.57}$$

Thus, the dissimilarity from the geometric process is in the absence of the scale parameter $a^{N(t)}$ in the argument of the failure rate function $\lambda(t)$. But the presence of this parameter, in fact, enables the development of the corresponding renewal-type theory for geometric processes. Unfortunately this is not possible now for the defined geometric failure rate reduction model.

Remark 5.10 The dissimilarity between geometric age reduction and failure rate reduction models is similar to that between the proportional and accelerated life models, as the failure rate for the ALM is $a\lambda(at)$ and $a\lambda(t)$ for the corresponding PH model.

The arithmetic failure rate reduction model was studied in a number of publications (Chan and Shaw, 1993; Doyen and Gaudoin, 2004, among others). The meaningful renewal-type theory cannot be developed in this case but some useful results for modelling and statistical inference can be obtained. According to Doyen and Gaudoin (2004), this model is based on two assumptions:

- Each repair action reduces the intensity process λ_t by an amount depending on the history of the imperfect repair process;
- Between consecutive imperfect repairs, realizations of the intensity process are vertically parallel to the initial (governing) failure rate $\lambda(t)$.

These assumptions lead to the following general form of the intensity process:

$$\lambda_t = \lambda(t) - \sum_{1}^{N(t)} \vartheta_i(\vartheta_1,\ldots,\vartheta_{i-1},S_1,\ldots,S_i), \tag{5.58}$$

where the function ϑ_i models the reduction of the intensity process that results from the i th imperfect repair, $i = 1,2,\ldots$.

Equation (5.58) can be simplified for specific settings. Assume that

$$\vartheta_i(\vartheta_1,\ldots,\vartheta_{i-1},S_1,\ldots,S_i) = \lambda_{S_i} - a\lambda_{S_i} = (1-a)\lambda_{S_i}, \tag{5.59}$$

where a is a reduction factor, $0 \le a \le 1$, that is constant for all cycles. Therefore, the intensity process in the first interval $[0, S_1)$ is $\lambda(t)$. In the second interval $[S_1, S_2)$, it is $\lambda(t) - a\lambda(S_1)$. The intensity process in the third interval is (Rausandt and Hoylandt, 2004)

$$\lambda(t) - a\lambda(S_1) - a(\lambda(S_2) - a\lambda(S_1))$$
$$= \lambda(t) - a[(1-a)^0 \lambda(S_2) + (1-a)^1 \lambda(S_1) .$$

Similarly, it can be shown that the general form of the intensity process in this special case is

$$\lambda_t = \lambda(t) - a\sum_{i=0}^{N(t)} (1-a)^i \lambda(S_{N(t)-i}) . \tag{5.60}$$

The structure of this equation has a certain similarity with Equation (5.29), which defines the intensity process for the Kijima II model. Another model suggested by Doyen and Gaudoin (2004) resembles the Kijima I model (5.33) for age reduction when only the 'input' of the last cycle is reduced. The intensity process for this model is obviously defined as

$$\lambda_t = \lambda(t) - a\lambda(S_{N(t)}) . \tag{5.61}$$

The intermediate cases between (5.60) and (5.61) can also be considered.

We end this section with a short summary comparing the properties of the two considered approaches to imperfect repair modelling. It seems that age reduction models are better motivated as they have a clear interpretation *via* the 'reduction of degradation principle' (*e.g.*, the reduction of the cumulative failure rate or of the cumulative wear). They also usually allow derivation of the renewal-type equations, which can be important in certain applications (*e.g.*, involving spare parts assessment). Although the failure rate itself can still be considered as a characteristic of degradation, its reduction as a model for degradation reduction looks rather formal. The vertical shift in the failure rate is also less motivated than a horizontal shift. The latter implies a clearly understandable shift in the corresponding distribution function and a convenient form of the MRL function in age reduction models.

5.7 Imperfect Repair *via* Direct Degradation

As most of the imperfect repair models considered in this chapter can be interpreted in terms of degradation and its reduction, it is reasonable to discuss, at least in general, an approach that is directly based on reduction of some cumulative degradation. In this section, we will consider only some initial reasoning in this direction.

Assume that an item's degradation at each cycle of the corresponding repair process is described by an increasing stochastic process $W_t,\ t \ge 0, W_0 = 0$ with independent increments. A failure occurs when this process reaches a predetermined (deterministic) level r. The corresponding distribution of the hitting time X_1 for this process is the Cdf of the time to failure in this case, *i.e.*,

$$F_1(t) = \Pr[W_t \geq r] = \Pr[X_1 \leq t].$$

Thus, the duration of the first cycle of the repair process is distributed in accordance with the Cdf $F_1(t)$. Perfect repair results in the restart of this process after the repair. Imperfect repair means that not all deterioration has been eliminated by the repair action. In line with the models of the previous sections, assume that the first imperfect repair action results in reducing degradation to the level $q_1 r$, $0 \leq q_1 \leq 1$. The perfect repair action in this case corresponds to $q_1 = 0$, whereas minimal repair is defined by $q_1 = 1$. In accordance with the independent increments property of the underlying stochastic process $W_t, t \geq 0$, $W_0 = 0$ the Cdf of the second cycle duration is

$$F_2(t) = \Pr[W_t \geq r - q_1 r] = \Pr[X_2 \leq t].$$

If all reduction factors on all subsequent cycles are equal to q_1, then we do not have deterioration in cycle durations starting with the third cycle. In this case, the repair process is described by the renewal process with delay (all cycles, except the first one, are i.i.d. distributed). Assume now that deterioration is modelled by the increasing sequence: $0 < q_1 < q_2 < q_3 < ... < 1$. Therefore,

$$F_{i+1}(t) = \Pr[W_t \geq r - q_i r] > F_i(t) = \Pr[W_t \geq r - q_{i-1} r], \ i = 1,2,..., \quad (5.62)$$

or, equivalently,

$$X_{i+1} <_{st} X_i, \ i = 1.2....,$$

which means that the cycle durations are ordered in the sense of usual stochastic ordering (3.40). Thus, the history of the corresponding imperfect repair process at time t is defined by the time elapsed since the last repair and the number of this repair. An obvious special case is the following geometric-type setting

$$F_{i+1}(t) = \Pr[W_t \geq r - q^i r], \ i = 1,2,... . \quad (5.63)$$

As in the case of the geometric process, it can be proved under the 'natural' assumptions on the process $W_t, t \geq 0$ that the expectation of the waiting time

$$S_n = \sum_1^n X_i$$

is converging when $n \to \infty$.

A suitable candidate for $W_t, t \geq 0$ is the gamma process. The gamma process is a stochastic process with independent, non-negative increments having a gamma distribution with identical scale parameters. It is often used to model gradual damage monotonically accumulating over time, such as wear, fatigue and corrosion (Abdel–Hammed, 1975, 1987; van Noortwijk *et al.*, 2007). The stochastic differential equation, from which the gamma process follows, is given by Wenocur (1989). An advantage of modelling deterioration processes using gamma processes is that the required mathematical calculations are relatively straightforward. In mathematical terms, the gamma process is defined as follows. Equation (2.22) defines the gamma probability density function with the shape parameter α and the scale parameter λ as

$$Ga(x \mid \alpha, \lambda) = f(t) = \frac{\lambda^{\alpha} x^{\alpha-1}}{\Gamma(\alpha)} \exp\{-\lambda x\}.$$

The following definition derives from this.

Definition 5.9. The gamma process with the shape function $\alpha(t) > 0$ and the scale parameter $\lambda > 0$ is the continuous time stochastic process $W_t, t \geq 0$ such that

- $W_0 = 0$ with probability 0;
- Independent increments $W(t_2) - W(t_1)$ in the interval $[t_1, t_2) \in [0, \infty)$ are gamma distributed as $Ga(x \mid \alpha(t_2) - \alpha(t_1), \lambda)$, where $\alpha(t)$ is a non-decreasing right-continuous function with $\alpha(0) = 0$.

As follows from this definition, the accumulated (in accordance with the gamma process) deterioration in $[0, t)$ is described by the pdf $Ga(x \mid \alpha(t), \lambda)$. From the properties of the gamma distribution:

$$E[W_t] = \frac{\alpha(t)}{\lambda}, \quad Var(W_t) = \frac{\alpha(t)}{\lambda^2}.$$

A special case of the increasing power function as a model for $\alpha(t)$ is often used for describing deterioration in structures and other mechanical units (see, *e.g.*, Elingwood and Mori, 1993). Note that the gamma process with stationary increments is defined by the linear shape function αt and the scale parameter λ. The gamma process with $\alpha = \lambda = 1$ is usually called the *standardized gamma process*. Although realizations of the Wiener process with drift (Definition 10.1) are not monotone, this process is sometimes also used in degradation modelling (Kahle and Wendt, 2004) as its mean is increasing.

An important property of the gamma process is that it is a jump process. The number of jumps in any time interval is infinite with probability one. Nevertheless, $E[W_t]$ is finite, as the majority of jumps are 'extremely small'. Dufresne *et al.* (1991) showed that the gamma process can be regarded as the limit of a compound Poisson process. The compound Poisson process is another possibility for the deterioration process $W_t, t \geq 0$. It is defined as the following random sum:

$$W_t = \sum_{1}^{N(t)} W_i, \tag{5.64}$$

where $N(t)$ is the NHPP and $W_i > 0,\ i = 1,2,...$ are i.i.d. random variables, which are independent of the process $N(t)$.

Note that for a compound Poisson process, the number of jumps in any time interval is finite with probability one. Because deterioration should preferably be monotone, we can choose the best deterioration process to be a compound Poisson process or a gamma process. In the presence of observed data, however, the advantage of the gamma process over the compound Poisson process is evident: discrete measurements usually consist of deterioration increments rather than of jump intensities and jump sizes (van Noortwijk *et al.*, 2007).

Combining our imperfect repair model (5.63) with the relationship for the distribution of hitting time for the gamma process (Noortwijk *et al.*, 2007) results in the following cycle-duration distributions for $i = 1,2,...$:

$$\begin{aligned} F_{i+1}(t) &= \Pr[W_t \geq r - q^i r] \\ &= \int_{r-q^i r}^{\infty} Ga(x \mid \alpha(t), \lambda)dx , \\ &= \frac{\Gamma(\alpha(t), (r - q^i r)\lambda)}{\Gamma(\alpha(t))} , \end{aligned} \qquad (5.65)$$

where $\Gamma(b, x)$ is an incomplete gamma function for $x \geq 0, b > 0$ defined as

$$\Gamma(b, x) = \int_x^{\infty} t^{b-1} \exp\{-t\}dt .$$

Relationship (5.65) is an approximate one, as the gamma process, being a jump process, does not reach the level r 'exactly' but attains it with a random *overshoot*. In fact, it is more appropriate to describe this model equivalently in terms of imperfect maintenance rather than in terms of imperfect repair (Nicolai, 2008). Consider, for example, the first cycle. The process value just before the repair (maintenance) action is $r + w_r$, where w_r denotes the value of the defined overshoot. Therefore, in accordance with the model, the next cycle should start with deterioration level $q \cdot (r + w_r)$ and not with qr as in (5.65). As the expected value of the overshoot in practice is usually negligible in comparison with r, (5.65) can be considered practically exact.

The considered degradation-based model of imperfect repair is the simplest one. There can be some other relevant settings. For example, the threshold r can be a random variable R. In this case, Equation (5.63) becomes

$$F_{i+1}(t) = \Pr[W_t \geq R - q^i R], \; i = 1,2,... \qquad (5.66)$$

and therefore can be viewed as a special case of the random resource approach of Section 10.2 (Equation (10.9)). Some technical matters arising from the fact that the gamma process is a jump process can be resolved by considering this model in a more mathematically detailed way as in Nicolai (2008) and in Nicolai *et al.* (2008).

5.8 Chapter Summary

The notion of virtual age, as opposed to calendar age, is indeed appealing. The virtual age is an indicator of the current state of an object. In this way, it is an aggregated, overall characteristic. A similar notion (biological age) is often used in life sciences, but without a proper mathematical formalization. If, for example, someone has vital characteristics (blood pressure, cholesterol level, *etc.*) as those of a younger person, then the state of his health definitely corresponds to some younger age. On the other hand, there are no justified ways to make this statement precise, as the state of health of an individual is defined by numerous parameters. However, the corresponding formalization can be performed for some simple, ageing engineering items. In this chapter, we developed the virtual age theory for repairable and non-repairable items.

We consider two non-repairable identical items operating in different environments. The first one operates in a baseline (reference) environment, whereas the second item operates in a more severe environment. We define the virtual age of the second item *via* a comparison of its level of deterioration with the deterioration level of the first item. If the baseline environment is 'equipped' with the calendar age, then the virtual age of an item in the second environment, which was operating for the same time as the first one, is larger than the corresponding calendar age. In Section 5.1, we developed formal models for the described age correspondence using the accelerated life model and its generalizations.

Various models can be suggested for defining the corresponding virtual age of an imperfectly repaired item. The term *virtual age* was suggested by Kijima (1989). An important feature of this model is the assumption that the repair action does not change the baseline Cdf $F(x)$ (or the baseline failure rate $\lambda(x)$) and only the starting time t changes after each repair. Therefore, the Cdf of a lifetime after repair in Kijima's model is defined as the remaining lifetime distribution $F(x \mid t)$. We developed the renewal theory for this setting and also considered asymptotic properties of the corresponding imperfect repair process. We proved in Section 5.3 that, as $t \to \infty$, this process converges to an ordinary renewal process.

Other types of imperfect repair were discussed in Sections 5.5 and 5.6. Specifically, we considered an imperfect repair model with the underlying gamma process of deterioration. The repair action decreases the accumulated deterioration to some intermediate level between the perfect and the minimal repair. The gamma process is often used to model gradual damage monotonically accumulating over time. An advantage of modelling deterioration processes using gamma processes is that the required mathematical calculations are relatively straightforward.

6

Mixture Failure Rate Modelling

6.1 Introduction – Random Failure Rate

The main definitions and properties of the failure rate and related characteristics were considered in Chapter 2. A natural generalization of the notion of a classical failure rate is a failure rate that is itself random (see Section 3.1 for a general discussion).

As was mentioned in Section 3.1, the usual source of a possible randomness in the failure rate of a non-repairable item is a random environment (*e.g.*, temperature, mechanical or electrical load, *etc.*), which in the simplest case is modelled by a single random variable (Example 3.1). A popular interpretation is also a subjective one, when we consider a lifetime and an associated non-observable parameter with the assigned set of conditional distributions (Shaked and Spizzichino, 2001). On the other hand, repairable items can also be characterized by a random failure rate, as instants of repair are random in time. A random failure rate of this kind was considered in Chapters 4 and 5.

Let the failure rate of a non-repairable item now be a stochastic process $\lambda_t, t \geq 0$. As in the specific case of Section 3.1.1, where this process was induced by some covariate process, we will call it the *hazard (failure) rate* process. One of the first publications to address the issue of a random failure rate was the paper by Gaver (1963). A number of interesting models for specific hazard rate processes were considered in Lemoine and Wenocur (1985), Wenocur (1989), Kebir (1991), and Singpurwalla and Yongren (1991), to mention a few. Recall that the corresponding stochastic process for repairable systems is called the *intensity process* (Chapters 4 and 5).

Our goal in this chapter is to analyse the simplest model for the hazard rate process when it is defined by a random variable Z (Example 3.1) in the following way:

$$\lambda_t = \lambda(t, Z)\,. \tag{6.1}$$

It turns out that this formally simple model is meaningful for theoretical studies and for practical applications as well. Consider a lifetime T with failure rate (6.1) defined for each realization $Z = z$. In accordance with exponential representation (2.5), we can formally write

$$\bar{F}(t,Z) = \exp\left\{-\int_0^t \lambda(u,Z)du\right\}, \tag{6.2}$$

meaning that this equation holds for each realization $Z = z$.

For the sake of presentation, we briefly repeat the reasoning of Section 3.1 and use the general Equations (3.3)–(3.7) for this specific case of the hazard rate process (6.1).

Applying the operation of expectation with respect to Z to both sides of (6.2) results in

$$\bar{F}(t) = \Pr[T > t] = E[\bar{F}(t,Z)] = E\left[\exp\left\{-\int_0^t \lambda(u,Z)du\right\}\right].$$

We will call $F(t)$ and $\bar{F}(t)$ the observed (marginal) distribution and survival functions, respectively. It follows from this equation that the corresponding observed failure rate $\bar{\lambda}(t) = f(t)/\bar{F}(t)$ is not equal to the expectation of the random failure rate $\lambda(u,Z)$, *i.e.*,

$$\bar{\lambda}(t) \neq E[\lambda(t,Z)].$$

Assume for simplicity that $\lambda(t,z) = z\lambda(t)$, where $\lambda(t)$ is a failure rate for some lifetime distribution. In this case, $\bar{F}(t,z)$ is a strictly convex function with respect to z and Jensen's inequality can be applied ($E[g(X)] > g(E[X])$ for some strictly convex function g and a random variable X). Therefore, using the Fubini's theorem and assuming that $E[Z] < \infty$ (see also Equations (3.5)–(3.7)) we obtain

$$\bar{F}(t) > \exp\left\{-\int_0^t E[\lambda(u,Z]du\right\}, t > 0. \tag{6.3}$$

It can be proved that

$$\bar{\lambda}(t) < E[\lambda(u,Z)] = \lambda(t)E[Z], \ t > 0.$$

Thus, the observed failure rate is smaller than the expectation of the failure rate process for the specific case considered. In Section 6.5 we will show explicitly that this inequality is true for a more general form of $\lambda(t,Z)$. Some other useful orderings will also be considered later in this chapter. On the other hand, owing to Jensen's inequality, (6.3) always holds if the finite expectation is obtained with respect to $\lambda(t,Z)$.

The described mathematical setting can be interpreted in terms of mixtures of distributions. The term "mixture" in this context will be used interchangeably with the terms "observed" or "marginal". This interpretation will be crucial for what follows in this and the following chapter. Mixtures of distributions play an important role in various disciplines.

Assume that in accordance with Equation (6.2), the Cdf $F(t)$ is indexed by a random variable Z in the following sense:

$$\Pr[T \le t \mid Z = z] \equiv \Pr[T \le t \mid z] = F(t,z) .$$

The corresponding failure rate $\lambda(t,z)$ is $f(t,z)/\bar{F}(t,z)$. Let Z be interpreted as a continuous non-negative random variable with support in $[a,b]$, $a \ge 0, b \le \infty$ and the pdf $\pi(z)$. Thus, the mixture Cdf is defined by

$$F_m(t) = \int_a^b F(t,z)\pi(z)dz , \tag{6.4}$$

where the subscript m stands for "mixture". As in (3.8) and (3.9), the mixture failure rate $\lambda_m(t)$ is defined in the following way:

$$\lambda_m(t) = \frac{f_m(t)}{\bar{F}_m(t)} = \frac{\int_a^b f(t,z)\pi(z)dz}{\int_a^b \bar{F}(t,z)\pi(z)dz} = \int_a^b \lambda(t,z)\pi(z \mid t)dz , \tag{6.5}$$

where the conditional pdf $\pi(z \mid t)$ is given by Equation (3.10). The probability $\pi(z \mid t)dz$ can be interpreted as the probability that $Z \in (z, z+dz]$ on condition that $T > t$. Note that, this interpretation *via* the conditional pdf is just a useful reasoning, whereas formally $\lambda_m(t)$ is defined by Equation (6.5).

Our main focus will be on continuous mixtures, but some results on discrete mixtures will be also discussed. Similar to (6.4), the discrete mixture Cdf can be defined as the following finite or infinite sum (see also Example 3.3):

$$F_m(t) = \sum_k F(t, z_k)\pi(z_k) , \tag{6.6}$$

where $\pi(z_k)$ is the probability mass of z_k. The corresponding pdf and the failure rate are then defined in a similar way to the continuous case.

In Section 3.1, some results on the shape of the failure rate were already discussed. The shape of the failure rate is very important in reliability analysis as, among other things, it describes the ageing properties of the corresponding lifetime distribution. Why is the understanding of the properties and the shape of the mixture failure rate so important? Apart from a purely mathematical interest, there are many applications where these issues become pivotal.

Our main interest here is in lifetime modelling for heterogeneous populations (Aalen, 1988). One can hardly find homogeneous populations in real life, although most of the studies on failure rate modelling deal with a homogeneous case. Neglecting existing heterogeneity can lead to substantial errors and misconceptions in stochastic analysis in reliability, survival and risk analysis as well as other disciplines. Some results on minimal repair modelling in heterogeneous populations were presented in Section 4.7.

Mixtures of distributions usually present an effective tool for modelling heterogeneity. The origin of mixing in practice can be 'physical' when, for example, a number of devices of different (heterogeneous) types, performing the same function and not distinguishable in operation, are mixed together. This occurs when we have 'identical' items of different makes. A similar situation arises when data from different distributions are pooled to enlarge the sample size (Gurland and Sethuraman, 1995).

It is well known that mixtures of DFR distributions are always DFR (Barlow and Proschan, 1975). On the other hand, mixtures of increasing failure rate (IFR) distributions can decrease, at least in some intervals of time, which means that the IFR class of distributions is not closed under the operation of mixing (Lynch, 1999). IFR distributions usually model lifetimes governed by ageing processes, which means that the operation of mixing can dramatically change the pattern of ageing, *e.g.*, from positive ageing (IFR) to negative ageing (DFR) (Example 3.2). A gamma mixture of Weibull distributions with increasing failure rates was considered in this example. As follows from Equation (3.11), the resulting mixture failure rate initially increases to a single maximum and then decreases asymptotically, converging to 0 as $t \to \infty$ (Figure 3.1). This fact was experimentally observed in Finkelstein (2005c) for a heterogeneous sample of miniature light bulbs, as illustrated by Figure 6.1. It should be noted, however, that change in ageing patterns often occurs in practice at sufficiently large ages of items, as in the case of human mortality. Therefore, the role of asymptotic methods in analysis is evident and the next chapter will be devoted to mixture failure rate modelling for large t. Thus, the discussed facts and other implications of heterogeneity should be taken into account in applications.

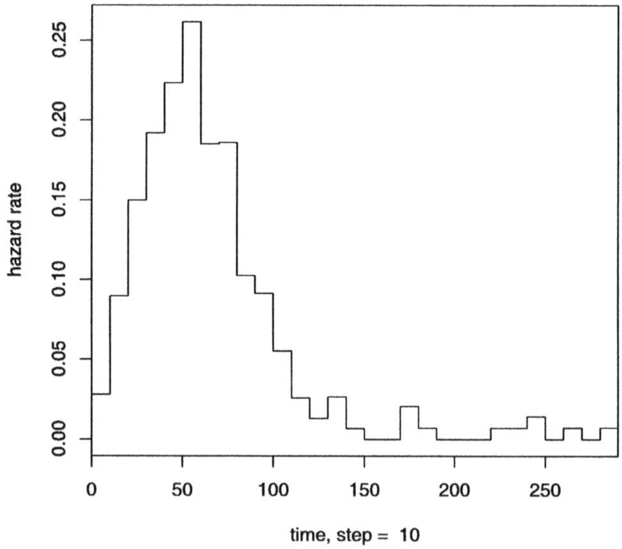

Figure 6.1. Empirical failure (hazard) rate for miniature light bulbs

Another equivalent interpretation of mixing in heterogeneous populations is based on a notion of a non-negative random unobserved parameter (frailty) Z. The term "frailty" was suggested in Vaupel *et al.* (1979) for the gamma-distributed Z

and the multiplicative failure rate model of the form $\lambda(t,z)=z\lambda(t)$. Since that time, multiplicative frailty models have been widely used in statistical data analysis and demography (see, *e.g.*, Andersen *et al.*, 1993). It is worth noting, however, that the specific case of the gamma-frailty model was, in fact, first considered by the British actuary Robert Beard (Beard, 1959, 1971).

A convincing 'experiment' showing the deceleration in the observed failure (mortality) rate is performed by nature. It is well known that human mortality follows the Gompertz (1825) lifetime distribution with an exponentially increasing mortality rate. We briefly discussed this distribution in Section 2.3.9. Assume that heterogeneity for this baseline distribution is described by the multiplicative gamma-frailty model, *i.e.*,

$$\lambda(t,Z)=Za\exp\{bt\};\, t\geq 0,\, a,b>0\,.$$

Owing to its computational simplicity, the gamma-frailty model is practically the only one widely used in applications so far. It will be shown later that the mixture failure rate $\lambda_m(t)$, in this case, is monotone in $[0,\infty)$ and asymptotically tends to a constant as $t\to\infty$, although 'individual' failure rates increase sharply as exponential functions for all $t\geq 0$. The function $\lambda_m(t)$ is monotonically increasing for the real demographic values of parameters of this model. This fact explains the recently observed deceleration in human mortality at advanced age (human mortality plateau, as in Thatcher, 1999). Similar deceleration in mortality was experimentally obtained for populations of medflies by Carey *et al.* (1992). Interesting results were also obtained by Wang *et al.* (1998).

While considering heterogeneous populations in different environments, the problem of ordering mixture failure rates for stochastically ordered mixing random variables arises. Assume, for example, that one mixing variable is larger than the other one in the sense of the usual stochastic ordering defined by Equation (3.40). Will this guarantee that the corresponding mixture failure rates will also be ordered in the same direction? We will show in this chapter that this is not sufficient and another stronger type of stochastic ordering should be considered for this reason. Some specific results for the case of frailties with equal means and different variances will also be obtained.

There are many situations where the concept of mixing helps to explain results that seem to be paradoxical. A meaningful example is a Parondo paradox in game theory (Harmer and Abbot, 1999), which describes the dependent losing strategies which eventually win. Di Crescenzo (2007) presents the reliability interpretation of this paradox. This author compares pairs of systems with two independent components in each series. The i th component of the first system ($i=1,2$) is less reliable than the corresponding component of the second one (in the sense of the usual stochastic order (3.40)). The first system is modified by a random choice of its components. Each component is chosen randomly from a set of components identical to the previous ones, and the corresponding distribution of a new component is defined as a discrete mixture (with $\pi=1/2$) of initial distributions of components of the first system. Thus, the described randomization defines a new system that is shown to be more reliable (under suitable conditions) than the second one, although initial components are less reliable than those of the second system. A formal proof of this phenomenon is presented in this paper, but the result can easily be interpreted in terms of certain properties of mixture failure rates to be discussed in this chapter.

We start with some simple properties describing the shape of the failure rate for the discrete mixture of two distributions.

6.2 Failure Rate of Discrete Mixtures

Consider a mixture of two lifetime distributions $F_1(t)$ and $F_2(t)$ with pdfs $f_1(t)$ and $f_2(t)$ and failure rates $\lambda_1(t)$ and $\lambda_2(t)$, respectively. Although our interest is mostly in mixtures with one governing distribution defined by Equation (6.6), we will briefly discuss in this section a more general case of different distributions ($k = 2$).

Let the masses π and $1-\pi$ define the discrete mixture distribution. The mixture survival function and the mixture pdf are

$$\bar{F}_m(t) = \pi\bar{F}_1(t) + (1-\pi)\bar{F}_2(t),$$
$$f_m(t) = \pi f_1(t) + (1-\pi)f_2(t),$$

respectively. In accordance with the definition of the failure rate, the mixture failure rate in this case is

$$\lambda_m(t) = \frac{\pi f_1(t) + (1-\pi)f_2(t)}{\pi\bar{F}_1(t) + (1-\pi)\bar{F}_2(t)}.$$

As $\lambda_i(t) = f_i(t)/\bar{F}_i(t),\ i = 1,2,$ this can be transformed into

$$\lambda_m(t) = \pi(t)\lambda_1(t) + (1-\pi(t))\lambda_2(t)\,, \tag{6.7}$$

where the time-dependent probabilities are

$$\pi(t) = \frac{\pi\bar{F}_1(t)}{\pi\bar{F}_1(t) + (1-\pi)\bar{F}_2(t)}, \quad 1-\pi(t) = \frac{(1-\pi)\bar{F}(t)}{\pi\bar{F}_1(t) + (1-\pi)\bar{F}_2(t)},$$

which corresponds to the continuous case defined by Equation (6.5). It easily follows from Equation (6.7) (Block and Joe, 1997) that

$$\min\{\lambda_1(t), \lambda_2(t)\} \le \lambda_m(t) \le \max\{\lambda_1(t), \lambda_2(t)\}\,.$$

For example, if the failure rates are ordered as $\lambda_1(t) \le \lambda_2(t)$, then

$$\lambda_1(t) \le \lambda_m(t) \le \lambda_2(t)\,. \tag{6.8}$$

Now we can show directly that if both distributions are DFR, then the mixture Cdf is also DFR (Navarro and Hernandez, 2004), which is a well-known result for the general case. Differentiating (6.7) results in

$$\lambda'_m(t) = \pi(t)\lambda'_1(t) + (1-\pi(t))\lambda'_2(t) - \pi(t)(1-\pi(t)(\lambda_1(t)-\lambda_2(t))^2 .$$

Therefore, as $\lambda'_i(t) \le 0,\ i = 1,2$, the mixture failure rate is also decreasing. The proof of this fact for the continuous case can be found, *e.g.*, in Ross (1996).

It follows from (6.8) that the mixture failure rate is contained between $\lambda_1(t)$ and $\lambda_2(t)$. As $\bar{F}(0) = 1$, the initial value of the mixture failure rate is just the 'ordinary' mixture of initial values of the two failure rates, *i.e.*,

$$\lambda_m(0) = \pi\lambda_1(0) + (1-\pi)\lambda_2(0) .$$

When $t > 0$, the conditional probabilities $\pi(t)$ and $1-\pi(t)$ are not equal to π and $1-\pi$, respectively. Finally,

$$\lambda_m(t) < \pi\lambda_1(t) + (1-\pi)\lambda_2(t),\ t > 0, \tag{6.9}$$

which follows from Equation (6.3), where Z is a discrete random variable with masses π and $1-\pi$. Thus, $\lambda_m(t)$ is always smaller than the expectation $\pi\lambda_1(t) + (1-\pi)\lambda_2(t)$. We shall discuss this property and the corresponding comparison in more detail for the continuous case.

The next chapter will be devoted to the asymptotic behaviour of $\lambda_m(t)$ as $t \to \infty$. We will show under rather weak conditions that in both discrete and continuous cases the mixture failure rate tends to the failure rate of the strongest population. For the considered model, this means that

$$\lim_{t\to\infty}(\lambda_m(t) - \lambda_1(t)) = 0 . \tag{6.10}$$

It is worth noting that the shapes of mixture failure rates in the discrete case can vary substantially. Many examples of the possible shapes for different distributions are given in Jiang and Murthy (1995) and in Lai and Xie (2006). For example, the possible shape of the mixture failure rate for any two Weibull distributions can be one of eight different types including IFR, DFR, UBT, MBT (modified bathtub shape: the failure rate first increases and then follows the bathtub shape). It was proved, however, that there is no BT shape option in this case.

6.3 Conditional Characteristics and Simplest Models

Our main interest in these two chapters is in continuous mixtures, as they are usually more suitable for modelling heterogeneity in practical settings. In addition, the corresponding models represent our uncertainty about parameters involved, which is also often the case in practice.

Let the support of the mixing random variable Z be $[0,\infty)$ for definiteness. We shall consider the general case, $[a,b]$, where necessary. Using the definition of the conditional pdf in Equations (3.10) and (6.5), denote the conditional expectation of Z given $T > t$ by $E[Z \mid t]$, *i.e.*,

$$E[Z \mid t] = \int_0^\infty z\,\pi(z \mid t)dz\,.$$

An important characteristic for further consideration is $E'[Z \mid t]$, the derivative with respect to t, *i.e.*,

$$E'[Z \mid t] = \int_0^\infty z\,\pi'(z \mid t)dz\,,$$

where

$$\begin{aligned}\pi'(z \mid t) &= -\frac{f(t,z)\pi(z)}{\int_0^\infty \bar{F}(t,z)\pi(z)dz} + \frac{\bar{F}(t,z)\pi(z)\lambda_m(t)}{\int_0^\infty \bar{F}(t,z)\pi(z)dz} \\ &= \lambda_m(t)\pi(z \mid t) - \frac{f(t,z)\pi(z)}{\int_0^\infty \bar{F}(t,z)\pi(z)d\theta}.\end{aligned} \quad (6.11)$$

Equations (3.10) and (6.5) were used for deriving (6.11). After simple transformations, we obtain the following useful result.

Lemma 6.1. The following equation for $E'[Z \mid t]$ holds:

$$E'[Z \mid t] = \lambda_m(t)E[Z \mid t] - \frac{\int_0^\infty z\,f(t,z)\pi(z)dz}{\int_0^\infty \bar{F}(t,z)\pi(z)dz}\,. \quad (6.12)$$

We will now consider two specific cases where the mixing variable Z can be 'entered' directly into the failure rate model. These are the additive and multiplicative models widely used in reliability and lifetime data analysis. The third well-known case of the accelerated life model (ALM) cannot be studied in a similar way. However, asymptotic theory for the mixture failure rate for this and the first two models will be discussed in the next chapter.

6.3.1 Additive Model

Let $\lambda(t,z)$ be indexed by parameter z in the following way:

$$\lambda(t,z)=\lambda(t)+z\,, \tag{6.13}$$

where $\lambda(t)$ is a deterministic, continuous and positive function for $t>0$. It can be viewed as some baseline failure rate. Equation (6.13) defines for $z\in[0,\infty)$ a family of 'horizontally parallel' functions. We will mostly be interested in an increasing $\lambda(t)$. In this case, the resulting mixture failure rate can have different intuitively non-evident shapes, whereas, as was stated earlier, a mixture of DFR distributions is always DFR. Noting that $f(t,z)=\lambda(t,z)\overline{F}(t,z)$ and applying Equation (6.5) for this model results in

$$\lambda_m(t)=\lambda(t)+\frac{\int_0^\infty z\,\overline{F}(t,z)\pi(z)dz}{\int_0^\infty \overline{F}(t,z)\pi(z)d\theta}=\lambda(t)+E[Z\mid t]\,. \tag{6.14}$$

Using this relationship and Lemma 6.1, a specific form of $E'[Z\mid t]$ can be obtained:

$$E'[Z\mid t]=(\lambda(t)+E[Z\mid t])E[Z\mid t]-\frac{\int_0^\infty (z\,\lambda(t)\,\overline{F}(t,z)+z^2\overline{F}(t,z))\pi(z)dz}{\int_0^\infty \overline{F}(t,z)\pi(z)dz}$$

$$=[E[Z\mid t]]^2-\int_0^\infty z^2\pi(z\mid t)dz=-Var(Z\mid t)\,, \tag{6.15}$$

where $Var(Z\mid t)$ denotes the variance of Z given $T>t$. This result can be formulated in the form of:

Lemma 6.2. The conditional expectation of Z for the additive model is a decreasing function of $t\in[0,\infty)$, which follows from

$$E'[Z\mid t]=-Var(Z\mid t)<0\,.$$

Differentiating (6.14) and using Relationship (6.15), we immediately obtain the result that was stated in Lynn and Singpurwalla (1997).

Theorem 6.1. Let $\lambda(t)$ be an increasing, convex function in $[0,\infty)$. Assume that $Var(Z\mid t)$ is decreasing in $t\in[0,\infty)$ and

$$Var(Z\mid 0)>\lambda'(0)\,.$$

Then $\lambda_m(t)$ decreases in $[0,c)$ and increases in $[c,\infty)$, where c can be uniquely defined from the following equation:

$$Var(Z \mid t) = \lambda'(t).$$

It follows from this theorem that the corresponding model of mixing results in the BT shape of the mixture failure rate. Figure 6.2 illustrates this result for the case of linear baseline failure rate $\lambda(t) = ct,\ c > 0$. The initial value of the mixture failure rate is $\lambda_m(0) = E[Z]$. It first decreases and then increases, converging to the failure rate of the strongest population, which is ct in this case. The convergence to the failure rate of the strongest population in a general setting will be discussed in the next chapter.

In addition to Lynn and Singpurwalla (1997), we have included an assumption that $Var(Z \mid t)$ should decrease for $t \geq 0$. It seems that, similar to the fact that $E[Z \mid t]$ is decreasing in $[0,\infty)$, the conditional variance $Var(Z \mid t)$ should also decrease, as the "weak populations are dying out first" when t increases. It turns out that this intuitive reasoning is not true for the general case. The counter-example can be found in Finkelstein and Esaulova (2001), which shows that the conditional variance for some specific distribution of Z is increasing in the neighbourhood of 0. It is also shown that $Var(\theta \mid t)$ is decreasing in $[0,\infty)$ when Z is exponentially distributed.

It follows from the proof of this theorem that if $Var(Z \mid 0) \leq \lambda'(0)$, then $\lambda_m(t)$ is increasing in $[0,\infty)$ and the IFR property is preserved. We will discuss the IFR preservation property at the end of the next section.

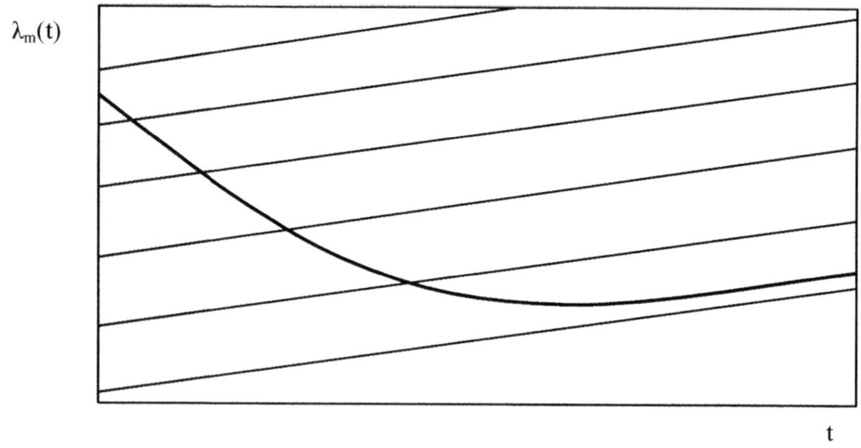

Figure 6.2. The BT shape of the mixture failure rate

6.3.2 Multiplicative Model

Let $\lambda(t, z)$ be now indexed by parameter z in the following multiplicative way:

$$\lambda(t, z) = z\,\lambda(t), \tag{6.16}$$

where, as previously, the baseline $\lambda(t)$ is a deterministic, continuous and positive function for $t > 0$. In survival analysis, Model (6.16) is usually called a proportional hazards (PH) model. The mixture failure rate (6.5) in this case reduces to

$$\lambda_m(t) = \int_0^\infty \lambda(t, z)\pi(z \mid t)dz = \lambda(t)E[Z \mid t]. \tag{6.17}$$

After differentiating:

$$\lambda'_m(t) = \lambda'(t)E[Z \mid t] + \lambda(t)E'[Z \mid t]. \tag{6.18}$$

It follows immediately from this equation that, when $\lambda(0) = 0$, the failure rate $\lambda_m(t)$ increases in the neighbourhood of $t = 0$. Further behaviour of this function depends on the other parameters involved. Example 3.2 shows that, *e.g.*, for the increasing baseline Weibull failure rate, the resulting mixture failure rate initially increases and then decreases converging to 0 as $t \to \infty$.

Substituting $\lambda_m(t)$ and the pdf

$$f(t, z) = \lambda(t, z)\bar{F}(t, z) = z\lambda(t)\bar{F}(t)$$

into Equation (6.12), similar to (6.15), the following result for the multiplicative model is obtained (Finkelstein and Esaulova, 2001):

Lemma 6.3. The conditional expectation of Z for the multiplicative model is a decreasing function of $t \in [0, \infty)$, as follows from

$$E'[Z \mid t] = -\lambda(t)Var(Z \mid t) < 0. \tag{6.19}$$

Equation (6.19) was also proved in Gupta and Gupta (1996) using the corresponding moment generating functions. Thus, it follows from Equation (6.17) and Lemma 6.3 that the function $\lambda_m(t)/\lambda(t)$ is a decreasing one. This property implies that $\lambda(t)$ and $\lambda_m(t)$ cross at most at only one point.

Example 6.1 Consider the specific case $\lambda(t) = const$. Then Equation (6.18) reduces to $\lambda'_m(t) = \lambda E'[Z \mid t]$. It follows from Lemma 6.3 that the mixture failure rate is decreasing. In other words, the mixture of exponential distributions is DFR. The foregoing can be considered as a new proof of this well-known fact. Other interesting proofs can be found in Barlow (1985) and Mi (1998). Note that the first paper describes this phenomenon from the 'subjective' point of view.

We end this section with some general considerations on the preservation of the mixture failure rate monotonicity property for the increasing family $\lambda(t,z)$, $z \in [0,\infty)$. As was stated in Barlow and Proschan (1975), this property is not preserved under the operation of mixing, although there are many specific cases when this preservation is observed. Example 3.2 shows that the Weibull-gamma mixture is not monotone. On the other hand, the Weibull-inverse Gaussian mixture is IFR for some values of parameters (Gupta and Gupta, 1996). The Gompertz-gamma mixture, as will be shown later in this chapter, is also IFR for certain values of parameters. Lynch (1999) had derived rather restrictive conditions for the preservation of the IFR property: the mixture failure rate $\lambda_m(t)$ is increasing if

- $\bar{F}(t,z)$ is log-concave in (t,z);
- $\bar{F}(t,z)$ is increasing in z for each $t > 0$;
- The mixing distribution is IFR.

The log-concavity property is a natural assumption because in the univariate case the IFR property is equivalently defined as $\bar{F}(t)$ being log-concave. This means that the derivative of $-\log \bar{F}(t)$, which, owing to the exponential representation, equals $\lambda(t)$, is positive. Therefore, the first condition seems also to be natural for $\bar{F}(t,z)$ as well. An important and rather stringent condition is, however, the second one. It is clear, *e.g.*, for the multiplicative model (6.16) that this condition does not hold, as the survival function

$$\bar{F}(t,z) = \exp\left\{-z\int_0^t \lambda(u)du\right\}$$

is decreasing in z for each $t > 0$. The same is true for the additive model (6.13). The choice of the IFR mixing distribution is not so important, and therefore the last assumption is not so restrictive. For the sake of computational simplicity, the gamma distribution is often chosen as the mixing one.

Example 6.2 Let the failure rate be given by the following linear function:

$$\lambda(t,z) = 2\frac{t}{z}.$$

Obviously, $\bar{F}(t,z)$ is increasing in z. It can be shown that $-\log \bar{F}(t,z)$ in this case is a concave function (Block *et al.*, 2003), but practical applications of this inverse variation law are not evident.

6.4 Laplace Transform and Inverse Problem

The Laplace transform methodology in multiplicative and additive models is usually very effective. It constitutes a convenient tool for dealing with mixture failure rates and corresponding conditional expectations especially when the Laplace transform of the mixing distribution can be obtained explicitly.

Consider now a rather general class of mixing distributions. Define distributions as belonging to the exponential family (Hougaard, 2000) if the corresponding pdf can be represented as

$$\pi(z) = \frac{\exp\{-\theta z\} g(z)}{\eta(\theta)}, \tag{6.20}$$

where $g(z)$ and $\eta(z)$ are some positive functions and θ is a parameter. The function $\eta(\theta)$ plays the role of a normalizing constant ensuring that the pdf integrates to 1. It is a very convenient representation of the family of distributions, as it allows for the Laplace transform to be easily calculated. The gamma, the inverse Gaussian and the stable (see later in this section) distributions are relevant examples of distributions in this family. The Laplace transform of $\pi(z)$ depends only on the normalizing function $\eta(z)$, which is quite remarkable (Hougaard, 2000). This can be seen from the following equation:

$$\begin{aligned}\pi^*(s) \equiv \int_0^\infty \exp\{-sz\}\pi(z)dz &= \frac{1}{\eta(\theta)}\int_0^\infty \exp\{-sz\}\exp\{-\theta z\}g(z)dz \\ &= \frac{\eta(\theta+s)}{\eta(\theta)}.\end{aligned} \tag{6.21}$$

A well-known fact from survival analysis states that the failure data alone do not uniquely define a mixing distribution and additional information (*e.g.*, on covariates) should be taken into account (a problem of non-identifiability, as, *e.g.*, in Tsiatis, 1974 and Yashin and Manton, 1997). On the other hand, with the help of the Laplace transform, the following inverse problem can be solved analytically at least for additive and multiplicative models of mixing (Finkelstein and Esaulova, 2001; Esaulova, 2006):

Given the mixture failure rate $\lambda_m(t)$ and the mixing pdf $\pi(z)$, obtain the failure rate $\lambda(t)$ of the baseline distribution.

This means that under certain assumptions any shape of the mixture failure rate can be constructed by the proper choice of the baseline failure rate.

Firstly, consider the additive model (6.13). The survival function and the pdf are

$$\bar{F}(t,z) = \exp\{-\Lambda(t) - zt\}, \quad f(t,z) = (\lambda(t) + z)\exp\{-\Lambda(t) - zt\},$$

respectively, where

$$\Lambda(t) = \int_0^\infty \lambda(u)du \tag{6.22}$$

is a cumulative baseline failure rate. Using Equation (6.4), the mixture survival function $\bar{F}_m(t)$ can be written via the Laplace transform as

$$\bar{F}_m(t) = \exp\{-\Lambda(t)\int_0^\infty \exp\{-zt\}\pi(z)dz = \exp\{-\Lambda(t)\}\pi^*(t), \tag{6.23}$$

where, as in (6.21), $\pi^*(t) = E[\exp\{-zt\}]$ is the Laplace transform of the mixing pdf $\pi(z)$. Therefore, using Equation (6.14):

$$\lambda_m(t) = \lambda(t) + \frac{\int_0^\infty z\exp\{-zt\}\pi(z)dz}{\int_0^\infty \exp\{-zt\}\pi(z)dz} = \lambda(t) - \frac{d}{dt}\log\pi^*(t). \tag{6.24}$$

It also follows from (6.14) that

$$E[Z \mid t] = -\frac{d}{dt}\log\pi^*(t).$$

It is worth noting that this conditional expectation does not depend on the baseline lifetime distribution and depends only on the mixing distribution.

The solution of the inverse problem for this special case is given by the following relationship:

$$\lambda(t) = \lambda_m(t) + \frac{d}{dt}\log\pi^*(t). \tag{6.25}$$

If the Laplace transform of the mixing distribution can be derived explicitly, then Equation (6.25) gives a simple analytical solution for the inverse problem. Assume, *e.g.*, that 'we want' the mixture failure rate to be constant, *i.e.*, $\lambda_m(t) = c$. Then the baseline failure rate is obtained as

$$\lambda(t) = c + E[Z \mid t].$$

At the end of this section some meaningful examples will be considered, whereas a simple explanatory one follows.

Example 6.3 Let $\pi(z)$ be uniformly distributed in $[0,b]$. Then the conditional expectation can be easily derived directly from (6.24) as

$$E[Z \mid t] = \frac{1}{t} - \frac{b}{\exp\{bt\} - 1}.$$

Obtaining the limit as $t \to 0$ results in the obvious $E[Z \mid 0] = b/2$. On the other hand, this function, in accordance with Lemma 6.1, is decreasing and converging to 0 as $t \to \infty$.

The corresponding survival function for the multiplicative model (6.16) is $\exp\{-z\Lambda(t)\}$. Therefore, the mixture survival function for this specific case, in accordance with Equation (6.4), is

$$\bar{F}_m(t) = \int_0^\infty \exp\{-z\Lambda(t)\}\pi(z)dz = \pi^*(\Lambda(t)). \tag{6.26}$$

As previously, it is written in terms of the Laplace transform of the mixing distribution, but this time as a function of the cumulative baseline failure rate $\Lambda(t)$. The mixture failure rate is given by

$$\lambda_m(t) = -\frac{\overline{F}'_m(t)}{\overline{F}_m(t)} = -\frac{d}{dt}\log \pi^*(\Lambda(t)). \tag{6.27}$$

It follows from Equations (6.17) and (6.27) that

$$\begin{aligned} E[Z \mid t] &= -\frac{\frac{d}{d\Lambda(t)}\pi^*(\Lambda(t))}{\pi^*(\Lambda(t))} \\ &= -\frac{d}{d\Lambda(t)}\log \pi^*(\Lambda(t)). \end{aligned} \tag{6.28}$$

The general solution to the inverse problem in terms of the Laplace transform is also simple in this case. From (6.27):

$$\pi^*(\Lambda(t)) = \exp\{-\Lambda_m(t)\},$$

where $\Lambda_m(t)$, similar to (6.22), denotes the cumulative mixture failure rate. Applying the inverse Laplace transform $L^{-1}(\cdot)$ to both sides of this equation results in

$$\lambda(t) = \Lambda'(t) = \frac{d}{dt}L^{-1}(\exp\{-\Lambda_m(t)\}). \tag{6.29}$$

Specifically, for the exponential family of mixing densities (6.20) and for the multiplicative model under consideration, the mixture failure rate is obtained from Equations (6.21) and (6.27) as

$$\begin{aligned} \lambda_m(t) &= -\frac{d}{dt}\log\frac{\eta(\theta+\Lambda(t))}{\eta(\theta)} \\ &= -\lambda(t)\frac{\frac{d}{d(\theta+\Lambda(t))}\eta(\theta+\Lambda(t))}{\eta(\theta+\Lambda(t))}, \end{aligned} \tag{6.30}$$

and, therefore, the conditional expectation is defined as

$$E[Z \mid t] = -\frac{\frac{d}{d(\theta+\Lambda(t))}\eta(\theta+\Lambda(t))}{\eta(\theta+\Lambda(t))}.$$

Using Equation (6.30), the solution to the inverse problem (6.29) can be obtained in this case as the derivative of the following function:

$$\Lambda(t) = \eta^{-1}(\exp\{-\lambda_m(t)\}\eta(\theta)) - \theta \,. \tag{6.31}$$

Example 6.4 Consider the special case defined by the gamma mixing distribution. This example is meaningful for the rest of this chapter and for the following chapter. We will derive an important relationship for the mixture failure rate, which is wellknown in the statistical and demographic literature. Thus, the mixing pdf $\pi(z)$ is defined as

$$\pi(z) = \frac{\beta^{\alpha} z^{\alpha-1}}{\Gamma(\alpha)} \exp\{-\beta z\}, \alpha, \beta > 0 \,. \tag{6.32}$$

In accordance with the definitions of the exponential family (6.20) and its Laplace transform (6.21),

$$\eta(\beta) = \frac{\Gamma(\alpha)}{\beta^{\alpha}}, \quad \pi^{*}(t) = \frac{\beta^{\alpha}}{(\beta + t)^{\alpha}} \,.$$

Therefore, from Equation (6.30):

$$\lambda_m(t) = \frac{\alpha\lambda(t)}{\beta + \Lambda(t)} \tag{6.33}$$

and

$$E[Z \mid t] = \frac{\alpha}{\beta + \Lambda(t)} \,.$$

Finally, differentiating Equation (6.31), the solution of the inverse problem is obtained as

$$\lambda(t) = \frac{\beta}{\alpha} \lambda_m(t) \exp\left\{\frac{\Lambda_m(t)}{\alpha}\right\}. \tag{6.34}$$

Assume that the mixture failure rate is constant, *i.e.*, $\lambda_m(t) = c$. It follows from (6.34) that for obtaining a constant $\lambda_m(t)$ the baseline $\lambda(t)$ should be exponentially increasing, *i.e.*,

$$\lambda(t) = \frac{\beta}{\alpha} c \exp\left\{\frac{ct)}{\alpha}\right\}.$$

This result is really striking: we are mixing the exponentially increasing family of failure rates and arriving at a constant mixture failure rate.

Equation (6.33) was first obtained by Beard (1959) and then independently derived by Vaupel *et al.* (1979) in the demographic context. In the latter paper the term 'frailty' was also first used for the mixing variable Z. Therefore, this model is usually called "the gamma-frailty model" in the literature. Owing to relatively simple computations, the gamma-frailty model is widely used in various applications.

Example 6.5 Let the mixing distribution follow the inverse Gaussian law. We will write the pdf of this distribution in the traditional parameterization as in Hougaard (2000) (compare with the pdf in Section 2.3.8), *i.e.*,

$$\pi(z) = (2\pi)^{1/2} z^{-3/2} \nu^{1/2} \exp\{\sqrt{\theta\nu}\} \exp\{-\theta z / 2 - \nu / 2z\}.$$

In accordance with Equation (6.20), the corresponding functions $\mu(z)$ and $\eta(\theta)$ for the exponential family are

$$\mu(z) = (2\pi)^{1/2} z^{-3/2} \nu^{1/2} \exp\{-\nu / 2z\}, \quad \eta(\theta) = \exp\{\sqrt{\theta\nu}\}.$$

Therefore, similar to the previous example,

$$\lambda_m(t) = \frac{\sqrt{\nu}\lambda(t)}{2\sqrt{\theta + \Lambda(t)}}, \quad E[Z \mid t] = \frac{\sqrt{\nu}}{2\sqrt{\theta + \Lambda(t)}}.$$

Finally, the solution to the inverse problem is given by

$$\lambda(t) = \frac{2}{\nu} \lambda_m(t)(\sqrt{\theta\nu} + \Lambda_m(t)).$$

The inverse problem for some other families of mixing densities can also be considered (Esaulova, 2006). For example, the positive stable distribution (Hougaard, 2000) has a Laplace transform that is convenient for computations (see Equation (6.68) of Example 6.8). On the other hand, the three-parameter power variance function (PVF) includes exponential family and positive stable distributions as specific cases (Hougaard, 2000).

6.5 Mixture Failure Rate Ordering

6.5.1 Comparison with Unconditional Characteristic

The 'unconditional mixture failure rate' was defined in Inequality (6.3) for the special case of the multiplicative model. Denote this characteristic by $\lambda_P(t)$. A generalization of Inequality (6.3) (to be formally proved by Theorem 6.2) can be formulated as

$$\lambda_m(t) < \lambda_P(t) \equiv \int_a^b \lambda(t,z)\pi(z)dz, \quad t > 0\,; \lambda_m(0) = \lambda_P(t). \tag{6.35}$$

Thus, owing to conditioning on the event that an item had survived in $[0,t]$, *i.e.*, $T > t$, the mixture failure rate is smaller than the unconditional one for each $t > 0$. Inequality (6.35) can be interpreted as: *"the weakest populations are dying out first"*. This interpretation is widely used in various special cases, *e.g.*, in the demographic literature. This means that as time increases, those subpopulations that have larger failure rates have higher chances of dying, and therefore the proportion of subpopulations with a smaller failure rate increases. This results in Inequality (6.35) and in a stronger property in the forthcoming Theorem 6.2.

Inequality (6.35) is written in terms of failure rate ordering. The usual stochastic order for two random variables X and Y was defined by Definition 3.4. The failure (hazard) rate order is defined in the following way.

Definition 6.1. A random variable X with a failure rate $\lambda_X(t)$ is said to be larger in terms of failure (hazard) rate ordering than a random variable Y with a failure rate $F_X(t)$ if

$$\lambda_X(t) \le \lambda_Y(t), \quad t \ge 0. \tag{6.36}$$

The conventional notation is $X \ge_{hr} Y$. It easily follows from exponential representation (2.5) that failure rate ordering is a stronger ordering, and therefore it implies the usual stochastic ordering (3.40).

The function $\lambda_P(t)$ in (6.35) is a supplementary one and it 'captures' the monotonicity pattern of the family $\lambda(t,z)$. Therefore, $\lambda_P(t)$ under certain conditions has a similar shape to individual $\lambda(t,z)$. If, *e.g.*, $\lambda(t,z),\ z \in [a,b]$ is increasing in t, then $\lambda_P(t)$ is increasing as well. By contrast, as was already discussed in this chapter, the mixture failure rate $\lambda_m(t)$ can have a different pattern: it can ultimately decrease, for instance, or preserve the property that it is increasing in t as in Lynch (1999). There is even a possibility of a number of oscillations (Block *et al.*, 2003). However, despite all possible patterns, Inequality (6.35) holds, and under some additional assumptions, the following difference can monotonically increase in time:

$$(\lambda_P(t) - \lambda_m(t)) \uparrow, t \ge 0. \tag{6.37}$$

Definition 6.2. (Finkelstein and Esaulova, 2006b). Inequality (6.35) defines a weak 'bending-down property' for the mixture failure rate, whereas (6.37) defines a strong 'bending-down property'.

The main additional assumption that will be needed for the following theorem is that the family of failure rates $\lambda(t,z),\ z \in [a,b]$ is ordered in z.

Theorem 6.2. Let the failure rate $\lambda(t,z)$ in the mixing model (6.4) and (6.5) be differentiable with respect to both arguments and be ordered as

$$\lambda(t,z_1) < \lambda(t,z_2), \quad z_1 < z_2, \forall z_1, z_2 \in [a,b], t \ge 0. \tag{6.38}$$

Then

- The mixture failure rate $\lambda_m(t)$ bends down with time at least in a weak sense, defined by (6.35);
- If, additionally, $\partial\lambda(t,z)/\partial z$ is increasing in t, then $\lambda_m(t)$ bends down with time in a strong sense, defined by (6.37).

Proof. Ordering (6.38) is equivalent to the condition that $\lambda(t,z)$ is increasing in z for each $t \geq 0$. In accordance with Equation (6.5), the definition of $\lambda_P(t)$ in (6.35) and integrating by parts:

$$\Delta\lambda(t) \equiv \int_a^b \lambda(t,z)[\pi(z) - \pi(z \mid t)]dz$$

$$= \lambda(t,z)[\Pi(z) - \Pi(z \mid t)]\,|_a^b - \int_a^b \lambda_z'(t,z)[\Pi(z) - \Pi(z \mid t)]dz$$

$$= \int_a^b -\lambda_z'(t,z)[\Pi(z) - \Pi(z \mid t)]dz > 0, \;\; t > 0, \tag{6.39}$$

where

$$\Pi(z) = \Pr[Z \leq z], \quad \Pi(z \mid t) = \Pr[Z \leq z \mid T > t]$$

are the corresponding conditional and unconditional distributions, respectively. Inequality (6.39) and the first part of the theorem follow from $\lambda_z'(t,z) > 0$ and from the following inequality:

$$\Pi(z) - \Pi(z \mid t) < 0, \;\; t > 0, z \in [a,b]. \tag{6.40}$$

To obtain (6.40), it is sufficient to prove that

$$\Pi(z \mid t) = \frac{\int_a^z \bar{F}(t,u)\pi(u)du}{\int_a^b \bar{F}(t,u)\pi(u)du}$$

is increasing in t. It is easy to see that the derivative of this function is positive if

$$\frac{\int_a^z \bar{F}_t'(t,u)\pi(u)du}{\int_a^z \bar{F}(t,u)\pi(u)du} > \frac{\int_a^b \bar{F}_t'(t,u)\pi(u)du}{\int_a^b \bar{F}(t,u)\pi(u)du}.$$

As $\overline{F}_t'(t,z) = -\lambda(t,z)\overline{F}(t,z)$, it is sufficient to show that (Finkelstein and Esaulova, 2006b)

$$\lambda(t,z)\int_a^z \overline{F}(t,u)\pi(u)du > \int_a^z \lambda(t,u)\overline{F}(t,u)\pi(u)du\,,$$

which follows from (6.38). Therefore, as the functions $\partial\lambda(t,z)/\partial z$ and $\Pi(z \mid t)$ are increasing in t, the final integrand in (6.39) is also increasing in t. Thus, the difference $\Delta\lambda(t)$ is also increasing, which immediately leads to the strong bending-down property (6.37). ■

It is worth noting that the decreasing of $\Pi[Z \mid t]$ in t can also be interpreted *via "the weakest populations are dying out first"* principle, as this distribution tends to be more concentrated around small values of $Z \geq a$ as time increases.

The light bulb example of Section 6.1 (Figure 6.1) shows the strong bending-down property for the mixture failure rate in practice. It was conducted by the author at the Max Planck Institute for Demographic Research (Finkelstein, 2005c). We recorded the failure times for a population of 750 miniature lamps and constructed the empirical failure rate function (in relative units) for the time interval 250 h. The results were convincing: the failure rate initially increased (a tentative fit showed the Weibull law) and then decreased to a very low level. The pattern of the observed failure rate is similar to that in Figure 3.1.

6.5.2 Likelihood Ordering of Mixing Distributions

We will show now that a natural ordering for our mixing model is the likelihood ratio ordering. For brevity, the terms "smaller" or "decreasing" are used and the evident symmetrical "larger" or "increasing" are omitted or *vice versa.* A similar reasoning can be found in Block *et al.* (1993) and Shaked and Spizzichino (2001).

Let Z_1 and Z_2 be continuous non-negative random variables with the same support and densities $\pi_1(z)$ and $\pi_2(z)$, respectively.

Definition 6.3. Z_2 is smaller than Z_1 in the sense of the likelihood ratio ordering:

$$Z_1 \geq_{lr} Z_2 \tag{6.41}$$

if $\pi_2(z)/\pi_1(z)$ is a decreasing function (Ross, 1996).

Definition 6.4. Let $Z(t), t \in [0,\infty)$ be a family of random variables indexed by a parameter t (*e.g.*, time) with probability density functions $p(z,t)$. We say that $Z(t)$ is decreasing in t in the sense of the likelihood ratio (the decreasing likelihood ratio (DLR) class) if

$$L(z,t_1,t_2) = \frac{p(z,t_2)}{p(z,t_1)}$$

is decreasing in z for all $t_2 > t_1$.

This property can also be formulated in terms of log-convexity of Glazer's function defined by Equation (2.36), as in Navarro (2008). It can be proved (Ross, 1996) that the likelihood ratio ordering implies the failure rate ordering. Therefore, it is the strongest of the three types of ordering considered so far. Thus, in accordance with Equations (3.40), (6.36) and (6.41), we have

$$Z_1 \geq_{lr} Z_2 \Rightarrow Z_1 \geq_{hr} Z_2 \Rightarrow Z_1 \geq_{st} Z_2 . \tag{6.42}$$

The following simple result states that the family of conditional mixing random variables $Z \mid t,\ t \in [0, \infty]$ forms the DLR class.

Theorem 6.3. Let the family of failure rates $\lambda(t, z)$ in mixing model (6.5) be ordered as in (6.38).

Then the family of random variables $Z \mid t \equiv Z \mid T > t$ is DLR in $t \in [0, \infty)$.

Proof. In accordance with the definition of the conditional mixing distribution (3.10) in the mixing model (6.5), the ratio of the densities for different instants of time is

$$L(z, t_1, t_2) = \frac{\pi(z \mid t_2)}{\pi(z \mid t_1)} = \frac{\bar{F}(t_2, z) \int_a^b \bar{F}(t_1, z)\pi(z)dz}{\bar{F}(t_1, z) \int_a^b \bar{F}(t_2, z)\pi(z)dz} . \tag{6.43}$$

Therefore, monotonicity in z of $L(z, t_1, t_2)$ is defined by the function

$$\frac{\bar{F}(t_2, z)}{\bar{F}(t_1, z)} = \exp\left\{ - \int_{t_1}^{t_2} \lambda(u, z)du \right\},$$

which, owing to Ordering (6.38), is decreasing in z for all $t_2 > t_1$. ■

Consider now two different mixing random variables Z_1 and Z_2 with probability density functions $\pi_1(z)$, $\pi_2(z)$ and the corresponding cumulative distribution functions $\Pi_1(z), \Pi_2(z)$, respectively. Intuition suggests that if Z_1 is larger than Z_2 in some stochastic sense to be defined, then the corresponding mixture failure rates should be ordered accordingly: $\lambda_{m1}(t) \geq \lambda_{m2}(t)$. The question is what type of ordering will guarantee this inequality? Simple examples show (Esaulova, 2006) that usual stochastic ordering is too weak for this purpose. It was stated already that the likelihood ratio ordering is a natural one for the family of random variables $Z \mid t$ in our mixing model. Therefore, it seems reasonable to order Z_1 and Z_2 in this sense, and see whether this ordering will lead to the desired ordering of the corresponding mixture failure rates or not.

The following lemma states that the likelihood ratio ordering is stronger than the usual stochastic ordering (3.40). This well-known fact is already indicated by Relationship (6.42), but we need a new proof to be used later.

Lemma 6.4. Let

$$\pi_2(z) = \frac{g(z)\pi_1(z)}{\int_a^b g(z)\pi_1(z)dz}, \tag{6.44}$$

where $g(z)$ is a continuous, decreasing function and the integral is a normalizing constant (integration of $\pi_1(z)$ should result in 1).

Then Z_1 is stochastically larger than Z_2.

Proof. Indeed,

$$\Pi_2(z) = \frac{\int_a^z g(u)\pi_1(u)du}{\int_a^b g(u)\pi_1(u)du} = \frac{\int_a^z g(u)\pi_1(u)du}{\int_a^z g(u)\pi_1(u)du + \int_z^b g(u)\pi_1(u)du}$$

$$= \frac{g^*(a,z)\int_a^z \pi_1(u)du}{g^*(a,z)\int_a^z \pi_1(u)du + g^*(z,b)\int_z^b \pi_1(u)du} \geq \int_a^z \pi_1(u)du = \Pi_1(z), \tag{6.45}$$

where $g^*(a,z)$ and $g^*(z,b)$ are the mean values of the function $g(z)$ for the corresponding integrals. As this function decreases, $g^*(z,b) \leq g^*(a,z)$ and the inequality in (6.45) follows. ■

Now we are able to prove the main ordering theorem (Finkelstein and Esaulova, 2006), showing that under certain assumptions the mixture failure rates for different mixing distributions are ordered in the sense of the failure rate ordering (6.36). A similar result is stated by Theorem 1.C.17 in Shaked and Shanthikumar (2007). Using general results on the totally positive functions (Karlin, 1968), these authors under more stringent conditions prove that the corresponding mixture random variables are ordered in a stronger sense of the likelihood ratio ordering. Our approach, by contrast, is based on direct reasoning and can also be used for 'deriving' the likelihood ratio ordering of mixing distributions as the necessary condition for the corresponding failure (hazard) rate ordering (see Equation 6.49).

Theorem 6.4. Let Equation (6.44) hold, where $g(z)$ is a decreasing function, which means that Z_1 is larger than Z_2 in the sense of the likelihood ratio ordering. Assume also that Ordering (6.38) holds.

Then the following inequality holds for $\forall t \in [0,\infty)$:

$$\lambda_{m1}(t) \equiv \frac{\int_a^b f(t,z)\pi_1(z)dz}{\int_a^b \bar{F}(t,z)\pi_1(z)dz} \geq \frac{\int_a^b f(t,z)\pi_2(z)dz}{\int_a^b \bar{F}(t,z)\pi_2(z)dz} \equiv \lambda_{m2}(t). \qquad (6.46)$$

Proof. Inequality (6.46) means that the mixture failure rate, which is obtained for a stochastically larger mixing distribution (in the likelihood ratio ordering sense), is larger for $\forall t \in [0,\infty)$ than the one obtained for the stochastically smaller mixing distribution. Therefore, the corresponding (mixture) random variables are ordered in the sense of the failure (hazard) rate ordering.

We shall prove, first, that

$$\Pi_1(z \mid t) = \frac{\int_a^z \bar{F}(t,u)\pi_1(u)du}{\int_a^b \bar{F}(t,u)\pi_1(u)du} \leq \frac{\int_a^z \bar{F}(t,u)\pi_2(u)du}{\int_a^b \bar{F}(t,u)\pi_2(u)du} \equiv \Pi_2(z \mid t). \qquad (6.47)$$

Indeed, using Equation (6.44):

$$\frac{\int_a^z \bar{F}(t,u)\pi_2(u)du}{\int_a^b \bar{F}(t,u)\pi_2(u)du} = \frac{\int_a^z \bar{F}(t,u)\dfrac{g(u)\pi_1(u)}{\int_a^b g(u)\pi_1(u)du}du}{\int_a^b \bar{F}(t,u)\dfrac{g(u)\pi_1(u)}{\int_a^b g(u)\pi_1(u)du}du}$$

$$= \frac{\int_a^z g(u)\bar{F}(t,u)\pi_1(u)du}{\int_a^b g(u)\bar{F}(t,u)\pi_1(u)du} \geq \frac{\int_a^z \bar{F}(t,u)\pi_1(u)du}{\int_a^b \bar{F}(t,u)\pi_1(u)du},$$

where the last inequality follows using exactly the same argument as in Inequality (6.45) of Lemma 6.4. Performing integration by parts as in (6.39) and taking into account Inequality (6.47) results in

$$\lambda_{m1}(t) - \lambda_{m2}(t) = \int_a^b \lambda(t,z)[\pi_1(z \mid t) - \pi_2(z \mid t)]dz$$

$$= \int_a^b -\lambda'_z(t,z)[\Pi_1(z \mid t) - \Pi_2(z \mid t)]dz \geq 0, \; t > 0 \; . \qquad (6.48)$$

Thus, when the mixing distributions are ordered in the sense of the likelihood ordering, the mixture failure rates are ordered as $\lambda_{m1}(t) \geq \lambda_{m2}(t)$. ■

A starting point for Theorem 6.4 is Equation (6.44) with the crucial assumption of a decreasing function $g(z)$ defining, in fact, the likelihood ratio ordering. This was our reasonable guess, as the usual stochastic order was not sufficient for the desired mixture failure rate ordering and a stronger ordering had to be considered. But this guess can be justified directly by considering the difference $\Delta\lambda(t) = \lambda_{m1}(t) - \lambda_{m2}(t)$ and using Equations (6.5) and (3.10). The corresponding numerator (the denominator is positive) is transformed into a double integral in the following way:

$$\int_a^b \lambda(t,z)\bar{F}(t,z)\pi_1(z)dz \int_a^b \bar{F}(t,z)\pi_2(z)dz$$

$$-\int_a^b \lambda(t,z)\bar{F}(t,z)\pi_2(z)dz \int_a^b \bar{F}(t,z)\pi_1(z)dz$$

$$= \int_a^b\int_a^b \bar{F}(t,u)\bar{F}(t,s)[\lambda(t,u)\pi_1(u)\pi_2(s) - \lambda(t,s)\pi_1(u)\pi_2(s)]duds$$

$$= \underset{u>s}{\int_a^b\int_a^b} \bar{F}(t,u)\bar{F}(t,s)(\lambda(t,u) - \lambda(t,s))(\pi_1(u)\pi_2(s) - \pi_1(s)\pi_2(u))duds\,. \quad (6.49)$$

Therefore, the final double integral is positive if Ordering (6.38) in the family of failure rates holds and $\pi_2(z)/\pi_1(z)$ is decreasing. Thus, the likelihood ratio ordering is derived as a necessary condition for the corresponding ordering of mixture failure rates.

What happens when Z_1 and Z_2 are ordered only in the sense of usual stochastic ordering: $Z_1 \geq_{st} Z_2$? As was already mentioned, this ordering is not sufficient for the mixture failure rate ordering (6.46). However, it is sufficient for the ordinary stochastic order of the corresponding random variables (Shaked and Shanthikumar, 2007). Indeed, similar to (6.48), it can be seen integrating by parts and taking into account that $F_z'(t,z) > 0$ and that $\Pi_1(z) - \Pi_2(z) \leq 0$:

$$F_{m1}(t) - F_{m2}(t) = \int_a^b F(t,z)[\pi_1(z) - \pi_2(z)]dz$$

$$= \int_a^b - F_z'(t,z)[\Pi_1(z) - \Pi_2(z)]dz \geq 0,\ t > 0\,.$$

Denote the corresponding mixture random variables by Y_1 and Y_2, respectively. Thus, the assumed ordering $Z_1 \geq_{st} Z_2$ results in the following stochastic ordering for Y_1 and Y_2:

$$Y_1 \leq_{st} Y_2\,,$$

which is evidently weaker than Inequality (6.46). Note that the latter inequality can equivalently be written as $Y_1 \leq_{hr} Y_2$.

6.5.3 Mixing Distributions with Different Variances

If mixing variables are ordered in the sense of the likelihood ratio ordering, then automatically

$$E[Z_1] \geq E[Z_2], \tag{6.50}$$

which obviously holds for the weaker (usual) stochastic ordering (3.40) as well. Inequality (6.50), in fact, can be considered as a definition of a very weak ordering of random variables Z_1 and Z_2.

Let $\Pi_1(z)$ and $\Pi_2(z)$ now be two mixing distributions with equal means. It follows from Equation (6.17) that for the multiplicative model, which will be considered in this section, the initial values of the mixture failure rates are equal in this case:

$$\lambda_{m1}(0) = \lambda_{m2}(0).$$

Intuitive considerations and general reasoning based on the principle "the weakest populations are dying out first" suggest that, unlike (6.46), the mixture failure rates will be ordered as

$$\lambda_{m1}(t) < \lambda_{m2}(t), \; t > 0 \tag{6.51}$$

if the variance of Z_1 is larger than the variance of Z_2. It will be shown, however, that this is true only for a special case and that for the general multiplicative model this ordering holds only for a sufficiently small time t.

Example 6.6 For a meaningful example, consider a multiplicative frailty model (6.17), where Z has a gamma distribution:

$$\pi(z) = \frac{\beta^{\alpha} z^{\alpha-1}}{\Gamma(\alpha)} \exp\{-\beta z\}, \lambda, \beta > 0.$$

Substituting this density into (3.8) and taking into account the multiplicative form of the failure rate,

$$\lambda_m(t) = \frac{\lambda(t) \int_0^{\infty} \exp\{-z\Lambda(t)\} z\pi(z)dz}{\int_0^{\infty} \exp\{-z\Lambda(t)\}\pi(z)dz},$$

where $\Lambda(t)$, as previously, denotes a cumulative baseline failure rate.

It follows from Example 6.4 that the mixture failure rate in this case is

$$\lambda_m(t) = \frac{\alpha\lambda(t)}{\beta + \Lambda(t)}.$$

As $E[Z] = \alpha / \beta$ and $Var(Z) = \alpha / \beta^2$, this equation can now be written in terms of $E[Z]$ and $Var(Z)$ in the following way:

$$\lambda_m(t) = \lambda(t)\frac{E^2[Z]}{E[Z] + Var(Z)\Lambda(t)}, \tag{6.52}$$

which, for the specific case $E[Z] = 1$, gives the result of Vaupel *et al.* (1979) that is widely used in demography:

$$\lambda_m(t) = \frac{\lambda(t)}{1 + Var(Z)\Lambda(t)}. \tag{6.53}$$

Using Equation (6.52), we can compare mixture failure rates of two populations with different Z_1 and Z_2 on condition that $E[Z_2] = E[Z_1]$. Therefore, the comparison is straightforward, *i.e.*,

$$Var(Z_1) \geq Var(Z_2) \Rightarrow \lambda_{m1}(t) \leq \lambda_{m2}(t). \tag{6.54}$$

Intuitively it can be expected that this result could be valid for arbitrary mixing distributions, at least for the multiplicative model. However, the mixture failure rate dynamics in time can be much more complicated even for this special case. The following theorem shows that ordering of variances is a sufficient and necessary condition for ordering of mixture failure rates, but only for the initial time interval.

Theorem 6.5. Let Z_1 and Z_2 be two mixing distributions with equal means in the multiplicative model (6.16) and (6.17).

Then ordering of variances

$$Var(Z_1) > Var(Z_2) \tag{6.55}$$

is a sufficient and necessary condition for ordering of mixture failure rates in the neighbourhood of $t = 0$, *i.e.*,

$$\lambda_{m1}(t) < \lambda_{m2}(t);\ t \in (0, \varepsilon), \tag{6.56}$$

where $\varepsilon > 0$ is sufficiently small.

Proof. Sufficient condition:
From Equation (6.17) we have

$$\Delta\lambda(t) = \lambda_{m1}(t) - \lambda_{m2}(t) = \lambda(t)(E[Z_1 \mid t] - E[Z_2 \mid t). \tag{6.57}$$

Equation (6.19) reads:

$$E'[Z_i \mid t] = -\lambda(t)Var(Z_i \mid t) < 0, i = 1,2, t \geq 0, \tag{6.58}$$

where

$$E[Z_i \mid 0] \equiv E[Z_i], \quad Var(Z_i \mid 0) \equiv Var(Z_i). \tag{6.59}$$

Thus, if Ordering (6.55) holds, Ordering (6.56) follows immediately after showing that the derivative of the function

$$\frac{\lambda_{m1}(t)}{\lambda_{m2}(t)} = \frac{E[Z_1 \mid t]}{E[Z_2 \mid t]}$$

at $t = 0$ is negative. This follows from Equation (6.58). Finally, the equation $\lambda_{m1}(0) = \lambda_{m2}(0)$ for the case of equal means is also taken into account.

Necessary condition:
The corresponding proof is rather technical (see Finkelstein and Esaulova, 2006 for details) and is based on considering the numerator of the difference $\Delta\lambda(t)$, which is

$$\lambda(t)\int_a^b\int_a^b [\exp\{-\Lambda(t)(u+s)\}](u-s)\pi_1(u)\pi_2(s)duds .$$

6.6 Bounds for the Mixture Failure Rate

In this section, we are mostly interested in simple bounds for the mixture failure rate for the multiplicative model of mixing. The obtained bounds can be helpful in various applications, *e.g.*, for mortality rate analysis in heterogeneous populations. We show that when the failure rates of subpopulations follow the proportional hazards (PH) model with the multiplicative frailty Z and the common proportionality factor k, the resulting mixture failure rate has a strict upper bound $k\lambda_m(t)$, where $\lambda_m(t)$ has a meaning of the mixture failure rate in a heterogeneous population without a proportionality factor ($k \equiv 1$). Furthermore, this result presents another explicit justification of the fact that the PH model in each realization does not result in the PH model for the corresponding mixture failure rates.

It is well known that the PH model is a useful tool, *e.g.*, for modelling the impact of environment on lifetime random variables. It is widely used in survival analysis. Combine the multiplicative model (6.16) with the PH model in the following way:

$$\lambda(t,z,k) = zk\lambda(t) \equiv z_k\lambda(t), \tag{6.60}$$

where z, as previously, comes from the realization of an unobserved random frailty Z and k is a proportional factor from the 'conventional' PH model. For the sake of modelling, this factor is written in an 'aggregated' form and not *via* a vector of explanatory variables, as is usually done in statistical inference. Therefore, the baseline $F(t)$ is indexed by the random variable $Z_k = kZ$. Equivalently, Equation (6.60) can be interpreted as a frailty model with a mixing random variable Z and a baseline failure rate $k\lambda(t)$. These two simple equivalent interpretations will help us in what follows. Without losing generality, assume that the support for Z is $[0,\infty)$. Similar to (6.17), the mixture failure rate $\lambda_{mk}(t)$ for the described case is defined as

$$\lambda_{mk}(t) = k\lambda(t)\int_0^\infty z\pi_k(z \mid t)dz \equiv \lambda(t)E[Z_k \mid t]. \tag{6.61}$$

As $Z_k = kZ$, its pdf is

$$p_k(z) = \frac{1}{k}\pi\left(\frac{z}{k}\right).$$

Theorem 6.6. Let the mixture failure rates for the multiplicative models (6.16) and (6.60) be given by Equations (6.17) and (6.61) respectively and let $k > 1$. Assume that the following quotient increases in z:

$$\frac{\pi_k(z)}{\pi(z)} = \frac{\pi\left(\frac{z}{k}\right)}{k\pi(z)} \uparrow. \tag{6.62}$$

Then:

$$\lambda_{mk}(t) > \lambda_m(t),\ \forall t \in [0,\infty). \tag{6.63}$$

Proof. Although Inequality (6.63) seems trivial at first sight, it is valid only for some specific cases of mixing (*e.g.*, for the multiplicative model, which is considered now). Denote

$$\Delta\lambda_m(t) = \lambda_{mk}(t) - \lambda_m(t). \tag{6.64}$$

Similar to (6.49) and using Equation (6.5), it can be seen that the sign of this difference is defined by the sign of the following difference:

$$\begin{aligned}
&\int_0^\infty z\bar{F}(t,z)\pi_k(z)dz\int_0^\infty \bar{F}(t,z)\pi(z)dz - \int_0^\infty z\bar{F}(t,z)\pi_k(z)dz\int_0^\infty \bar{F}(t,z)\pi(z)dz \\
&= \int_0^\infty\int_0^\infty \bar{F}(t,u)\bar{F}(t,s)[u\pi_k(u)\pi(s) - s\pi_k(u)\pi(s)]duds \\
&= \underset{u>s}{\int_0^\infty\int_0^\infty} \bar{F}(t,u)\bar{F}(t,s)(u-s)(\pi_k(u)\pi(s) - \pi_k(s)\pi(u))duds.
\end{aligned} \tag{6.65}$$

Therefore, the sufficient condition for Inequality (6.63) is Relationship (6.62). It is easy to verify that this condition is satisfied, *e.g.*, for the gamma and the Weibull densities, which are often used for mixing. In fact, while deriving Equation (6.65), the multiplicative form of the model was not used. Thus, Theorem 6.6 is valid for the general mixing model (6.5), although the proportionality $Z_k = kZ$ has a clear meaning only for the multiplicative model. ■

Example 6.7 Consider the multiplicative gamma-frailty model of Example 6.6. The mixture failure rate $\lambda_m(t)$ in this case is given by Equation (6.52). The mixture failure rate $\lambda_{mk}(t)$ is

$$\lambda_{mk}(t) = \lambda(t)\frac{E^2[Z_k]}{E[Z_k] + Var(Z_k)\Lambda(t)}. \tag{6.66}$$

Let $k > 1$. Then

$$\lambda_{mk}(t) = \lambda(t)\frac{k^2E^2[Z]}{kE[Z] + k^2Var(Z)\Lambda(t)} > \lambda_m(t),$$

which is a direct proof of Inequality (6.63) in this special case.

The upper bound for $\lambda_{mk}(t)$ is given by the following theorem.

Theorem 6.7. Let the mixture failure rates for multiplicative models (6.16) and (6.60) be given by Equations (6.17) and (6.61) respectively and let $k > 1$.
Then

$$\lambda_{mk}(t) < k\lambda_m(t),\ t > 0. \tag{6.67}$$

Proof. As $Z_k = kZ$, it is clear that $\lambda_{mk}(0) = k\lambda_m(0)$. Consider the difference in (6.64) in a slightly different way than in the previous theorem. The mixture failure rate $\lambda_{mk}(t)$ will be defined equivalently by the baseline failure rate $k\lambda(t)$ and the mixing variable Z. This means that

$$\lambda_{mk}(t) - k\lambda_m(t) = k\lambda(t)(\hat{E}[Z \mid t] - E[Z \mid t]),$$

where conditioning in $\hat{E}[Z \mid t]$ is different from that in $E[Z \mid t]$ in the described sense. Denote

$$\bar{F}_k(t,z) = \exp\{-zk\Lambda(t)\}.$$

Similar to (6.65), $sign[\lambda_{mk}(t) - k\lambda_m(t)]$ is defined by

$$sign \int_0^\infty \int_0^\infty_{u>s} \pi(u)\pi(s)(u-s)(\bar{F}_k(t,u)\bar{F}(t,s) - \bar{F}(t,u)\bar{F}_k(t,s))duds,$$

which is negative for all $t > 0$, as

$$\frac{\bar{F}_k(t,z)}{\bar{F}(t,z)} = \exp\{-(k-1)z\Lambda(t)\}$$

is decreasing in z. ■

It is worth noting that we do not need additional conditions for this bound as in the case of Theorem 6.5. An obvious but meaningful consequence of (6.67) is

$$\lambda_{mk}(t) \neq k\lambda_m(t), t > 0.$$

Therefore, this theorem gives another explicit justification of a well-known fact:

The PH model in each realization does not result in the PH model for the corresponding mixture failure rates.

Example 6.7 (continued). The gamma-frailty model is a direct illustration of Inequality (6.67), which can be seen in the following way:

$$\lambda_{mk}(t) = \lambda(t)\frac{k^2E^2[Z]}{kE[Z]+k^2Var(Z)\Lambda(t)}$$

$$< \lambda(t)\frac{kE^2[Z]}{E[Z]+Var(Z)\Lambda(t)} = k\lambda_m(t).$$

Example 6.8 In this example, we will consider the stable frailty distributions. A distribution is strictly stable (Feller, 1971) if the sum of independent random variables described by this distribution follows the same distribution, *i.e.*,

$$c(n)Z_1 =_D Z_1 + Z_2 + ... + Z_n,$$

where $=_D$ denotes "the same distributions". The function $c(n)$ has the form $n^{1/\alpha}$, where α is between 0 and 2. The normal distribution results from $\alpha = 2$ and the degenerate distribution is defined by $\alpha = 1$. It follows from Hougaard (2000) that the Laplace transform of a stable distribution with a positive support is given by

$$L(s) = \exp\left\{-\frac{\beta s^{\alpha}}{\alpha}\right\}, \tag{6.68}$$

where β is a positive parameter and $\alpha \in (0,1]$ for a positive stable distribution. Applying Equation (6.27) to Model (6.16) results in

$$\lambda_m(t) = \beta\lambda(t)(\Lambda(t))^{\alpha-1}. \tag{6.69}$$

On the other hand, applying Equation (6.27) to (6.60) gives

$$\lambda_{mk}(t) = k^{\alpha}\beta\lambda(t)(\Lambda(t))^{\alpha-1} = k^{\alpha}\lambda_m(t). \tag{6.70}$$

Therefore, we observe proportionality in this setting but with the changing coefficient of proportionality (from k to k^{α}, respectively). It is clear that this specific result does not contradict Theorems 6.6 and 6.7, as it follows from (6.69) and (6.70) that for positive stable distributions ($\alpha \in (0,1)$) and $k > 1$, the following inequalities hold:

$$\lambda_m(t) < \lambda_{mk}(t) < k\lambda_m(t), \ t > 0.$$

6.7 Further Examples and Applications

6.7.1 Shocks in Heterogeneous Populations

Consider the general mixing model (6.4) and (6.5) for a heterogeneous population and assume that at time $t = t_1$ an instantaneous shock had occurred that affects the whole population. With the corresponding complementary probabilities it either kills (destroys) an item or 'leaves it unchanged'. Without losing generality, let $t_1 = 0$; otherwise a new initial mixing variable should be defined and the corresponding procedure can easily be adjusted to this case. It is natural to suppose that the frailer (with larger failure rates) the items are, the more susceptible they are to failure. This means that the probability of a failure (death) from a shock is an increasing function of the value of the failure rate of an item at $t = 0$. Therefore a shock performs a kind of a burn-in operation (see, *e.g.*, Block *et al.*, 1993; Mi, 1994; Clarotti and Spizzichino, 1999; Cha, 2000, 2006).

The initial pdf of a frailty Z before the shock is $\pi(z)$. After a shock the frailty and its distribution change to Z_1 and $\pi_1(z)$, respectively. As previously, let the mixture failure rate for a population without a shock be $\lambda_m(t), t \geq 0$ and denote the corresponding mixture failure rate for the same population after a shock at $t = 0$ by $\lambda_{ms}(t), t \geq 0$. We want to compare $\lambda_{ms}(t)$ and $\lambda_m(t)$. It is reasonable to suggest that $\lambda_{ms}(t) < \lambda_m(t)$, as the items with higher failure rates are more likely to be eliminated. As was already mentioned, the natural ordering for mixing distributions is the ordering in the sense of the likelihood ratio defined by Inequality (6.41). In accordance with this definition, assume that

$$Z \geq_{lr} Z_1, \tag{6.71}$$

which means that $\pi_1(z)/\pi(z)$ is a decreasing function. Now we are able to formulate the following result, which is proved in a way similar to Theorems 6.6 and 6.7.

Theorem 6.8. Let the mixing variables before and after a shock at $t = 0$ be ordered in accordance with (6.71). Assume that $\lambda(t,z)$ is ordered in z, *i.e.*,

$$\lambda(t,z_1) < \lambda(t,z_2), \quad z_1 < z_2, \forall z_1, z_2 \in [0,\infty], \ t \geq 0. \tag{6.72}$$

Then

$$\lambda_{ms}(t) < \lambda_m(t), \forall t \geq 0 . \qquad (6.73)$$

Proof. Inequality (6.72) is a natural ordering for the family of failure rates $\lambda(t,z), z \in [0,\infty)$ and trivially holds, *e.g.*, for the specific multiplicative model. Conducting all steps as when obtaining Equation (6.65) finally results in the following relationship:

$$sign[\lambda_{ms}(t) - \lambda_m(t)]$$

$$= sign \int_a^b \int_{\substack{a \\ u>s}}^b \overline{F}(t,u)\overline{F}(t,s)(\lambda(t,u) - \lambda(t,s))(\pi_1(u)\pi(s) - \pi_1(s)\pi(u))duds ,$$

which is negative due to (6.71) and (6.72). ■

In accordance with Inequality (6.73), $\lambda_{ms}(t) < \lambda_m(t)$ for $t \geq 0$. This fact seems intuitively evident, but it is valid only owing to the rather stringent conditions of this theorem. It can be shown, for example, that replacing (6.71) with a weaker condition of usual stochastic ordering $Z \geq_{st} Z_1$ does not guarantee Ordering (6.73) for all t.

6.7.2 Random Scales and Random Usage

Consider a system with a baseline lifetime Cdf $F(x)$ and a baseline failure rate $\lambda(x)$. Let this system be used intermittently. A natural model for this pattern is, *e.g.*, an alternating renewal process with periods when the system is 'on' followed by periods when the system is 'off'. Assume that the system does not fail in the 'off' state. If chronological (calendar) time t is sufficiently large, the process can be considered stationary. The proportion of time when the system is operating in $[0,t)$ is approximately $zt,\ 0 < z \leq 1$ in this case. Thus the relationship between the usage scale x and the chronological time scale t is

$$x = zt,\ \ 0 < z \leq 1 . \qquad (6.74)$$

Equation (6.74) defines a scale transformation for the lifetime random variable in the following way:

$$F(t,z) \equiv F(zt) .$$

Along with time scales x and t there can be other usage scales. For instance, in the automobile reliability application, the cumulative mileage y can play the role of this scale (Finkelstein, 2004a).

Let parameter z turn into a random variable Z with the pdf $\pi(z)$, which describes a random usage. In our terms, this is a mixture, *i.e.*,

$$F_m(t) = E[F(Zt)] = F(t_u) = \int_0^1 F(zt)\pi(z)dz ,$$

where t_u is an equivalent (deterministic) usage scale, which can also be helpful in modelling. Using the definition of the failure rate $\lambda(t,z) = f(t,z)/\overline{F}(t,z)$ for this specific case

$$\lambda(t,z) = z\lambda(zt). \tag{6.75}$$

The mixture failure rate is defined as

$$\lambda_m(t) = \int_0^1 z\lambda(zt)\pi(z \mid t)dz. \tag{6.76}$$

Equation (6.75) defines the failure rate for a well-known accelerated life model (ALM) to be studied in the next chapter. It seems that there is only a slight difference in comparison with the multiplicative model (6.16), *i.e.*, the multiplier z in the argument of the baseline failure rate $\lambda(t)$, but it turns out that this difference makes modelling much more difficult.

Example 6.9 Let the baseline failure rate be constant: $\lambda(t) = \lambda$. Then $\lambda(t,z) = z\lambda$. Assume that the mixing distribution is uniform: $\pi(z) = 1,\ z \in [0,1]$. Direct computation (Finkelstein, 2004a) results in

$$\lambda_m(t) = \frac{(1 - \exp\{-\lambda t\}) - \lambda t \exp\{-\lambda t\}}{t(1 - \exp\{-\lambda t]} \to \frac{1}{t}$$

as $t \to \infty$. Thus, the failure rate in the calendar time scale is decreasing in $[0, \infty)$ and is asymptotically approaching t^{-1}, whereas the baseline failure rate in the usage scale x is constant. This means that a random usage can dramatically change the shape of the corresponding failure rate.

Let the baseline failure rate be an increasing power function (the Weibull law): $\lambda(t) = \lambda t^{\gamma-1}; \lambda > 0, \gamma > 1$. Equation (6.75) becomes $\lambda(t,z) = z^{\gamma}\lambda$. Assume for simplicity that the mixing random variable Z^{γ} is also uniformly distributed in $[0,1]$. Direct integration in (6.76) (Finkelstein, 2004a) gives

$$\lambda_m(t) = \frac{\gamma[(1 - \exp\{-\lambda_b t^{\gamma}\}) - \lambda_b t \exp\{-\lambda_b t^{\gamma}\}]}{t(1 - \exp\{-\lambda_b t^{\gamma}]} \to \frac{\gamma}{t} \text{ as } t \to \infty,$$

where $\lambda_b = \lambda(\gamma)^{-1}$. The shape of $\lambda_m(t)$ is similar to the shape that was discussed while deriving Relationship (3.11) for the gamma-Weibull mixture in the multiplicative model. But this is not surprising at all, because for the baseline Weibull distribution only, the accelerated life model can be reparameterized to result in the multiplicative model (Cox and Oakes, 1984). As in Equation (3.11), $\lambda_m(t)$ in this case asymptotically tends to 0, although the baseline failure rate is increasing.

6.7.3 Random Change Point

In reliability analysis, it is often reasonable to assume that early failures follow one distribution (infant mortality), whereas after some time another distribution with another pattern comes into play. Alternatively, a device starting to function at some small level of stress can experience an increase of this stress at some instant of time $t = z$. Most often a change in the original pattern of the failure rate is caused by some external factors (*e.g.*, a change in environment). The simplest failure rate change point model (Patra and Dey, 2002) is defined as

$$\lambda(t,z) = \lambda_1(t)I(t < z) + \lambda_2(t)I(t \geq z),\ t \geq 0, \tag{6.77}$$

where $\lambda_1(t)$ is the failure rate before the change point, $\lambda_2(t)$ is the failure rate after it and $I(t < z), I(t \geq z)$ are the corresponding indicators.

Denote the Cdfs that correspond to $\lambda_1(t), \lambda_2(t)$ and $\lambda(t,z)$ by $F_1(t), F_2(t)$ and $F(t,z)$, respectively. The survival function corresponding to the failure rate $\lambda(t,z)$ is

$$\bar{F}(t,z) = \bar{F}_1(t)I(t < z) + \bar{F}_1(z)\frac{\bar{F}_2(t)}{\bar{F}_2(z)}I(t \geq z),$$

where the definition of the mean remaining lifetime (2.3) is used. Assume now that the change point Z is a random variable. It is clear that this is a mixing model and we can use our expressions for $\pi(z \mid t)$ and $\lambda_m(t)$, *i.e.*,

$$\pi(z \mid t) = \frac{\pi(z)}{\int_0^\infty \bar{F}(t,z)\pi(z)dz}\begin{cases} \bar{F}_1(t), & t < z, \\ \dfrac{\bar{F}_1(z)}{\bar{F}_2(z)}\bar{F}_2(t), & t \geq z. \end{cases}$$

Eventually,

$$\lambda_m(t) = \frac{\lambda_1(t)\bar{F}_1(t)\int_t^\infty \pi(z)dz + \lambda_2(t)\bar{F}_2(t)\int_0^t \dfrac{\bar{F}_1(z)}{\bar{F}_2(z)}\pi(z)dz}{\bar{F}_1(t)\int_t^\infty \pi(z)dz + \bar{F}_2(t)\int_0^t \dfrac{\bar{F}_1(z)}{\bar{F}_2(z)}\pi(z)dz}. \tag{6.78}$$

Let specifically $\lambda_1(t) = \lambda_1, \lambda_2(t) = \lambda_2$ and $\pi(z)$ also be an exponential distribution with parameter λ_c. Equation (6.78) simplifies to

$$\lambda_m(t) = \frac{\lambda_1 + \dfrac{\lambda_2\lambda_c}{\lambda_2 - \lambda_1 - \lambda_c}(1 - \exp\{-(\lambda_2 - \lambda_1 - \lambda_c\}t)}{1 + \dfrac{\lambda_c}{\lambda_2 - \lambda_1 - \lambda_c}(1 - \exp\{-(\lambda_2 - \lambda_1 - \lambda_c\}t)}. \tag{6.79}$$

It is clear that $\lambda_m(0) = \lambda_1$. Let $\lambda_2 > \lambda_1 + \lambda_c$. Then

$$\lim_{t\to\infty} \lambda_m(t) = \lambda_1 + \lambda_c.$$

It can be shown that $\lambda'(t) > 0, \forall t \geq 0$, which means that $\lambda(t)$ monotonically increases from λ_1 to $\lambda_1 + \lambda_c$ as $t \to \infty$. Let $\lambda_1 < \lambda_2 < \lambda_1 + \lambda_c$. It follows from Equation (6.79) that

$$\lim_{t\to\infty} \lambda_m(t) = \lambda_2 . \tag{6.80}$$

Finally, (6.80) also holds for $\lambda_2 < \lambda_1$. Therefore, $\lim_{t\to\infty} \lambda(t)$ in this special case depends on the relationships between λ_1, λ_2 and λ_c.

6.7.4 MRL of Mixtures

The MRL function was defined by Equation (2.7). Along with the failure rate, this is also the most important characteristic of a lifetime random variable. The MRL function can constitute a convenient and reasonable model of mixing in applications, although we think that this approach has not received the proper attention in the literature so far. In accordance with (2.7), the MRL can be defined for each value of z via the corresponding survival function as

$$m(t,z) = \frac{\int_t^\infty \bar{F}(u,z)du}{\bar{F}(t,z)} . \tag{6.81}$$

Substitution of the mixture survival function $\bar{F}_m(t)$ instead of $\bar{F}(t)$ in the right-hand side of Equation (2.7) results in the following formal definition of the mixture MRL function:

$$m_m(t) = \frac{\int_t^\infty \bar{F}_m(u)du}{\bar{F}_m(t)} = \frac{\int_t^\infty \int_0^\infty \bar{F}(u,z)\pi(z)d\theta du}{\int_0^\infty \bar{F}(t,z)\pi(\theta)dz} . \tag{6.82}$$

Assuming that the integrals in (6.82) are finite, we can transform this equation by changing the order of integration, *i.e.*,

$$m_m(t) = \frac{\int_t^\infty \int_0^\infty \bar{F}(u,z)\pi(z)dzdu}{\int_0^\infty \bar{F}(t,z)\pi(z)dz} = \int_0^\infty m(t,z)\pi(z \mid t)dz , \tag{6.83}$$

where, in accordance with Equation (3.10), the conditional density $\pi(z \mid t)$ of the mixing variable Z (on the condition that $T > t$) is

$$\pi(z \mid t) = \frac{\pi(z)\bar{F}(t,z)}{\int_0^\infty \bar{F}(t,z)\pi(z)dz}.$$

Therefore, formal definition (6.82) is equivalent to a self-explanatory mixing rule (6.83). Equation (6.83) enables us to analyse the shape of $m_m(t)$. It can also be done directly *via* Equation (6.82) or *via* the corresponding mixture failure rate $\lambda_m(t)$, because sometimes it is more convenient to define $\lambda_m(t)$ from the very beginning. It is clear that if $\lambda_m(t)$ is increasing (decreasing) in $[0,\infty)$, then $m_m(t)$ is decreasing (increasing) in $[0,\infty)$. It also follows from the results of Section 2.4 that if, for example, $\lambda_m(t)$ has a bathtub shape and condition $m'_m(0) < 0$ takes place, then the MRL function $m_m(t)$ is decreasing in $[0,\infty)$. It can be shown that under some assumptions mixtures of increasing MRL distributions also have increasing MRL functions.

Mixing in Equations (6.82) and (6.83) is defined by the 'ordinary' mixture of the corresponding distribution. The model of mixing, however, can be defined directly by $m(t,z)$ as well. The simplest natural model of this kind is

$$m(t,z) = \frac{m(t)}{z}, \; z > 0, \tag{6.84}$$

which is similar to the multiplicative model of mixing for the failure rate. This model was considered in Zahedi (1991) for modelling the impact of an environment as an alternative to the Cox PH model. Some ageing properties of mixtures, defined by Relation (6.84), were described by Badia *et al.* (2001). Properties of the mixture MRL function were also analysed in Mi (1999) and Finkelstein (2002a), among others.

6.8 Chapter Summary

The mixture failure rate $\lambda_m(t)$ is defined by Equation (6.5) as a conditional expectation of a random failure rate $\lambda(t,Z)$. A family of failure rates of subpopulations $\lambda(t,z),\ z \in [a,b]$ describes heterogeneity of a population itself. Our main interest in this chapter is in failure rate modelling for heterogeneous populations. One can hardly find homogeneous populations in real life, although most studies on failure rate modelling deal with a homogeneous case. Neglecting existing heterogeneity can lead to substantial errors and misconceptions in stochastic analysis in reliability, survival and risk analysis and other disciplines.

It is well known that mixtures of DFR distributions are always DFR. On the other hand, mixtures of increasing failure rate (IFR) distributions can decrease at least in some intervals of time, which means that the IFR class of distributions is not closed under the operation of mixing. As IFR distributions usually model lifetimes governed by ageing processes, the operation of mixing can dramatically change the pattern of ageing, *e.g.*, from positive ageing (IFR) to negative ageing (DFR).

The mixture failure rate is bent down due to "the weakest populations are dying out first" effect. This should be taken into account when analysing the failure data for heterogeneous populations. If mixing random variables are ordered in the sense of the likelihood ratio, the mixture failure rates are ordered accordingly. Mixing distributions with equal expectations and different variances can lead to the corresponding ordering for mixture failure rates in some special cases. For the general mixing distribution in the multiplicative model, however, this ordering is guaranteed only for a sufficiently small amount of time.

The problem with random usage of engineering devices can be reformulated in terms of mixtures. This is done for the automobile example in Section 6.7.2, where the behaviour of the mixture failure rate was analysed for this special case.

The mixture MRL function $m_m(t)$ is defined by Equation (6.83) and can be studied in a similar way to $\lambda_m(t)$, but this topic needs further attention. Alternatively, it can be defined in a direct way, *e.g.*, as in an inverse-proportional model (6.84).

7

Limiting Behaviour of Mixture Failure Rates

7.1 Introduction

In this chapter, we obtain explicit asymptotic results for the mixture failure rate $\lambda_m(t)$ as $t \to \infty$. A general class of distributions is suggested that contains as special cases additive, multiplicative and accelerated life models that are widely used in practice. Although the accelerated life model (ALM) is the main tool for modelling and statistical inference in accelerated life testing (Bagdonavicius and Nikulin, 2002), there are practically no results in the literature on the mixture failure rate modelling for this model. One could mention some initial descriptive findings by Anderson and Louis (1995) and analytical derivation of bounds for the distance of a mixture from a parental distribution in Shaked (1981).

The approach developed in this chapter allows for the *asymptotic analysis* of the mixture failure rates for the ALM and, in fact, results in some counterintuitive conclusions. Specifically, when the support of the mixing distribution is $[0, \infty)$, the mixture failure rate in this model converges to 0 as $t \to \infty$ and does not depend on the baseline distribution. On the other hand, the ultimate behaviour of $\lambda_m(t)$ for other models depends on a number of factors, and specifically on the baseline distribution. Depending on the parameters involved, it can converge to 0, tend to ∞ or exhibit some other behaviour.

There are many applications where the behaviour of the failure rate at relatively large values of t is really important. In the previous chapter, the example of the oldest-old mortality was discussed when the exponentially increasing Gompertz mortality curve is bent down for advanced ages (mortality plateau). As we already stated, owing to the principle "the weakest populations are dying out first", many mixtures with the IFR baseline failure rate exhibit (at least ultimately) a decreasing mixture failure rate pattern. This change of the ageing pattern should definitely be taken into account in many engineering applications as well. For instance, what is the reason for the preventive replacement of an ageing item if, owing to heterogeneity, the 'new' item can have a larger failure rate and therefore be less reliable? In spite of the mathematically intensive contents, this chapter presents a number of clearly formulated results that can be used in practical analysis.

The developed approach is different from that described in Block *et al.* (1993, 2003) and Li (2005) and, in general, follows Finkelstein and Esaulova (2006a). On one hand, we obtain *explicit* asymptotic formulas in a direct way; on the other hand, we are also able to analyse some useful general asymptotic properties of the models. In Section 7.5, we discuss the multivariate frailty in the competing risks framework. This discussion is based on the generalization of the univariate approach to the bivariate case.

The presentation of this chapter is rather technical. Therefore, the sketches of the proofs are deferred to Section 7.7 and can be skipped by the reader who is uninterested in mathematical details.

First, we turn to some introductory results for the limiting behaviour of discrete mixtures that will help in understanding the nature of the limiting behaviour, when $\lambda_m(t)$ tends to the failure rate of the strongest population.

7.2 Discrete Mixtures

Let the frailty (unobserved random parameter) Z for the lifetime T be a discrete random variable taking values in a set $z_1, z_2, ..., z_n$ with probabilities $\pi_i(z_i)$, $i = 1,2,...,n$, respectively. This discrete case can be very helpful for understanding certain basic issues for a more 'general' continuous setting. Some initial properties for discrete mixtures were already discussed in Section 6.2. In this section, the mixture of two distributions will be considered and it will be shown under some weak assumptions that the corresponding mixture failure rate is converging to the failure rate of the strongest population. This result is obviously important from both a theoretical and a practical point of view, as it explains certain facts that were already observed for various special cases.

Similar to the continuous case, the mixture failure rate can be defined as

$$\lambda_m(t) = \sum_1^n \lambda(t, z_i)\pi(z_i \mid t), \tag{7.1}$$

where conditional probabilities $\pi(z_i \mid t)$ of $Z = z_i$ given $T > t, i = 1,2,...,n$ are

$$\pi(z_i \mid t) = \frac{\pi(z_i)\bar{F}(t, z_i)}{\sum_1^n \bar{F}(t, z_i)\pi(z_i)}. \tag{7.2}$$

Note that Equations (7.1) and (7.2) define the mixing model governed by the distribution $F(t, z_i)$ indexed by the discrete random variable Z. This setting is basic and is suitable for describing heterogeneity *via* the unobserved parameter Z. The multiplicative model (6.16), which will be studied in this section, is defined for the discrete case in a similar way as

$$\lambda(t, z_i) = z_i\lambda(t), \tag{7.3}$$

where $\lambda(t)$ is a baseline failure rate. Therefore, as in (6.17), Equation (7.1) reads

$$\lambda_m(t) = \lambda(t)E[Z \mid t].$$

Let, for simplicity, $n = 2$. The following results can easily be adjusted to the general case. Denote $\pi(z_1) = \pi_1$, $\pi(z_2) = \pi_2 = 1 - \pi_1$ and let $z_2 > z_1 > 0$.

Then

$$\lambda_m(t) = \lambda(t, z_1)\pi(z_1 \mid t) + \lambda(t, z_2)\pi(z_2 \mid t), \tag{7.4}$$

where

$$\pi(z_i \mid t) = \frac{\pi_i \bar{F}(t, z_i)}{\pi_1 \bar{F}(t, z_1) + \pi_2 \bar{F}(t, z_2)}, \ i = 1,2 . \tag{7.5}$$

Example 7.1 Consider the Weibull distribution of the following form:

$$\bar{F}(t, z_i) = \exp\{-z_i t^b\} \, ; \, \lambda(t, z_i) = z_i b t^{b-1}; \ b > 1, \ i = 1,2 .$$

Thus, in accordance with (7.4) and (7.5), the corresponding mixture failure rate for the multiplicative model is

$$\lambda_m(t) = z_1 b t^{b-1} \frac{\pi_1 \exp\{-z_1 t^b\}}{\pi_1 \exp\{-z_1 t^b\} + \pi_2 \exp\{-z_2 t^b\}}$$

$$+ z_2 b t^{b-1} \frac{\pi_2 \exp\{-z_2 t^b\}}{\pi_1 \exp\{-z_1 t^b\} + \pi_2 \exp\{-z_2 t^b\}} .$$

These equations suggest that when $t \to \infty$,

$$\lambda_m(t) - \lambda(t, z_1) = (z_2 - z_1) b t^{b-1} \frac{\pi_2}{\pi_1} \exp\{-(z_2 - z_1) t^b\}(1 + o(1)) \to 0 , \tag{7.6}$$

and the mixture failure rate, as $t \to \infty$, converges to the failure rate of the strongest population from above. When $b = 1$, the setting reduces to the well-known exponential case (Barlow and Proschan, 1975).

Although the failure rate $\lambda(t, z_1)$ in this example is increasing as a power function, the distance between it and the mixture failure rate $\lambda_m(t)$ tends to 0 as $t \to \infty$. For the general setting (7.4), when $\lambda(t, z_1) \to \infty$, we will distinguish between the convergence

$$\lambda_m(t) - \lambda(t, z_1) \to 0 \text{ as } t \to \infty \tag{7.7}$$

and the following asymptotic equivalence:

$$\lambda_m(t) = \lambda(t, z_1)(1 + o(1)) \text{ as } t \to \infty , \tag{7.8}$$

which will mostly be used in this chapter in the following alternative notation (Relationship (2.54)):

$$\lambda_m(t) \sim \lambda(t, z_1) \text{ as } t \to \infty .$$

It is obvious that when $\lambda(t, z_1)$ has a finite limit, then (7.7) and (7.8) coincide. The main limiting results of this chapter will be asymptotic in the sense of Relationship (7.8), but the following theorem refers to both notions.

Theorem 7.1. Consider the mixture model (7.4) and (7.5). Let

$$\lambda(t, z_1) = z_1\lambda(t),\ \lambda(t, z_2) = z_2\lambda(t);\ z_2 > z_1 > 0 ,$$

where $\lambda(t) \to \infty$ as $t \to \infty$.

Then

- Relationship (7.8) holds;
- Relationship (7.7) holds if

$$\lambda(t)\exp\{-(z_2 - z_1)\int_0^t \lambda(u)du\} \to 0 \text{ as } t \to \infty . \tag{7.9}$$

Proof. Denote $c \equiv z_2 / z_1 > 1$. Using simple transformations, similar to Block and Joe (1997):

$$\frac{\lambda_m(t)}{\lambda(t, z_1)} = 1 + \frac{\pi_1(c-1)}{\pi_1 + \pi_2(\bar{F}(t, z_1))^{1-c}} .$$

As $\bar{F}(t, z_1) \to 0$ for $t \to \infty$, we immediately arrive at (7.8), whereas the condition

$$\lambda(t, z_1)(\bar{F}(t, z_1))^{c-1} \to 0 \text{ as } t \to \infty ,$$

which is equivalent to (7.9), leads to the convergence result (7.7). ■

It is clear that this theorem holds when $\lim \lambda(t)$ is constant as $t \to \infty$. In this case (7.7) and (7.8) coincide. Similar results for the discrete mixture of different distributions can be found in Block and Joe (1997) and Block *et al.* (2003). See also Vaupel and Yashin (1985) for some meaningful illustrative graphs.

Remark 7.1 Condition (7.9) is a rather weak one. In essence, it states that the pdf of a distribution with an ultimately increasing failure rate tends to 0 as $t \to \infty$. All distributions that are typically used in lifetime data analyses meet this requirement. But one can consider some 'bizarre' distributions, for which Condition (7.9) does not hold. Let, for instance,

$$\lambda(t) = \beta_{n+1},\ t \in [n, n+1),\ n = 0,1,2,.. ,$$

where

$$\beta_1 = 1,\ \beta_{n+1} = \exp\left\{\sum_{i=1}^{n} \beta_i\right\};\ n = 1,2,\dots\ .$$

The failure rate $\lambda(t)$ defined in this way is a piecewise continuous function. It is easy to verify in this case that (7.9) does not hold. Therefore, there is no convergence defined by Relationship (7.7). The author would like to thank Professors Henry Block and Thomas Savits for this example.

The rest of this chapter is devoted to a much more general continuous mixing model, which includes, as already mentioned, the additive, the multiplicative and the accelerated life models as special cases.

7.3 Survival Model

We will define now a rather general class of lifetime distributions $F(t,z)$ and we will study the asymptotic behaviour of the corresponding mixture failure rate $\lambda_m(t)$ in the mixing model (6.4) and (6.5). It is more convenient from the start to give this definition in terms of the cumulative failure rate $\Lambda(t,z)$ rather than in terms of the failure rate $\lambda(t,z)$. The basic model is given by the following general equation:

$$\Lambda(t,z) = A(z\phi(t)) + \psi(t),\quad \Lambda(t,z) \equiv \int_0^t \lambda(t,z)\,. \tag{7.10}$$

The natural properties of the cumulative failure rate of the absolutely continuous distribution $F(t,z)$ (for $\forall z \in [0,\infty)$) imply that the functions $A(s), \phi(t)$ and $\psi(t)$ are differentiable, that the right-hand side of (7.10) is non-decreasing in t and tends to infinity as $t \to \infty$, and that

$$A(z\phi(0)) + \psi(0) = 0\,.$$

Therefore, these properties will be assumed throughout the chapter, although some of them will not be needed in the formal proofs.

An important additional simplifying assumption is that the functions

$$A(s),\ s \in [0,\infty);\ \phi(t),\ \ t \in [0,\infty)$$

are increasing functions of their arguments, and therefore we will view

$$1 - \exp\{-A(z\phi(t)),\ \ z \neq 0$$

as a lifetime Cdf. The failure rate $\lambda(t,z)$ is obtained by differentiation of the cumulative failure rate $\Lambda(t,z)$, *i.e.*,

$$\lambda(t,z) = z\phi'(t)A'(z\phi(t)) + \psi'(t)\,. \tag{7.11}$$

We are now able to explain why we start with the cumulative failure rate rather than with the failure rate itself, as is often done in lifetime modelling. The reason is that we can easily suggest intuitive interpretations for (7.10), whereas it is certainly not so simple to interpret the failure rate structure of the form (7.11) without stating that it follows from the structure of the cumulative failure rate.

Relationship (7.10) defines a rather broad class of survival models that can be used, *e.g.*, for modelling the impact of environment on characteristics of survival. The obvious restriction of the model is the multiplicative nature of the argument $z\phi(t)$, which can easily be generalized to $g(z)\phi(t)$, but not to a general function of t and z. From a practical point of view, Relationship (7.10) is general enough and, as was already mentioned, includes as specific cases the proportional hazards (PH), additive hazards (AH) and accelerated life (ALM) models that are widely used in reliability, survival and risk analysis.

PH (multiplicative) Model: Let

$$A(u) \equiv u,\ \phi(t) = \Lambda(t),\ \psi(t) \equiv 0\,.$$

Then

$$\lambda(t,z) = z\lambda(t), \quad \Lambda(t,z) = z\Lambda(t) = z\int_0^t \lambda(u)du\,. \tag{7.12}$$

ALM: Let

$$A(u) \equiv \Lambda(u),\ \phi(t) = t,\ \psi(t) \equiv 0\,.$$

Then

$$\Lambda(t,z) = \int_0^{tz} \lambda(u)du = \Lambda(tz), \quad \lambda(t,z) = z\lambda(tz)\,. \tag{7.13}$$

AH Model: Let

$$A(u) \equiv u,\ \phi(t) = t,\ \psi(t) \text{ is increasing},\ \psi(0) = 0\,.$$

Then

$$\lambda(t,z) = z + \psi'(t), \quad \Lambda(t,z) = zt + \psi(t)\,. \tag{7.14}$$

Equations (7.12)–(7.14) show that even the simplest forms of the functions involved result in a number of well-known models. The functions $\lambda(t)$ and $\psi'(t)$ play the role of baseline failure rates in Equations (7.12)–(7.13) and (7.14). Note that, in all these models, the functions $\phi(t)$ and $A(s)$ are monotonically increasing.

The asymptotic behaviour of mixture failure rates for the PH and AH models was studied for some specific mixing distributions in, *e.g.*, Gurland and Sethuraman (1995) and Finkelstein and Esaulova (2001). Theorem 6.1 of the previous chapter also describes the ultimate behaviour of the mixture failure rate in the AH model.

7.4 Main Asymptotic Results

In this section, the main asymptotic results are formulated and discussed. The proofs are rather technical and cumbersome. Therefore, the corresponding sketches are deferred to the last section of this chapter. As the methodology of the proofs is innovative and important for the developed approach, we feel that the reader should have an opportunity to follow the abridged versions. The full text of the proofs and some additional results can be found in Finkelstein and Esaulova (2006a).

The following theorem derives an asymptotic formula for the mixture failure rate $\lambda_m(t)$ under rather mild assumptions.

Theorem 7.2. Let the cumulative failure rate $\Lambda(t,z)$ be given by Model (7.10) and let the mixing pdf $\pi(z),\ z\in[0,\infty)$ be defined as

$$\pi(z)=z^{\alpha}\pi_1(z)\,, \tag{7.15}$$

where $\alpha>-1$ and $\pi_1(z),\pi_1(0)\neq 0$ is a function bounded in $[0,\infty)$ and continuous at $z=0$. Assume also that

$$\phi(t)\to\infty \quad \text{as } t\to\infty \tag{7.16}$$

and that $A(s)$ satisfies

$$\int_0^\infty \exp\{-A(s)\}s^{\alpha}ds<\infty\,. \tag{7.17}$$

Then

$$\lambda_m(t)-\psi'(t)\sim(\alpha+1)\frac{\phi'(t)}{\phi(t)}. \tag{7.18}$$

Specifically, if the additive term is equal to zero (this term is, in fact, not important, as all conditions are formulated in terms of the functions $A(s),\phi(t)$ and $\pi_1(z)$), Equation (7.18) reduces (as $t\to\infty$) to

$$\lambda_m(t)\sim(\alpha+1)\frac{\phi'(t)}{\phi(t)}. \tag{7.19}$$

Remark 7.2 Assumption (7.15) holds for the main lifetime distributions, such as Weibull, gamma, lognormal, *etc*. Assumption (7.16) states a natural condition for the function $\phi(t)$, which can often be viewed as a scale transformation. Condition (7.17) means that the Cdf $1-\exp\{-A(s)\}$ should not be 'too heavy-tailed' (as, *e.g.*, is the Pareto distribution $1-s^{-\beta}$, for $s\geq 1,\beta-\alpha>1$) and is equivalent to the condition of the existence of the moment of order $\alpha+1$ for this Cdf. Examples in the next section will clearly show that these conditions are not at all stringent and can easily be met in most practical situations.

A crucial feature of this result is that the asymptotic behaviour of the mixture failure rate depends only (omitting an obvious additive term) on the behaviour of the mixing distribution in the neighbourhood of 0 and on the derivative of the logarithm of the scale function $\phi(t):(\log\phi(t))'=\phi'(t)/\phi(t)$. When $\pi(0)\neq 0$ and $\pi(z)$ is bounded in $[0,\infty)$, the result does not depend on the mixing distribution at all, as $\alpha=0$ in this case.

The following example shows that the asymptotic relationship for $\lambda_m(t)$ could be different if the behaviour of the mixing distribution at $z=0$ does not comply with Condition (7.15).

Example 7.2 Consider the multiplicative model (7.12) and let the mixing density be given by

$$\pi(z)=\frac{1}{\sqrt{\pi z}\sqrt{z}}\exp\{-1/z\}.$$

It can be shown by direct integration in (6.5) that

$$\lambda_m(t)=\frac{\lambda(t)}{\sqrt{\Lambda(t)}},$$

which definitely differs from (7.19). It will be shown in the next section that for the multiplicative model, Relationship (7.19) implies that $\lambda_m(t)\sim(\alpha+1)\lambda(t)/\Lambda(t)$.

Theorem 7.2 deals with the case when the support of a mixing distribution includes 0, *i.e.*, $z\in[0,\infty)$. In this case, the strongest population for $\psi'(t)=0$ is not usually defined as, *e.g.*, in multiplicative and accelerated life models. For a specific additive model and for Model (7.11) with $\psi'(t)\neq 0$, the strongest population can formally be defined as, *e.g.*, $\lambda(t,0)=\psi'(t)$. If, however, the support is separated from 0, the situation changes significantly and the mixture failure rate can tend to the failure rate of the strongest population as $t\to\infty$, even when the additive term in (7.11) is 0. The following theorem states reasonable conditions for that, and we assume, for simplicity, that $\psi(t)=0$.

Theorem 7.3. Let, as in Theorem 7.2, the class of lifetime distributions be defined by Equation (7.10), where $\phi(t)\to\infty$, and let $A(s)$ be twice differentiable.

Assume that, as $s\to\infty$

$$\frac{A''(s)}{(A'(s))^2}\to 0 \tag{7.20}$$

and

$$sA'(s)\to\infty. \tag{7.21}$$

Also assume that for all $b, c > a, b < c$, the quotient $A'(bs)/A'(cs)$ is bounded as $s \to \infty$. Finally, let the mixing pdf $\pi(z)$ be defined in $[a, \infty)$, $a > 0$, bounded in this interval and continuous at $z = a$ and let it satisfy $\pi(a) \neq 0$.

Then

$$\lambda_m(t) \sim a\phi'(t)A'(a\phi(t)). \tag{7.22}$$

Remark 7.3 There are many assumptions in this theorem, but they are rather natural and hold at least for the specific models under consideration. It can easily be verified that Conditions (7.20) and (7.21) trivially hold for the specific multiplicative and additive models of the previous section. We will discuss these conditions later within the framework of the ALM. More generally, the conditions of the theorem hold if $A(s)$ belongs to a rather wide class of functions of smooth variation (Bingham *et al.*, 1987).

Assume additionally that the family of failure rates (7.11) is ordered in z, at least ultimately (for large t), *i.e.*,

$$\lambda(t, z_1) < \lambda(t, z_2), \quad z_1 < z_2, \forall z_1, z_2 \in [z_0, \infty], \; z_0 \geq 0, t \geq 0. \tag{7.23}$$

Then, as mentioned, Theorem 7.3 can be interpreted *via* the principle that *the mixture failure rate converges to the failure rate of the strongest population.* The right-hand side of (7.22) can also be interpreted in this case as the failure rate of the strongest population for a survival model, defined by a random variable with the Cdf $1 - \exp\{A(z\phi(t)\}$).

7.5 Specific Models

7.5.1 Multiplicative Model

The general definition of the mixture failure rate in (6.5) for the multiplicative model (7.12) reduces to

$$\lambda_m(t) = \frac{\int_0^\infty z\lambda(t)\exp\{-z\Lambda(t)\}\pi(z)dz}{\int_0^\infty \exp\{-z\Lambda(t)\}\pi(z)dz}. \tag{7.24}$$

As $A(u) \equiv u, \phi(t) = \Lambda(t), \psi(t) \equiv 0$ in this specific case, Theorems 7.2 and 7.3 simplify to the following corollaries.

Corollary 7.1. Assume that the mixing pdf $\pi(z), z \in [0, \infty)$ in Model (7.12) can be written as

$$\pi(z) = z^\alpha \pi_1(z), \tag{7.25}$$

where $\alpha > -1$ and $\pi_1(z)$ is bounded in $[0, \infty)$, continuous at $z = 0$, and satisfies $\pi_1(0) \neq 0$.

Then the mixture failure rate for Model (7.12) has the following asymptotic behaviour:

$$\lambda_m(t) \sim \frac{(\alpha+1)\lambda(t)}{\int_0^t \lambda(u)du}. \tag{7.26}$$

Corollary 7.2. Assume that the mixing pdf $\pi(z),\ z \in [a,\infty)$ in (7.12) (we can define $\pi(z)=0$ for $z \in [0, a\,))$ is bounded, right semicontinuous at $z = a$, and satisfies $\pi(a) \neq 0$.

Then, in accordance with (7.22), the mixture failure rate for Model (7.12) has the following asymptotic behaviour:

$$\lambda_m(t) \sim a\lambda(t). \tag{7.27}$$

Corollary 7.1 states a remarkable fact: the asymptotic behaviour of the mixture failure rate $\lambda_m(t)$ depends only on the behaviour of the mixing pdf in the neighbourhood of $z = 0$ and the baseline failure rate $\lambda(t)$.

Corollary 7.2 describes the convergence of a mixture failure rate to the mixture failure rate of the strongest population. In this simple multiplicative case, the family of the failure rates is trivially ordered in z and the strongest population has the failure rate $a\lambda(t)$.

The next theorem generalizes the result of Corollary 7.2.

Theorem 7.4. Assume that the mixing pdf $\pi(z)$ in (7.12) has a support in $[a,b]$, $a > 0, b \leq \infty$, and, for $z \geq a$, can be defined as

$$\pi(z) = (z-a)^{\alpha}\pi_1(z-a), \tag{7.28}$$

where $\alpha > -1$ and $\pi_1(z-a)$ is bounded in $z \in [a,b]$, with $\pi_1(0) \neq 0$.

Then

$$\lambda_m(t) \sim a\lambda(t). \tag{7.29}$$

as $t \to \infty$.

It is quite remarkable that this asymptotic result does not depend on a mixing distribution (even in the case of a singularity at $z = a$). Relationship (7.29) also describes the convergence to the failure rate of the strongest population, which differs dramatically from the convergence described by (7.26). The explanation of this difference is quite obvious: owing to the multiplicative nature of the model, the behaviour of $z\lambda(t)$ in the neighbourhood of $z = 0$ (for Density (7.25)) is different from the behaviour of this product in the neighbourhood of $z = a$ (for Density (7.28)). Thus, the effect of a multiplier $z \to 0$ is a decisive factor for the shape of the function on the right-hand side of (7.26).

Example 7.3 Let the mixing distribution be the gamma distribution with the pdf

$$\pi(z) = \frac{\beta^c z^{c-1}}{\Gamma(c)} \exp\{-\beta z\};\ c, \beta > 0,$$

where the notation for the shape parameter was changed to c in order to avoid a confusion with parameter α in (7.26). Owing to the results of Section 6.4, the exact formula for the mixture failure rate in this case is given by

$$\lambda_m(t) = \frac{c\lambda(t)}{\beta + \Lambda(t)} = \frac{c\lambda(t)}{\beta + \int_0^t \lambda(u)du}. \tag{7.30}$$

It is clear that, as $\Lambda(t) \to \infty$ for $t \to \infty$ and $c = \alpha + 1$ for the gamma pdf, this formula perfectly complies with the general asymptotic result (7.26).

It follows from Equation (3.11) that, as $t \to \infty$, the gamma mixture for the baseline Weibull distribution is asymptotically proportional to $1/t$, which also complies with (7.26).

Example 7.4 Consider the gamma mixture for the baseline Gompertz distribution with the failure rate $\lambda(t) = a\exp\{bt\}$, $a, b > 0$. Computation in accordance with Equation (7.30) for this baseline failure rate results in

$$\lambda_m(t) = \frac{bc\exp\{bt\}}{\exp\{bt\} + \left(\dfrac{b\beta}{a} - 1\right)}. \tag{7.31}$$

If $b\beta = a$, then $\lambda_m(t) \equiv bc$. However, if $b\beta > a$, then $\lambda_m(t)$ increases to bc and if $b\beta < a$, it decreases to bc. The corresponding graph is given in Figure 7.1.

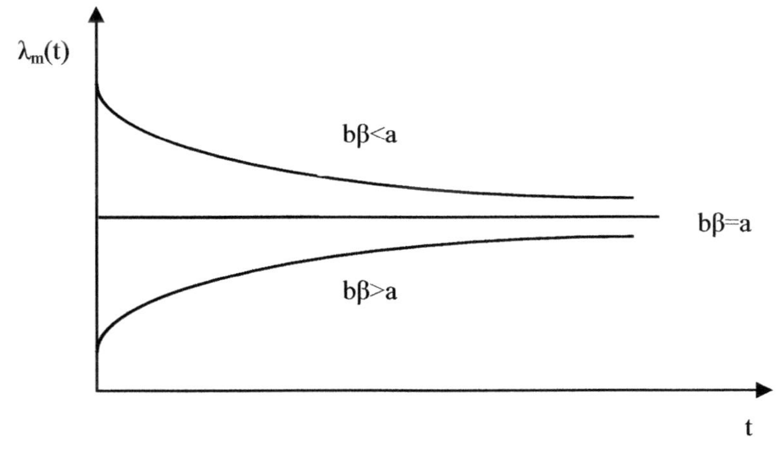

Figure 7.1. Gamma-Gompertz mixture failure rate

Obviously, Relationship (7.26) gives the same asymptotic value bc for $t \to \infty$ as the exact Equation (7.31). Thus, we are mixing exponentially increasing failure rates and as a result obtaining a slowly increasing (decreasing) mixture failure rate, which converges to a constant value. We already mentioned that this important example models the deceleration of human mortality at advanced ages (the mortality plateau).

7.5.2 Accelerated Life Model

Although Equation (7.13) is also simple in this case, the presence of a mixing parameter z in the arguments makes the corresponding analysis of the mixture failure rate more complex than for the multiplicative model. Similar to (7.24), the mixture failure rate in this specific case is defined as

$$\lambda_m(t) = \frac{\int_0^\infty z\lambda(tz)\exp\{-\Lambda(tz))\pi(z)dz}{\int_0^\infty \exp\{-\Lambda(tz))\pi(z)dz} . \tag{7.32}$$

The asymptotic behaviour of the mixture failure rate $\lambda_m(t)$ can be described as a specific case of Theorem 7.2 with $A(s) = \Lambda(s)$, $\phi(t) = t$ and $\psi(t) \equiv 0$.

Corollary 7.3. Assume that the mixing pdf $\pi(z), z \in [0,\infty)$ can be defined as $\pi(z) = z^{\alpha}\pi_1(z)$, where $\alpha > -1$ and $\pi_1(z)$ is continuous at $z = 0$ and bounded in $[0,\infty)$, $\pi_1(0) \neq 0$. Let the baseline distribution with the cumulative failure rate $\Lambda(t)$ have a moment of order $\alpha + 1$.

Then the asymptotic behaviour of the mixture failure rate for the ALM (7.13) is given by

$$\lambda_m(t) \sim \frac{\alpha + 1}{t} \tag{7.33}$$

as $t \to \infty$.

The conditions of Corollary 7.3 are not that strong and are relatively natural. Most of the widely used lifetime distributions have moments of all orders. The Pareto distribution will be discussed in the next example. As stated, the conditions on the mixing distribution hold for, *e.g.*, the gamma and the Weibull distributions, which are commonly used as mixing distributions.

Note that Relationship (7.33) is a surprising result, at least at first sight, as it does not depend on the baseline distribution. It is also dramatically different from the multiplicative case (7.26).

The following example shows other possibilities for the asymptotic behaviour of $\lambda_m(t)$ when one of the conditions of Corollary 7.3 does not hold.

Example 7.5 Consider the gamma mixing distribution with the scale parameter equal to 1, *i.e.*, $\pi(z) = z^{\alpha}\exp\{-x\}/\Gamma(\alpha+1)$. Let the baseline distribution be the Pareto distribution with the pdf $f(t) = \beta / t^{\beta+1}$ $t \geq 1,\ \beta > 0$.

For $\beta > \alpha + 1$, the conditions of Corollary 7.3 and Relationship (7.33) hold. Let $\beta \le \alpha + 1$, which means that the baseline distribution does not have an $(\alpha + 1)$ th moment. Then one of the conditions of Corollary 7.3 is violated. In this case, it can be shown by direct derivation (Finkelstein and Esaulova, 2006a) that, as $t \to \infty$

$$\lambda_m(t) \sim \frac{\beta}{t}. \tag{7.34}$$

Although the dependence on time in (7.34) is the same as in (7.33), the mixture failure rate in this case depends on the baseline distribution *via* the parameter β. Both relationships can be combined as

$$\lambda_m(t) \sim \frac{\min(\beta, \alpha + 1)}{t}.$$

It can be shown that the same asymptotic relationship holds not only for the gamma distribution, but also for any other mixing distribution $\pi(z)$ of the form $\pi(z) = z^\alpha \pi_1(z)$. If $\beta > \alpha + 1$, the function $\pi_1(z)$ should be bounded and $\pi_1(0) \ne 0$.

As $A(s) = \Lambda(s)$ and $\phi(t) = t$, Theorem 7.3 simplifies to the following corollary.

Corollary 7.4. Assume that the mixing pdf $\pi(z),\ z \in [a, \infty)$ is bounded, continuous at $z = a$ and satisfies $\pi(a) \ne 0$. Let

$$\frac{\lambda'(t)}{(\lambda(t))^2} \to 0, \quad t\lambda(t) \to \infty \tag{7.35}$$

as $t \to \infty$. Assume also that for all $b, c > 0,\ b < c$, the quotient $\lambda(bx)/\lambda(cx)$ is bounded as $x \to \infty$.

Then, in accordance with (7.22), the mixture failure rate for Model (7.13) has the following asymptotic behaviour:

$$\lambda_m(t) \sim a\lambda_0(at). \tag{7.36}$$

Condition (7.35) is rather weak. For example, in the marginal case of the Pareto distribution (with the baseline failure rate of the form $\lambda(t) = ct^{-1},\ c > 0,\ t \ge 1$), this condition is not satisfied, but in mixing we are primarily interested in the increasing baseline failure rates.

7.5.3 Proportional Hazards and Other Possible Models

Owing to its simplicity, the asymptotic behaviour of $\lambda_m(t)$ in the additive hazards model (7.14) does not deserve special attention. As $A(s) = s$ and $\phi(t) = t$, Conditions (7.16) and (7.17) of Theorem 7.2 hold and the asymptotic result (7.18) simplifies to

$$\lambda_m(t) \sim \frac{\alpha + 1}{t} + \psi'(t).$$

as $t \to \infty$.

Thus, even in the case where the support of the mixing variable is $[0,\infty)$, the function $\psi'(t)$ can be formally interpreted as the failure rate of the strongest population. Then Theorem 7.2, where the baseline failure rate $\lambda(t)=\psi'(t)$ is an increasing convex function, can be 'completed' by stating that $\lambda_m(t)$ is ultimately increasing in such a way that

$$\lim_{t\to\infty}(\lambda_m(t)-\lambda(t))=0\,.$$

Theorem 7.3 can also be generalized in an obvious way.

Some combinations of the specific models considered can be analysed using the asymptotic approach developed in this chapter. For instance, the generalized 'proportional-accelerated life-additive' model

$$\lambda(t,z)=z^k\lambda(z^m t)+\psi'(t),\;\; k,m>0$$

can be formally investigated after a suitable adjustment of Model (7.10) and of the corresponding asymptotic Theorem 7.2, although the practical usefulness of this model is not clear so far.

Esaulova (2006) generalized the results of this section to a lifetime model of the following form:

$$\Lambda(t,z)=A(\eta(z)\phi(t))+\psi(t),$$

where $\eta(z)$ is a differentiable and strictly monotone function for all $z\geq 0$.

7.6 Asymptotic Mixture Failure Rates for Multivariate Frailty

7.6.1 Introduction

In the previous sections, we considered a lifetime random variable T indexed by a frailty parameter Z. The next obvious step of generalization is to study a multivariate frailty. This means that there can be several unobserved parameters (independent or dependent), which is often the case in practice. Note that T, as previously, is the univariate lifetime random variable. The simplest model is the bivariate multiplicative model, which is an obvious generalization of the univariate multiplicative model (7.12), *i.e.*,

$$\lambda(t,z_1,z_2)=z_1,z_2\lambda(t)\,. \tag{7.37}$$

We shall consider this specific model in Section 7.6.4.

Let Z_1 and Z_2 be interpreted as non-negative random variables with supports in $[0,\infty)$. Similar to Section 6.1,

$$\Pr[T\leq t\mid Z_1=z,Z_2=z_2]\equiv\Pr[T\leq t\mid z_1,z_2]=F(t,z_1,z_2)$$

and

$$\lambda(t,z_1,z_2)=\frac{f(t,z_1,z_2)}{\bar{F}(t,z_1,z_2)}\,.$$

Assume that Z_1 and Z_2 have a bivariate joint pdf $\pi(z_1, z_2)$. The mixture failure rate is defined in this case as

$$\lambda_m(t) = \frac{f_m(t)}{F_m(t)} = \frac{\int_0^\infty \int_0^\infty f(t, z_1, z_2)\pi(z_1, z_2)dz_1 dz_2}{\int_0^\infty \int_0^\infty \overline{F}(t, z_1, z_2)\pi(z_1, z_2)dz_1 dz_2}$$

$$= \int_0^\infty \int_0^\infty \lambda(t, z_1, z_2)\pi(z_1, z_2 \mid t)dz_1 dz_2 , \tag{7.38}$$

where the conditional pdf, similar to Equations (3.10) and (6.5), is

$$\pi(z_1, z_2 \mid t) = \pi(z_1, z_2)\frac{\overline{F}(t, z_1, z_2)}{\int_0^\infty \int_0^\infty \overline{F}(t, z_1, z_2)\pi(z_1, z_2)dz_1 dz_2} . \tag{7.39}$$

In what follows in this section, we consider two specific bivariate frailty models. Our goal is to apply the developed asymptotic methodology to the bivariate setting.

7.6.2 Competing Risks for Mixtures

Consider firstly, a system of two statistically independent components in series with lifetimes $T_1 \geq 0, T_2 \geq 0$ and distribution functions $F_1(t), F_2(t)$, respectively. As the system fails when the first failure of a component occurs, this setting can also be interpreted in terms of the corresponding competing risks. The Cdf function of a system is obviously

$$F_s(t) = 1 - \overline{F}_1(t)\overline{F}_2(t) .$$

Therefore, the competing risks setting reduces the bivariate problem to the univariate one.

As in the univariate case, assume that distributions $F_i(t), i = 1,2$ are indexed by random variables (frailties) Z_i with supports in $[0, \infty)$, *i.e.*,

$$\Pr[T_i \leq t \mid Z_i = z] \equiv \Pr[T_i \leq t \mid z] = F_i(t, z) .$$

The corresponding mixture failure rates, as in Equation (6.5), are defined in the following way:

$$\lambda_{m,i}(t)=\frac{f_{m,i}(t)}{\overline{F}_{m,i}(t)}=\frac{\int_0^\infty f_i(t,z)\pi_i(z)dz}{\int_0^\infty \overline{F}_i(t,z)\pi_i(z)dz}$$

$$=\int_0^\infty \lambda_i(t,z)\pi_i(z\mid t)dz,\quad i=1,2, \tag{7.40}$$

where the conditional pdf $\pi_i(z\mid t)$ is given by Equation (3.10).

Assume now that the components of our system are conditionally independent given $Z_1=z_1, Z_2=z_2$. This is an important assumption. Then

$$F_s(t,z_1,z_2)=1-\overline{F}_1(t,z_1)\overline{F}_2(t,z_2), \tag{7.41}$$

and the corresponding pdf is

$$f_s(t,z_1,z_2)=f_1(t,z_1)\overline{F}(t,z_2)+f_2(t,z_2)\overline{F}_1(t,z_1). \tag{7.42}$$

Thus the components are dependent only *via* the possible dependence between Z_1 and Z_2, which is described by the joint pdf $\pi(z_1,z_2)$. The mixture failure rate of this system $\lambda_{m,s}(t)$ is given by Equation (7.38), where $\lambda_m(t), f(t,z_1,z)$ and $\overline{F}(t,z_1,z_2)$ are substituted by $\lambda_{m,s}(t), f_s(t,z_1,z)$ and $\overline{F}_s(t,z_1,z_2)$, respectively. Obviously, the failure rate of the system is the sum of the components' failure rates, *i.e.*,

$$\lambda_s(t,z_1,z_2)=\lambda_1(t,z_1)+\lambda_2(t,z_2). \tag{7.43}$$

If Z_1 and Z_2 are independent, which means that

$$\pi(z_1,z_2)=\pi_1(z_1)\pi_2(z_2),$$

then the bivariate conditional density (7.39) is also a product of the corresponding univariate conditional densities. This can be seen using Equations (7.39)–(7.41):

$$\pi(z_1,z_2\mid t)=\pi_1(z_1)\pi_2(z_2)\frac{\overline{F}_1(t,z_1)\overline{F}_2(t,z_2)}{\int_0^\infty\int_0^\infty \overline{F}_1(t,z_1)\overline{F}_2(t,z_2)\pi_1(z_1),\pi_2(z_2)dz_1dz_2}$$

$$=\frac{\pi_1(z_1)\overline{F}_1(t,z_1)\pi_2(z_2)\overline{F}_2(t,z_2)}{\int_0^\infty \overline{F}_1(t,z_1)\pi_1(z_1)dz_1\int_0^\infty \overline{F}_2(t,z_2)\pi_2(z_2)dz_2}$$

$$=\pi_1(z_1\mid t)\pi_2(z_2\mid t). \tag{7.44}$$

Therefore, when the components of the system are conditionally independent and Z_1 and Z_2 are independent, the mixture failure rate of the system is the sum of the mixture failure rates of individual components, which, taking into account Equations (7.38),(7.43) and (7.44), is clearly seen from the following:

$$\lambda_{m,s}(t) = \int_0^\infty \int_0^\infty \lambda(t, z_1, z_2)\pi(z_1, z_2 \mid t)dz_1 dz_2$$

$$= \int_0^\infty \int_0^\infty [\lambda_1(t, z_1) + \lambda_2(t, z_2)]\pi(z_1, z_2 \mid t)dz_1 dz_2$$

$$= \int_0^\infty \lambda_1(t, z_1)\pi_1(z_1 \mid t)dz_1 + \int_0^\infty \lambda_2(t, z_2)\pi_1(z_2 \mid t)dz_2$$

$$= \lambda_{m,1}(t) + \lambda_{m,2}(t). \tag{7.45}$$

Note that this result does not hold for the case of shared frailty for $Z_1 \equiv Z_2 \equiv Z$, which can be shown directly by similar integration. Mixture failure rates for some specific mixing distributions and shared frailties were considered by Yashin and Iachine (1999).

7.6.3 Limiting Behaviour for Competing Risks

Now we turn to a study of the asymptotic behaviour of a mixture failure rate of a system for the case when frailties Z_1 and Z_2 are correlated. The method is based on the approach of Section 7.3 developed for the univariate case.

Assume that survival functions for the components are given by (7.10), where the non-important additive term is set to be 0, *i.e.*,

$$F_i(t, z_i) = \exp\{-A_i(z_i \phi_i(t)), \; i = 1,2. \tag{7.46}$$

The following theorem generalizes Theorem 7.2 to the bivariate case. Its proof can be found in Finkelstein and Esaulova (2008).

Theorem 7.5. Let the components' survival functions in the competing risks model (7.41) be defined by Equation (7.46), where the mixing variables Z_1 and Z_2 have the joint pdf $\pi(z_1, z_2)$. Let the following hold:

- $\pi(z_1, z_2) = z_1^{\alpha_1} z_2^{\alpha_2} \pi_0(z_1, z_2)$, where the function $\pi_0(z_1, z_2)$ is continuous at $(0,0)$ and bounded in $[0, \infty) \times [0, \infty)$, $\pi(0,0) \neq 0$ and $\alpha_1, \alpha_2 > -1$.
- The increasing functions $\phi_i(t), i = 1,2$ tend to infinity as $t \to \infty$.
- The increasing functions $A_i(s), i = 1,2$ are differentiable and satisfy

$$\int_0^\infty \exp\{-A_i(s)\} s^{\alpha_i} ds < \infty.$$

Then the asymptotic mixture failure rate of the system is given by the following asymptotic relationship:

$$\lambda_{m,s}(t) \sim (\alpha_1 + 1)\frac{\phi_1'(t)}{\phi_1(t)} + (\alpha_2 + 1)\frac{\phi_2'(t)}{\phi_2(t)}. \tag{7.47}$$

It follows from the additive nature of (7.47) and Equation (7.19) that the asymptotic mixture failure rate in our model can be viewed as the sum of univariate mixture failure rates of each component with its own independent frailty. Therefore, taking into account Equation (7.45), we can interpret Theorem 7.5 in the following way:

The mixture failure rate $\lambda_{m,s}(t)$ in the correlated frailty model with conditionally independent components is asymptotically equivalent to the corresponding mixture failure rate in the independent frailty model.

Therefore, this theorem describes some 'vanishing dependence' as $t \to \infty$.

The first assumption of Theorem 7.5 imposes certain restrictions on the mixing distribution. In the univariate case, Equation (7.15) holds, *e.g.*, for gamma, Weibull and lognormal distributions. In the bivariate case, all mixing densities that are positive and continuous at the origin are obviously admissible.

Example 7.6 Gumbel Bivariate Exponential Distribution
The survival function of this distribution is (Equation 3.26)

$$S(z_1, z_2) = \exp\{-z_1 - z_2 - \delta z_1 z_2\},$$

where $0 \le \delta \le 1$. The mixing pdf is

$$\pi(z_1, z_2) = \exp\{-z_1 - z_2 - \delta z_1 z_2\}\{(1 + \delta z_1)(1 + \delta z_2) - \delta\}.$$

This pdf is bounded and continuous in $[0, \infty)^2$ and $\pi(0,0) = 1 - \delta$. Thus, for $0 \le \delta < 1$, the mixing density satisfies the conditions of Theorem 7.5 and Relationship (7.47) holds.

It can easily be checked that the Farlie–Gumbel–Morgenstern distribution defined in Example 3.9 also meets the requirements for the admissible mixing distribution. Other distributions of this class are the Dirichlet distribution, the inverted Dirichlet distribution, some types of multivariate logistic distributions (Kotz *et al.*, 2000), *etc.*

There are also examples of when conditions of Theorem 7.5 do not hold. The Marshall–Olkin bivariate exponential distribution defined by Equation (3.30) depends on $\max(z_1, z_2)$ and therefore is not absolutely continuous.

Finally, in order to illustrate the result of Theorem 7.5, as in Section 7.4, consider the specific cases. Assume that the model for each component is a multiplicative one, *i.e.*,

$$\lambda_i(t, z_i) = z_i \lambda_i(t),\ i = 1,2,$$

and that $\alpha_1 = \alpha_2 = 0$ in the joint mixing pdf $\pi(z_1, z_2)$. Then, in accordance with

(7.47), as $t \to \infty$,

$$\lambda_{m,s}(t) \sim \frac{\lambda_1(t)}{\int_0^t \lambda_1(u)du} + \frac{\lambda_2(t)}{\int_0^t \lambda_2(u)du}.$$

In a similar way, for the ALM $\lambda_i(t, z_i) = z_i \lambda(tz_i)$, we get

$$\lambda_{m,s}(t) \sim \frac{2}{t}.$$

Both of these formulas show that the asymptotic behaviour of mixture failure rates does not depend on the mixing distribution.

7.6.4 Bivariate Frailty Model

In this section, we will briefly discuss another bivariate frailty model, which is defined for a single lifetime random variable T. It is a generalization of the simplest multiplicative model (7.37):

$$\lambda(t, z_1, z_2) = G(z_1, z_2)\lambda(t), \tag{7.48}$$

where $G(z_1, z_2)$ is some positive bivariate function. The corresponding survival and the probability density functions are

$$\bar{F}(t, z_1, z_2) = \exp\{-G(z_1, z_2)\Lambda(t)\},$$

$$f(t, z_1, z_2) = G(z_1, z_2)\lambda(t)\exp\{-G(z_1, z_2)\Lambda(t)\},$$

respectively.

Let the function $G(z_1, z_2)$ be invertible with respect to z_1, and denote by $B(z_1, z_2)$ the corresponding inverse function, *i.e.*,

$$B(G(z_1, z_2), z_2) \equiv z_1, \quad G(B(z_1, z_2), z_2) \equiv z_1.$$

Changing the variable of integration in the equation for the mixture (marginal) survival function $\bar{F}_m(t)$ to $s = G(z_1, z_2)$ gives

$$\begin{aligned}\bar{F}_m(t) &= \int_0^\infty\int_0^\infty \exp\{-G(z_1, z_2)\Lambda(t)\}\pi(z_1, z_2)dz_1dz_2 \\ &= \int_0^\infty \exp\{-\Lambda(t)s\}\int_0^\infty \frac{\partial B(s, z_2)}{\partial s}\pi(B(s, z_2), z_2)dz_2ds \\ &= \int_0^\infty \exp\{-\Lambda(t)s\}g(s)ds,\end{aligned}$$

where the function $g(s)$ is defined as

$$g(s) = \int_0^\infty \frac{\partial B(s, z_2)}{\partial s} \pi(B(s, z_2), z_2) dz_2 \ .$$

Similarly, the corresponding pdf is

$$f_m(t) = \lambda(t) \int_0^\infty \exp\{-\Lambda(t)s\} s g(s) \, ds \ .$$

It can be seen that, as

$$\int_0^\infty g(s) ds = \int_0^\infty \int_0^\infty \pi(z_1, z_2) dz_1 dz_2 = 1 ,$$

the function $g(s)$ can be interpreted as a pdf.

Equation (7.48) defines a multiplicative model. Therefore, the corresponding results of Section 7.4.1 can be applied with the obvious substitution of $\pi(z)$ by $g(z)$.

We will consider now an example where the asymptotic relationship for the mixture failure rate $\lambda_m(t)$ is obtained *via* direct integration.

Example 7.7 Let the function $G(z_1, z_2)$ in (7.48) be defined as in (7.37), *i.e.*,

$$G(z_1, z_2) = z_1 z_2 \ .$$

Obviously,

$$B(s, z_2) = \frac{s}{z_2}, \quad \frac{\partial B(s, z_2)}{\partial s} = \frac{1}{z_2}, \quad \frac{\partial G(z_1, z_2)}{\partial z_1} = z_2 \ .$$

Assume that the mixing distribution is uniform in $[0, b] \times [0, b]$ for some $b > 0$.

$$\pi(z_1, z_2) = \begin{cases} 1/b^2, & 0 \le z_1, z_2 \le b, \\ 0, & \textit{otherwise}. \end{cases}$$

Then

$$\bar{F}_m(t) = \frac{1}{b^2} \int_0^b \int_0^b \exp\{-\Lambda(t) xy\} dx dy = \frac{1}{b^2} \int_0^b \frac{1}{\Lambda(t) y} (1 - \exp\{-\Lambda(t) by\}) dy$$

$$= \frac{1}{\Lambda(t) b^2} \int_0^{\Lambda(t) b^2} \frac{1}{u} (1 - \exp\{-u\}) du \ .$$

It can be seen that, as $v \to \infty$,

$$\int_0^v \frac{1}{u} (1 - \exp\{-u\}) du \sim \log v \ .$$

Therefore, finally,

$$\bar{F}_m(t) \sim \frac{\log \Lambda(t)}{\Lambda(t)b^2} \tag{7.49}$$

as $t \to \infty$. Similarly (Esaulova, 2006),

$$f_m(t) \sim \frac{\lambda(t)\log \Lambda(t)}{(\Lambda(t))^2 b^2} \tag{7.50}$$

and

$$\lambda_m(t) = \frac{f_m(t)}{\bar{F}_m(t)} \sim \frac{\lambda(t)}{\Lambda(t)}, \quad t \to \infty, \tag{7.51}$$

which is a remarkably simple asymptotic relation that is similar to (7.26) for the univariate model.

Example 7.8 Consider another special case where the function $G(z_1, z_2)$ in (7.48) is additive, *i.e.*,

$$G(z_1, z_2) = z_1 + z_2 .$$

Then

$$B(s, z_2) = s - z_2, \quad \frac{\partial B(s, z_2)}{\partial s} \equiv \frac{\partial G(z_1, z_2)}{\partial z_1} = 1 .$$

Assume that the function

$$g(s) = \int_0^s \pi(s - z_2, z_2)dz_2 \tag{7.52}$$

can be written as

$$g(s) = s^\alpha g_1(s) ,$$

where $\alpha > -1$ and $g_1(z)$ is bounded in $[0, \infty)$, continuous at $z = 0$ and $g_1(0) \neq 0$. Then, in accordance with Corollary 7.1, the asymptotic formula ($t \to \infty$) for the mixture failure rate is

$$\lambda_m(t) \sim \frac{(\alpha + 1)\lambda(t)}{\Lambda(t)} . \tag{7.53}$$

On the other hand, it is more relevant to formulate the result explicitly in terms of the initial bivariate mixing pdf $\pi(z_1, z_2)$. Assume that $\pi(z_1, z_2)$ satisfies the first condition of Theorem 7.5. It follows from Equation (7.52) that, as $s \to 0$,

$$g(s) \sim s^{\alpha_1+\alpha_2+1}\pi(0,0).$$

Therefore, (7.53) can be transformed into

$$\lambda_m(t) \sim \frac{(\alpha_1+\alpha_2+2)\lambda(t)}{\Lambda(t)}, \tag{7.54}$$

which is the same result that can be obtained directly from (7.47). This is not surprising, because the considered model can also be interpreted as the mixture failure rate model for a series system with conditionally independent components and dependent frailties.

The next section contains the brief proofs of the theorems of this chapter. The full text of the proofs can be found in Finkelstein and Esaulova (2006a, 2008).

7.7 Sketches of the Proofs

Proof of Theorem 7.2. We start with a simple lemma.

Lemma 7.1. Let $g(z)$ and $h(z)$ be non-negative functions in $[0,\infty)$ satisfying the following conditions:

$$\int_0^\infty g(z)dz < \infty,$$

and let $h(z)$ be bounded and continuous at $z = 0$.

Then, as $t \to \infty$,

$$t\int_0^\infty g(tz)h(z)dz \to h(0)\int_0^\infty g(z)dz. \tag{7.55}$$

Proof. Substituting $u = tz$ gives

$$t\int_0^\infty g(tz)h(z)dz = \int_0^\infty g(u)h(u/t)du.$$

The function $h(u)$ is bounded and $h(u/t) \to 0$ as $t \to \infty$; thus convergence (7.55) holds by the dominated convergence theorem. ■

Now we can proceed with the proof of Theorem 7.2. The survival function, which corresponds to (7.10), is

$$\bar{F}(t,z) = \exp\{-(A(z\phi(t))\},$$

where we assume that the non-important additive term is zero: $\psi(t) \equiv 0$. Taking into account that $\phi(t) \to \infty$ as $t \to \infty$, and applying Lemma 7.1 to the function $g(u) = \exp\{-A(u)\}u^{\alpha}$, gives

$$\int_0^\infty \overline{F}(t,z)\pi(z)dz = \int_0^\infty \exp\{-(A(z\phi(t))\}z^\alpha \pi_1(z)dz$$

$$\sim \frac{\exp\{-\psi(t)\}\pi_1(0)}{\phi(t)^{\alpha+1}} \int_0^\infty \exp\{-A(s)\}s^\alpha \, ds\,, \tag{7.56}$$

where the integral is finite owing to (7.17).

Similarly, applying Lemma 7.1 to the corresponding pdf:

$$\int_0^\infty f(t,z)\pi(z)dz = \phi'(t)\int_0^\infty A'(z\phi(t))\exp\{-A(z\phi(t))\}z^{\alpha+1}\pi_1(z)dz$$

$$\sim \frac{\phi'(t)\pi_1(0)}{\phi(t)^{\alpha+2}} \int_0^\infty A'(s)\exp\{-A(s)\}s^{\alpha+1} ds\,. \tag{7.57}$$

It can be shown with the help of (7.17) that

$$\exp\{-A(s)\}s^{\alpha+1} \to 0 \quad \text{as} \quad t \to \infty.$$

Using this fact and integrating by parts yields

$$\int_0^\infty A'(s)\exp\{-A(s)\}s^{\alpha+1}ds = (\alpha+1)\int_0^\infty \exp\{-A(s)\}s^\alpha ds\,. \tag{7.58}$$

Combining Equations (7.56)–(7.58) finally results in

$$\frac{\int_0^\infty f(t,z)\pi(z)dz}{\int_0^\infty \overline{F}(t,z)\pi(z)dz} \sim (\alpha+1)\frac{\phi'(t)}{\phi(t)}.$$

Proof of Theorem 7.3. This theorem is rather technical and we must use three supplementary lemmas that present consecutive steps on the way to (7.22). We state these lemmas without proofs (Finkelstein and Esaulova, 2006a).

Lemma 7.2. Let $h(x)$ be a twice-differentiable function with an ultimately positive derivative, such that

$$\int_0^\infty \exp\{-h(y)\}dy < \infty\,.$$

Also let $h''(x)/(h'(x))^2 \to 0$ as $x \to \infty$.

Then

$$\int_x^\infty \exp\{-h(y)\}dy \sim \exp\{-h(x)\}\frac{1}{h'(x)}$$

as $x \to \infty$.

We use this lemma to obtain the following one.

Lemma 7.3. Let the assumptions of Lemma 7.2 hold. Assume additionally that

$$xh'(x) \to \infty, x \to \infty.$$

Let $\mu(u)$ be a positive, bounded and locally integrable function defined in $[a,\infty)$, continuous at $u = a$. Assume that $\mu(a) \neq 0$.

Then

$$\int_a^\infty \exp\{-h(ux)\}\mu(u)du \sim \frac{\mu(a)\exp\{-h(ax)\}}{xh'(ax)}$$

as $x \to \infty$.

Lemma 7.4. Under the assumptions of Lemma 7.2, the following asymptotic relationship holds as $x \to \infty$:

$$\int_a^\infty h'(ux)\exp\{-h(ux)\}u\mu(u)du \sim \frac{a\mu(a)}{x}\exp\{-h(ax)\}.$$

Now we are ready to prove Theorem 7.3 itself. Applying Lemma 7.3 as $t \to \infty$ results in

$$\int_a^\infty \bar{F}(t,z)\pi(z)dz = \int_a^\infty \exp\{-(A(z\phi(t))\}\pi(z)dz$$

$$\sim \frac{\pi(a)}{A'(a\phi(t))\phi(t)}\exp\{-A(a\phi(t))\}.$$

Therefore,

$$\int_a^\infty f(t,z)\pi(z)dz = \phi'(t)\int_a^\infty A'(z\phi(t))\exp\{-A(z\phi(t))\}z\pi(z)dz.$$

Using Lemma 7.4 results in the following relationship:

$$\int_a^\infty A'(z\phi(t))\exp\{-A(z\phi(t))\}z\pi(z)dz \sim \frac{a\pi(a)}{\phi(t)}\exp\{-A(a\phi(t))$$

and finally, we arrive at (7.22), *i.e.*,

$$\lambda_m(t) = \frac{\int_0^\infty f(t,z)\pi(z)dz}{\int_0^\infty \bar{F}(t,z)\pi(z)dz}$$

$$\sim \frac{\phi'(t)a\pi(a)\exp\{-A(a\phi(t))\}}{\phi(t)} \cdot \frac{A'(a\phi(t))\phi(t)}{\pi(a)\exp\{-A(a\phi(t))\}}$$

$$= a\phi'(t)A'(a\phi(t)).$$

Proof of Theorem 7.4. We consider the numerator and the denominator in (7.24) separately. Changing variables and applying Lemma 7.1 we obtain

$$\int_0^\infty \bar{F}(t,z)\pi(z)dz = \int_a^\infty \exp\{-z\Lambda_0(t)\}(z-a)^\alpha \pi_1(z-a)dz$$

$$= \exp\{-a\Lambda_0(t)\int_0^\infty \exp\{-z\Lambda_0(t)\}z^\alpha \pi_1(z)dz$$

$$\sim \frac{\exp\{-a\Lambda_0(t)\}\pi_1(0)\Gamma(\alpha+1)}{(\Lambda_0(t))^{\alpha+1}}. \tag{7.59}$$

Similarly,

$$\int_0^\infty zf(t,z)\pi(z)dz = \lambda_0(t)\int_a^\infty z\exp\{-z\Lambda_0(t)\}(z-a)^\alpha \pi_1(z-a)dz$$

$$= \lambda_0(t)\exp\{-a\Lambda_0(t)\}\int_0^\infty \exp\{-z\Lambda_0(t)\}z^{\alpha+1}\pi_1(z)dz$$

$$+ a\lambda_0(t)\exp\{-a\Lambda_0(t)\}\int_0^\infty \exp\{-z\Lambda_0(t)\}z^\alpha \pi_1(z)dz\,.$$

As $t \to \infty$, the first integral on the right-hand side is equivalent to $\pi_1(0)\Gamma(\alpha+2)(\Lambda_0(t))^{-\alpha-2}$ and the second integral is equivalent to $\pi_1(0)\Gamma(\alpha+1)(\Lambda_0(t))^{-\alpha-1}$, which decreases more slowly than the first one. Thus,

$$\int_0^\infty zf(tz)\pi(z)dz \sim \frac{a\lambda_0(t)\exp\{-a\Lambda_0(t)\}\pi_1(0)\Gamma(\alpha+1)}{(\Lambda_0(t))^{\alpha+1}}. \tag{7.60}$$

Finally, substituting (7.59) and (7.60) into Equation (7.24), we arrive at (7.29).

7.8 Chapter Summary

A general class of distributions is discussed in this chapter. This class contains as special cases the additive, multiplicative and accelerated life models that are widely used in reliability practice. The corresponding asymptotic theory is developed and applied to deriving and analysing asymptotic failure rates. We also use the developed approach for obtaining asymptotic failure rates in the correlated competing risks setting. It turns out that as $t \to \infty$, the correlation can 'fade out'.

There are many applications where the behaviour of the failure rate at relatively large values of t is really important. In Chapter 6, the example of the oldest-old mortality was discussed when the exponentially increasing Gompertz mortality curve is 'bent down' for advanced ages (mortality plateau).

Some of the obtained results are very surprising. For example, when the support of the mixing distribution is $[0,\infty)$, the mixture failure rate in the accelerated life model converges to 0 as $t \to \infty$ and does not depend on the baseline distribution. Under reasonable assumptions, we prove that the asymptotic behaviour of the mixture failure rate for other models depends only on the behaviour of the mixing distribution in the neighbourhood of the left-hand endpoint of its support, and not on the whole mixing distribution.

The presentation of results in this chapter is rather technical. Therefore, sketches of the proofs of the main theorems are deferred to the last section.

8

‘Constructing’ the Failure Rate

In this chapter, we will consider several specific settings when the failure rate can be obtained (constructed) directly as an exact or an approximate relationship. Along with meaningful heuristic considerations, exact solutions and approaches will also be discussed where possible. Most examples to follow are based on the operation of thinning of the Poisson process (Cox and Isham, 1980) or on equivalent reasoning. In many instances this method can be very helpful and often results in significant simplifications. The choice of the problems to be considered is defined by the projects in which the author took part recently and by the corresponding publications.

A basic feature of the models to be discussed is defined by an underlying point process of events that can be terminated in some way. Termination of this process usually results in, *e.g.*, a mission failure or the failure of a system, *etc.* Most of the results are obtained for the underlying Poisson process (homogeneous or nonhomogeneous). In this case, the corresponding failure rate, and therefore the probability of termination can usually be obtained in an explicit form under reasonable assumptions. Termination of renewal processes, however, usually cannot be modelled explicitly and only bounds and approximations exist for reliability measures of interest.

In Section 8.3 we apply the developed approach to obtaining the survival probability of an object which is moving in a plane and encountering moving or (and) fixed obstacles. In the *safety at sea* application terminology, each foundering or collision results in a failure (accident) with a predetermined probability. It will be shown that this setting can be reduced to the one-dimensional case. In Section 8.4, the notion of multiple availability is discussed. The corresponding probabilities are also obtained using the operation of thinning of the Poisson process. By properly adjusting the term ‘failure’, other sections of this chapter can also be easily interpreted in terms of safety and risk analysis.

8.1 Terminating Poisson and Renewal Processes

Two equivalent interpretations for the termination of the Poisson process are usually considered in the literature. The first one is often referred to as a method of the *per demand failure rate* (Thompson, 1988). Its probabilistic description is simple, and therefore it is widely used in reliability practice. In accordance with the notation of Chapter 4, let λ_r be the rate of the homogeneous Poisson process $N(t),\ t \geq 0$, describing instantaneous demands of some kind. Assume that each demand is instantaneously serviced with probability $1-\theta$ and is not serviced with the complementary probability θ. Let T be the time to failure of this 'system' defined as the time to the first non-serviced demand or, equivalently, to the termination of our process. In accordance with the definition of the homogeneous Poisson process,

$$\Pr[N(t)=n]=\exp\{-\lambda_r t\}\frac{(\lambda_r t)^n}{n!}, \tag{8.1}$$

and the corresponding survival probability (the probability that all demands in $[0,t]$ have been serviced) can be obtained directly in the following way:

$$\begin{aligned}\Pr[T \geq t]=\bar{F}(t)&=\sum_0^{\infty}(1-\theta)^k\exp\{-\lambda_r t\}\frac{(\lambda_r t)^k}{k!}\\ &=\exp\{-\theta\,\lambda_r\, t\}.\end{aligned} \tag{8.2}$$

It follows from Equation (8.2) that the corresponding failure rate, which is defined by the distribution $F(t)$, is given by a simple and meaningful relationship:

$$\lambda(t)=\theta\,\lambda_r\,. \tag{8.3}$$

Thus, the rate of the underlying Poisson process λ_r is decreased by the factor $\theta \leq 1$.

On the other hand, the classical operation of thinning of the point process (Cox and Isham, 1980) means that each point of the process is deleted with probability θ or retained in the process with the complementary probability $1-\theta$. Therefore, the described thinned Poisson process has the rate $(1-\theta)\lambda_r$. It follows from the properties of the Poisson process that the time to the first deletion (failure) is described by the Cdf with the failure rate $\theta\,\lambda_r$, which is equal to (8.3). Note that the operation of thinning can be very effective in many applications. A number of problems in reliability, risk and safety analysis can be interpreted by means of the described model.

Similar to (8.2), the result can be generalized in a straightforward way to the case of the NHPP with rate $\lambda_r(t)$, *i.e.*,

$$\begin{aligned}\bar{F}(t)&=\exp\left(-\int_0^t\theta\,\lambda_r(u)du\right),\\ \lambda(t)&=\theta\,\lambda_r(t)\,,\end{aligned} \tag{8.4}$$

where the additional assumption that the distribution $F(t)$ should be a proper one ($F(\infty)=1$) is imposed, *i.e.*,

$$\int_0^{\infty}\theta\lambda_r(u)dx=\infty\,.$$

Another useful and widely used interpretation is *via* the process of shocks. In fact, a shock can also be considered as a demand of some kind. In Chapter 10, we consider the demands for energy. When these demands are 'non-serviced', the death of an organism occurs. We understand the term "shock" in a very broad sense as some instantaneous, potentially harmful event. Shock models are widely used in practical and theoretical reliability. For example, they can present a useful framework for studying ageing properties of distributions (Barlow and Proschan, 1975; Beichelt and Fatti, 2002).

Assume that a shock is the only cause of failure. This means that a system is 'absolutely reliable' in the absence of shocks. Assume now, similar to the "per demand" interpretation, that a shock affecting a system independently from the previous shocks results in its failure (and in the termination of the corresponding Poisson shock process) with probability θ and does not cause any changes in the system with the complementary probability $1-\theta$. It is obvious that the survival probability and the failure rate are defined in this case by Equations (8.2) and (8.3), respectively.

Note that the described setting is often referred to as an extreme shock model, as only the impact of the current shock is taken into account, whereas in cumulative shock models the impact of preceding shocks is accumulated (Sumita and Shanthikumar, 1985; Gut and Husler, 2005).

When the function $\theta(t)$ depends on time, other approaches should be used for deriving the following generalization to Equation (8.4):

$$\bar{F}(t) = \exp\left(-\int_0^t \theta(t)\lambda_r(u)du\right),$$

$$\lambda(t) = \theta(t)\lambda_r(t)\,. \tag{8.5}$$

This result was first proved in a direct way using cumbersome derivations in Beichelt and Fischer (1980) (see also Beichelt, 1981, and Block *et al.*, 1985). We will present now a non-technical proof of (8.5) based on the notion of the conditional intensity function (CIF) $\lambda(t \mid \mathrm{H}(t))$ described by Definition 4.2 and Equation (4.4). As in Chapter 4, $\lambda_r(t)$ denotes the rate of an orderly point process of shocks. In accordance with Definition 4.2 and using the independence of the previous shocks property, the following reasoning becomes straightforward:

$$\begin{aligned}
\tilde{\lambda}(t \mid T(\mathrm{H}(t)) \ge t)dt &= \Pr[T \in [t, t+dt) \mid T(\mathrm{H}(t)) \ge t] \\
&= \frac{\Pr[T \in [t, t+dt), T(\mathrm{H}(t) \ge t]}{\Pr[T(\mathrm{H}(t) \ge t]} \\
&= \frac{\theta(t)\lambda_r(t \mid H(t))\Pr[T(\mathrm{H}(t) \ge t]}{\Pr[T(\mathrm{H}(t) \ge t]}dt \\
&= \theta(t)\lambda_r(t \mid \mathrm{H}(t))dt\,,
\end{aligned}$$

where $\tilde{\lambda}(t \mid T(\mathrm{H}(t)) \ge t)dt$ is the conditional probability of the termination of our point process of shocks in $[t, t+dt)$ and $\lambda_r(t \mid \mathrm{H}(t))$ is the corresponding CIF. The condition $T(\mathrm{H}(t)) \geq t$ means that all shocks in $[0,t)$ for this realization were survived.

Note that the function $\tilde{\lambda}(t \mid T(\mathrm{H}(t)) \geq t)$ depends on the realization $\mathrm{H}(t)$. Therefore, in accordance with Definition 2.1, it cannot define the conventional failure rate $\lambda(t)$. On the other hand, it is well known (see Equation 4.5) that $\lambda_r(t \mid \mathrm{H}(t)) = \lambda_r(t)$ for the specific case of the homogeneous Poisson process, as this is the only process with the memoryless property. Finally, the failure rate $\lambda(t)$ that corresponds to the random time to termination is

$$\lambda(t) = \tilde{\lambda}(t \mid T(\mathrm{H}(t)) \geq t) = \theta(t)\lambda_r(t)$$

in each realization of the considered NHPP process of shocks. Therefore, it is clear that Equation (8.5) holds. A similar reasoning can be found, *e.g.*, in Finkelstein (1999a) and Nachlas (2005).

Unfortunately, a renewal process of shocks does not allow for similar meaningful, simple formulas, and, as we have already mentioned, bounds and approximations should be used for the probabilities of interest in this case. In the rest of this section, we will briefly describe some initial results for terminating renewal processes only.

Under the same general assumption as for the NHPP, consider the terminating renewal process with a constant probability θ of termination at each cycle. As previously, T denotes the time to termination of a process and let X, with the Cdf $G(t)$ and $E[X] < \infty$, be the underlying interarrival time. The corresponding survival probability can be written in the form of the following infinite series:

$$\Pr[T \geq t] = \bar{F}(t) = \sum_{k=1}^{\infty} (1-\theta)^{k-1} \bar{G}^{(k)}(t), \tag{8.6}$$

where, as in Section 4.3.2, $G^{(n)}(t)$ denotes the n-fold convolution of $G(t)$ with itself and $\bar{G}^{(n)}(t) = 1 - G^{(n)}(t)$. Note that the corresponding series for the Poisson process is given by Equation (8.2). Special numerical methods should be used for obtaining $\bar{F}(t)$ in this case. Therefore, it is important to have simple approximations and bounds for this probability. It is well known (see, *e.g.*, Kalashnikov, 1997) that, as $\theta \to 0$, the following convergence in distribution takes place:

$$\bar{F}(t) \to 1 - \exp\left\{-\frac{\theta t}{E[X]}\right\}. \tag{8.7}$$

Thus, the failure rate that corresponds to the Cdf $F(t)$ in this case is approximately constant for sufficiently small θ, *i.e.*,

$$\lambda(t) \approx \frac{\theta}{E[X\}}. \tag{8.8}$$

Relationship (8.8) becomes Equation (8.3) when interarrival times are distributed exponentially. In practice, parameter θ is not always sufficiently small for effectively using this approximation and therefore, the corresponding upper and lower bounds for $\bar{F}(t)$ can be very helpful. Assume that $G(t)$ satisfies the Cramer-Lundberg condition, stating the existence of a constant $k > 0$ such that

$$\bar{\theta}\int_0^\infty \exp\{ku\}dG(u) = 1,$$

where $\bar{\theta} \equiv 1-\theta$. Then $\bar{F}(t)$ has the following bounds:

$$\frac{\exp\{-kt\}}{\bar{\theta}}\left(1 - kE[\xi(t)]\right) \le \bar{F}(t) \le \frac{\exp\{-kt\}}{\bar{\theta}},$$

where $\xi(t)$ is the forward waiting time (the time since arbitrary t to the next moment of renewal) in the renewal process governed by the Cdf $G(t)$ (Kalashnikov, 1997).

Another bound that is useful in practice but rather crude (Finkelstein, 2003a) is based on the following identity:

$$E\left[\bar{\theta}^{N(t)}\right] = \sum_{k=0}^{\infty} \bar{\theta}^k (G^{(k)}(t) - G^{(k+1)}(t)) = \bar{F}(t),$$

which immediately follows after recalling that for the renewal process (Ross, 1996)

$$\Pr[N(t) = n] = G^{(n)}(t) - G^{(n+1)}(t).$$

As the power function is a convex one, Jensen's inequality can be used, *i.e.*,

$$\bar{F}(t) = E\left[\bar{\theta}^{N(t),}\right] \ge \bar{\theta}^{E[N(t),]} = \bar{\theta}^{H(t)},$$

where, as usual, $H(t) = E[N(t)]$ is the corresponding renewal function.

8.2 Weaker Criteria of Failure

8.2.1 Fatal and Non-fatal Shocks

In the previous section, a system could be 'killed' by a single shock or, equivalently, a shock process could be terminated at each step. An important assumption was that the probability of this termination did not depend on the history of the shock process. Assume now that we are looking at a shock process, where a shock is fatal for a system only if it is 'too close' to the previous shock; otherwise the shock is harmless. As previously, assume that a shock is the only possible cause of a system's failure. A possible interpretation of this setting is the following: when the time between the two consecutive shocks is too small, the system cannot recover from the consequences of the previous shock and this event results in a failure. Therefore, the time required for recovery should be taken into account. Note that the setting of the previous section can be considered as a model with an instantaneous recovery. It is natural to assume that the recovery time is a random variable. Denote this variable by τ with the Cdf $R(t)$. Thus, if the shock occurs while the system is still in the process of recovery, a failure (disaster, catastrophe) occurs.

Assume that shocks arrive in accordance with the non-homogeneous Poisson process with rate $\lambda_r(t)$. As previously, the survival function $\bar{F}(t)$ (the probability of a system's failure-free performance in $[0,t)$) is of interest. Consider the following integral equation for $\bar{F}(t)$ (Finkelstein, 2007c):

$$\bar{F}(t) = \exp\left\{-\int_0^t \lambda_r(u)du\right\}\left(1+\int_0^t \lambda_r(u)du\right)$$
$$+\int_0^t \lambda_r(x)\exp\left\{-\int_0^x \lambda_r(u)du\right\}\left[\int_0^{t-x} \lambda_r(y)\exp\left\{-\int_0^y \lambda_r(u)du\right\}R(y)\hat{F}(t-x-y)dy\right]dx, \quad (8.9)$$

where the first term in the right hand side is the probability that there was no more than one shock in $[0,t)$ and the integrand of the second term defines the joint probability of the following events:

- The first shock had occurred in $[x, x+dx)$;
- The second shock had occurred in $[x+y, x+y+dy)$;
- The time between two shocks y is sufficient for recovering (the probability of this event is $R(y)$);
- The system is functioning without failures in $[x+y, t)$.

By $\hat{F}(t)$ in (8.9) we denote the probability of the system's functioning without failures in $[0,t)$ when the first shock occurred at $t=0$. Similar to Equation (8.9), the following integral equation with respect to $\hat{F}(t)$ can be obtained:

$$\hat{F}(t) = \exp\left\{-\int_0^t \lambda_r(u)du\right\} + \int_0^t \lambda_r(x)\exp\left\{-\int_0^x \lambda_r(u)du\right\}R(x)\hat{F}(t-x)dx. \quad (8.10)$$

Simultaneous Equations (8.9) and (8.10) can be solved numerically. First, $\hat{F}(t)$ should be obtained from (8.10) and then substituted in (8.9).

For the homogeneous Poisson process $\lambda_r(t) = \lambda_r$, these equations can be explicitly solved *via* the Laplace transform. In accordance with our notation for the Laplace transform of Section 4.3.2, denote the Laplace transforms of $\bar{F}(t), \hat{F}(t)$ and $R(t)$ by

$$\bar{F}^*(s) = \int_0^\infty \exp\{-st)\bar{F}(t)dt, \; \hat{F}^*(s) = \int_0^\infty \exp\{-st)\hat{F}(t)dt,$$

$$R^*(s) = \int_0^\infty \exp\{-st)R(t)dt\,,$$

respectively. Applying the Laplace transform to both sides of Equations (8.9) and (8.10) and using the property that the Laplace transform of a convolution is equal to the product of the Laplace transforms of the corresponding integrand functions, $\hat{F}^*(s)$ can eventually be derived (Finkelstein, 2007c) as

$$\overline{F}^*(s) = \frac{s[1-\lambda R^*(s+\lambda_r)] - \lambda_r{}^2 R^*(s+\lambda_r) + 2\lambda_r}{(s+\lambda_r)^2[1-\lambda R^*(s+\lambda_r)]}\,. \tag{8.11}$$

In general, the corresponding inverse transform can be obtained numerically, whereas explicit solutions can be obtained only for simple cases.

Example 8.1 Let $R(t) = 1 - \exp\{-\mu t\}$. Then

$$R^*(s+h) = \frac{\mu}{(s+\lambda_r)(s+\lambda_r+\mu)}$$

and

$$\overline{F}^*(s) = \frac{s+2\lambda_r+\mu}{s^2 + s(2\lambda_r+\mu) + \lambda_r{}^2}\,. \tag{8.12}$$

The inverse Laplace transform results in

$$\overline{F}(t) = A_1 \exp\{s_1 t\} + A_2 \exp\{s_2 t\}\,, \tag{8.13}$$

where s_1, s_2 are the roots of the denominator in (8.12) given by

$$s_{1,2} = \frac{-(2\lambda_r+\mu) \pm \sqrt{(2\lambda_r+\mu)^2 - 4\lambda_r{}^2}}{2}$$

and

$$A_1 = \frac{s_1+2\lambda_r+\mu}{s_1-s_2}, \quad A_2 = -\frac{s_2+2\lambda_r+\mu}{s_1-s_2}\,.$$

Equation (8.13) defines the exact solution for $\overline{F}(t)$. In applications, it is convenient to use simple approximate formulas. Consider the following reasonable assumption:

$$\frac{1}{\lambda_r} >> \bar{\tau} \equiv \int_0^\infty (1-R(x))dx\,. \tag{8.14}$$

Inequality (8.14) means that the mean interarrival time in the shock process is much larger than the mean time of recovery $\bar{\tau}$, and this is often the case in practice. In the study of repairable systems, a similar case is usually called the *fast repair approximation.* The fast repair approximation in availability problems will be studied in Section 8.4. Using this assumption, Equation (8.13) can be written as the following approximate relationship:

$$\bar{F}(t) \approx \exp\{-\lambda_r^2 \bar{\tau} t\}, \tag{8.15}$$

and therefore, the corresponding failure rate is approximately constant, *i.e.*,

$$\lambda(t) \approx \lambda_r^2 \bar{\tau}. \tag{8.16}$$

On the other hand, using Assumption (8.14), Approximation (8.16) can be obtained *via* the per demand failure rate method (8.1)–(8.2). The probability that the next shock will occur earlier than the recovery completed is

$$\theta = \int_0^\infty \lambda_r \exp\{-\lambda_r x\} R(x)) dx,$$

which, for the case of exponential $R(t)$ and the corresponding fast repair condition $\mu >> \lambda_r$, results in $\theta = \lambda_r \mu / (\lambda_r + \mu)$. Therefore,

$$\bar{F}(t) \approx \exp\{-\theta \lambda_r t) \approx \exp\{-\lambda_r^2 \bar{\tau} t\}. \tag{8.17}$$

The first approximation in (8.17) is due to the fact that the Poisson process is 'stopped' for those periods of recovery that are small in accordance with (8.14). We will discuss the accuracy of approximations of this kind in Section 8.5.

Example 8.2 Let the recovery time be constant $\tau_a > 0$. In this case, straightforward reasoning defines the survival probability as the following sum (Finkelstein, 2007c):

$$\bar{F}(t) = \exp\{-\lambda h t\} \sum_{k=0}^{[t/\tau_a]} \frac{(h(t-(k-1)\tau_a))^k}{k!},$$

where $[\cdot]$ denotes the integer part.

Another possible generalization of the shock models is to consider two independent shock processes: a process of harmful shocks with rate λ_{rh} and a process of healing (repair) shocks with rate λ_{rr}. Failure of the system is defined as the occurrence of two harmful events in a row. Therefore, if a harmful shock is followed by a healing one, a failure does not occur. This problem can be described mathematically by equations similar to (8.9) and (8.10) and can be solved using the Laplace transforms. On the other hand, similar to (8.17), an approximate relationship for the corresponding survival probability is given by

$$\overline{F}(t) \approx \exp\left\{-\frac{\lambda_{rh}^2}{\lambda_{rh}+\lambda_{rr}}t\right\} \approx \exp\left\{-\frac{\lambda_{rh}^2}{\lambda_{rr}}t\right\},$$

where the analogue of the fast repair approximation in this case is understood as $\lambda_{rr} >> \lambda_{rh}$.

8.2.2 Fatal and Non-fatal Failures

The approach of the previous section can also be applied to obtaining reliability characteristics of repairable systems with a weaker criterion of failure. Assume that a repairable system's failure is not considered as such (from a quality of performance point of view) if the repair time does not exceed a constant time τ_a. To distinguish between these two types of malfunctions, let us call the first event a *breakdown*, reserving the term 'failure' for the final event. There are many examples of such systems. Performance of a marine navigation system, *e.g.*, is characterized by its accuracy in obtaining navigation parameters. If a breakdown is repaired sufficiently quickly, then the corresponding latitude (or longitude) does not noticeably change and the failure of a system does not occur. The operation failure in this case occurs only when the navigation error, which increases with time of repair, exceeds a predetermined level. The repair eliminates the cause of a breakdown and resets the navigational error to a minimal level. Sometimes the described systems are called the systems with *time redundancy* (Zarudnij, 1973).

A system with time redundancy can have the following states:

- E_0 – a system is operating;
- E_1 – a system is under repair, but its duration does not exceed τ_a;
- E_2 – a system is in the state of failure, as the repair duration exceeds τ_a;

Denote by $p_i(t,\tau_a),\ i=0,1$ the joint probability that the reparable system is in the state E_i at time t and that it did not fail before in $[0,t)$. In accordance with our criterion of failure, the corresponding survival function is

$$\overline{F}(t) = p_0(t,\tau_a) + p_1(t,\tau_a). \tag{8.18}$$

We can proceed further analytically only after some simplifying assumptions. Let the Cdf of the time to a breakdown be exponential with the failure rate λ and the repair time be arbitrary with the Cdf $G(t)$ and the pdf $g(t)$. Under these assumptions, using a similar reasoning to the previous section and deriving the corresponding simultaneous integral equations, it can be proved (Zarudnij, 1973) that the following equation for the Laplace transform of $\overline{F}(t)$ holds:

$$\overline{F}^*(s) = \frac{s+\lambda(1-\exp\{-s\tau_a[1-G(\tau_a)]-g^*(s,\tau_a))}{s[s+\lambda-\lambda g^*(s,\tau_a)]}, \tag{8.19}$$

where

$$g^*(s,\tau_a) = \int_0^{\tau_a} \exp\{-sx\}g(y)dy \tag{8.20}$$

is a 'truncated' Laplace transform.

The survival probability $\overline{F}(t)$ can be obtained using numerical methods for the inverse Laplace transform. Note that, as τ_a is a constant, the denominator in (8.19) has an infinite number of roots in the complex plane. Therefore, a solution can be obtained only as an infinite series. On the other hand, we will consider now an effective asymptotic approach to obtaining $\overline{F}(t)$ for the case of the fast repair. Therefore, assume that

$$\gamma \equiv 1 - G(\tau_a) << 1, \tag{8.21}$$

which is a definition of the fast repair in the case under consideration.

As previously, let T denote the lifetime of our system. Consider the random variable γT. From $\Pr[T > t] = \overline{F}(t)$ it follows that $\Pr[\gamma T > t] = \overline{F}(t/\gamma)$. The Laplace transform of this function is obtained directly from Equation (8.19). It can be shown after some simple transformations (Zarudnij, 1973) that as $\gamma \to 0$, uniformly in every finite interval,

$$\gamma \overline{F}^*(\gamma s) \to \frac{1}{s + \dfrac{\lambda}{1 + \lambda\tau'}}, \tag{8.22}$$

where $\tau' < \tau_a$. In order to proceed, another reasonable assumption should be imposed: $\lambda\tau_a << 1$, meaning that the mean time between the system's breakdowns is much larger than τ_a. Using this condition and the fast repair assumption (8.21), the Laplace transform (8.22) can be inverted as

$$\overline{F}(t) \to \exp\{-(1 - G(\tau_a)\lambda t\}, \quad \gamma \to 0, \tag{8.23}$$

which is a very useful approximate relationship.

This approximate result can be also derived *via* the method of the per demand failure rate (8.1) and (8.2), where $\lambda_r = \lambda$ and $\theta = \gamma$. In this case, the approximation error in (8.23) is due to the fact that the Poisson process with the rate λ is 'stopped' for periods of repair, which are small in accordance with our assumptions. When the repair time is exponential, *i.e.*, $G(t) = 1 - \exp\{-\mu t\}$, Relationship (8.23) becomes

$$\overline{F}(t) \approx \exp\{-(\exp\{-\mu t\})\lambda t\}.$$

In the rest of this chapter, we will see how the method of per demand failure rate (and its variations) effectively works in different reliability and safety assessment settings.

8.3 Failure Rate for Spatial Survival

In this section, we consider a model of spatial survival in plane (Finkelstein, 1998, 2003c). It is a generalization of a per demand failure rate method of Section 8.1 to a spatial Poisson process of shocks. Random obstacles along the route of some moving object will play a role of these shocks. Although the initial setting is bivariate, the resulting constructed failure rate is a univariate function, and therefore the results of Section 8.1 can be used.

The setting of the problem is as follows: a sufficiently small normally or tangentially oriented interval is moving along a fixed route in the plane, crossing points of spatial Poisson random processes. Each crossing leads to a termination of the process (failure, accident) with a predetermined probability. As previously, the probability of passing the route without termination is of interest. An immediate application of the method to be considered is the *safety at sea* assessment. Our approach takes into account the fixed obstacles (*e.g.*, shallows), which can lead to foundering and the moving obstacles (*e.g.*, other ships), which can lead to collisions.

The field of fixed obstacles is considered to be random. In this application, there are two types of fixed obstacles: obstacles with known coordinates, marked in the corresponding navigational sea charts (and therefore not random), and obstacles with unknown coordinates, which following the subjective approach can be considered random. It turns out that, owing to the accuracy of navigation and motion control systems of a ship, weather influences, currents, *etc.*, the obstacles with the known coordinates can also be modelled as random points in the plane (the corresponding randomization procedure is performed in Finkelstein, 1998). The 'geometric densities' of these obstacles, which can be obtained from the navigational charts, define the rates of the corresponding planar point processes to be used in the model.

The values of probabilities of accidents in "safety at sea" analysis are usually in the range 10^{-4} to 10^{-6}. Such estimates are often meaningless since there are not enough data to justify them. Therefore, simple relations for comparison of these probabilities can be very helpful in practice.

The developed approach can also be used for obtaining solutions that are optimal, for example, for finding a route with maximal probabilities of safe performance with or without specific restrictions (time on the route, fuel consumption, *etc.*). In what follows we consider the 2-dimensional setting, but the generalization to $n=3$ is straightforward and can be applied to assessing air traffic safety.

8.3.1 Obstacles with Fixed Coordinates

Denote by $\{N(B)\}$ an orderly point process in the plane, where $N(B)$ is a number of points in some domain $B \subset \Re^2$. We shall consider points of the process as prospective point influences (shocks) on our system (shallows for a ship, for instance). Similar to (4.1), the rate of this process $\lambda_f(\xi)$ can be formally defined as

$$\lambda_f(\xi) = \lim_{S(\delta(\xi))\to 0} \frac{E[N(\delta(\xi))]}{S(\delta(\xi))}, \tag{8.24}$$

where $B = \delta(\xi)$ is the neighbourhood of ξ with the area $S(\delta(\xi))$ and the diameter tending to zero. The subscript f stands for "fixed" obstacles.

Definition 8.1. The spatial non-homogeneous Poisson process is defined similar to the 1-dimensional case by the following relations (Cox and Isham, 1980):

$$\Pr[N(\delta(\xi)) = 1 \mid H_{\delta(\xi)}] = \lambda_f(\xi)S(\delta(\xi)) + o(S(\delta(\xi))),$$

$$\Pr[N(\delta(\xi)) > 1 \mid H_{\delta(\xi)}] = o(S(\delta(\xi))),$$

where $H_{\delta(\xi)}$ denotes the configuration of all points outside $\delta(\xi)$.

It can be shown for an arbitrary B that $N(B)$ has a Poisson distribution with mean

$$\int_B \lambda_f(\xi)d\xi$$

and that the numbers of points in non-overlapping domains are mutually independent random variables (Cox and Isham, 1980).

Our goal is to obtain a generalization of Equations (8.4) and (8.5) to the bivariate case. The idea of this generalization is in a suitable parameterization allowing us to reduce the problem to the 1-dimensional case. Assume for simplicity that $\lambda_f(\xi)$ is a continuous function of ξ in an arbitrary closed circle in $\Re^2$. Let R_{ξ_1,ξ_2} be a fixed continuous curve connecting two distinct points in the plane, ξ_1 and ξ_2. We will call R_{ξ_1,ξ_2} a route. A point (a ship in our application) is moving in one direction along the route. Every time it 'crosses the point' of the process $\{N(B)\}$ (see later the corresponding regularization), an accident (failure) can happen with a given probability. We are interested in assessing the probability of moving along R_{ξ_1,ξ_2} without accidents. Let r be the distance from ξ_1 to the current point of the route (coordinate) and $\lambda_f(r)$ denote the corresponding rate. Thus, the 1-dimensional parameterization is considered. For defining the corresponding Poisson measure, the dimensions of objects under consideration should be taken into account.

Let $(\gamma_n^+(r), \gamma_n^-(r))$ be a small interval of length $\gamma_n(r) = \gamma_n^+(r) + \gamma_n^-(r)$ in a normal direction to R_{ξ_1,ξ_2} at the point with the coordinate r, where the upper index denotes the corresponding direction ($\gamma_n^+(r)$ is on one side of R_{ξ_1,ξ_2}, whereas $\gamma_n^-(r)$ is on the other). Let $\overline{R} \equiv \mid R_{\xi_1,\xi_2} \mid$ be the length of R_{ξ_1,ξ_2} and assume that the interval is small compared with the length of the route, *i.e.*,

$$\overline{R} >> \gamma_n(r), \forall r \in [0, \overline{R}].$$

The interval $(\gamma_n^+(r), \gamma_n^-(r))$ is moving along R_{ξ_1,ξ_2}, crossing points of a random field. For "safety at sea" applications, it is reasonable to assume the symmetrical $(\gamma_n^+(r) = \gamma_n^-(r))$ structure of the interval with length $\gamma_n(r) = 2\delta_s + 2\delta_o(r)$, where $2\delta_s, 2\delta_o(r)$ are the diameters of the ship and of an obstacle, respectively. For simplicity, we assume that all obstacles have the same diameter. Thus, the ship's dimensions are already 'included' in the length of our equivalent interval. There can be other models as well, *e.g.*, the diameter of an obstacle can be considered a random variable.

Taking Equation (8.24) into account, the *equivalent rate* of occurrence of points, $\lambda_{e,f}(r)$ is defined as

$$\lambda_{ef}(r) = \lim_{\Delta r \to 0} \frac{E[N(B(r, \Delta r, \gamma_n(r))]}{\Delta r}, \tag{8.25}$$

where $N(B(r, \Delta r, \gamma_n(r))$ is the random number of points crossed by the interval $\gamma_n(r)$ when moving from r to $r + \Delta r$. Thus, the specific domain in this case is defined as an area covered by the interval moving from r to $r + \Delta r$.

When $\Delta r \to 0$, $\gamma_n(r) \to 0$, and taking into account that $\lambda_f(\xi)$ is a continuous function (Finkelstein, 2003),

$$\begin{aligned} E[N(B(r, \Delta r, \gamma_n(r))] &= \int_{B(r, \Delta r, \gamma_n(r))} \lambda_f(\xi) dS(\delta(\xi)) \\ &= \gamma_n(r) \lambda_f(r) dr[1 + o(1)], \end{aligned}$$

which leads to the expected relationship for the equivalent rate of the corresponding 1-dimensional non-homogeneous Poisson process, *i.e.*,

$$\lambda_{ef}(r) = \gamma_n(r) \lambda_f(r)[1 + o(1)], \quad \Delta r \to 0, \gamma_n(r) \to 0. \tag{8.26}$$

As the radius of curvature of the route $R_c(r)$ is sufficiently large compared with $\gamma_n(r)$, *i.e.*,

$$\gamma_n(r) << R_c(r),$$

the domain covered by the interval $(\gamma_n^+(r), \gamma_n^-(r))$ when it moves from r to $r + \Delta r$ along the route, is asymptotically ($\Delta r \to 0$) rectangular with area $\gamma_n(r)\Delta r$. Hence, the performed r-parameterization along the fixed route reduces the problem to the 1-dimensional setting of the previous section.

Assume now, as in the previous sections of this chapter, that the crossing of a point with a coordinate r leads to an accident with probability $\theta_f(r)$ and to the survival with the complementary probability $\bar{\theta}_f(r) = 1 - \theta_f(r)$. Denote by R the random distance from the initial point of the route ξ_1 to a point of the route where an accident has occurred. Similar to (8.5), the probability of passing the route R_{ξ_1, ξ_2} without accidents can be derived in the following way:

$$\Pr\{R > \bar{R}\} = \exp\left\{-\int_0^{\bar{R}} \lambda_{af}(r) dr\right\}, \tag{8.27}$$

where

$$\lambda_{af}(r) \equiv \theta_f(r) \lambda_{ef}(r) \tag{8.28}$$

is the corresponding failure (accident) rate. Thus, we have constructed the analogue of the per demand failure rate. As previously, Equations (8.27) and (8.28) constitute a simple and convenient tool for obtaining probabilities of safe (reliable) performance.

8.3.2 Crossing the Line Process

The content of this topic requires a more advanced mathematical background (spatial-temporal point processes and elements of stochastic geometry), and therefore this section may be omitted by the less mathematically oriented reader.

Consider a random process of continuous curves in the plane to be called paths. In the "safety at sea" application, the ship's routes in the sea chart represent paths, whereas the rate of stochastic processes to be defined represents the intensity of navigation in the given sea area. A specific case of stationary random lines in the plane to be considered as our model is called a *stationary line process.* Thus, for simplicity, the route of a ship will be modelled by a line in the plane.

It is convenient to characterize a line in the plane by its (ρ,ψ) coordinates, where ρ is a perpendicular distance from the line to a fixed origin and ψ is the angle between this perpendicular line and a fixed reference direction. The following observation is very helpful and connects a line process with a point process, which is important for our discussion. A random process of undirected lines can be defined as a point process on the cylinder $\Re_+ \times S$, where $\Re_+ = (0,\infty)$ and S denotes the interval $(0,2\pi]$. Therefore, each point on the cylinder is equivalent to the line in $\Re^2$. The following result is obtained in Daley and Vere-Jones (1988).

Theorem 8.1. Let V be a fixed line in $\Re^2$ with coordinates (ρ_v,α) and let N_V be a point process on V generated by its intersections with a stationary line process.

Then N_V is a stationary point process on V with rate λ_V given by

$$\lambda_V = \lambda \int_S |\cos(\psi - \alpha)| P(d\psi), \tag{8.29}$$

where λ is the constant rate of a stationary line process and $P(d\psi)$ is the probability that an arbitrary line has orientation in $[\psi, \psi + d\psi)$.

If the line process is isotropic, then $\lambda_V = 2\lambda / \pi$. The rate λ is induced by a random measure defined by the total length of lines inside any closed bounded convex set in $\Re^2$. One cannot define the corresponding measure as the number of lines intersecting the above-mentioned set, because in this case, it will not be additive, as the same line can intersect several domains in the set. The importance of this theorem is that it makes the useful connection between the line process and the corresponding point process on V.

Assume that a line process is a homogeneous Poisson process. This means that the point process N_V generated by its intersections with an arbitrary line V is a Poisson point process.

Consider now a stationary-temporal Poisson line process in the plane. Similar to N_V, the Poisson point process $\{N_V(t), t>0\}$ of its intersections with V in time can be defined. The constant rate of this process $\lambda_V(1)$ defines the probability of intersection (with a line from a temporal line process) of an interval of unit length in V during a unit interval of time (given these units are sufficiently small). As previously, $\lambda_V(1) = 2\lambda(1)/\pi$ for the isotropic case. Having defined all necessary notions, we can proceed now with obtaining the rate of intersections.

Let V_{ξ_1,ξ_2} be a finite line route, connecting ξ_1 and ξ_2 in $\Re^2$ and let r, as in the previous section, be the distance from ξ_1 to the current point of V_{ξ_1,ξ_2}. Then $\lambda_V(1)drdt$ can be interpreted as the (approximate) probability of intersecting V_{ξ_1,ξ_2} by the temporal line process in

$$(r, r+dr)\times(t,t+dt); \forall r \in (0,\overline{R}),\ t>0.$$

A point (a ship) starts moving along V_{ξ_1,ξ_2} at $\xi_1, t=0$ with a given speed $v(t)$. We assume that an accident happens with a given probability when it intersects the line from the temporal line process. Note that intersections in Section 8.4.1 were time-independent, as the obstacles were not moving. A regularization procedure, involving dimensions (of a ship, in particular) can be performed, *e.g.*, in the following way. Define the 'attraction interval'

$$(r-\gamma_{ta}^-, r+\gamma_{ta}^+) \subset V_{\xi_1,\xi_2},\ \gamma_{ta}^+, \gamma_{ta}^- \geq 0,\ \gamma_{ta}(r) = \gamma_{ta}^+ + \gamma_{ta}^- << \overline{R},$$

where the subscript "ta" stands for tangential. Therefore, the attraction interval, which can be specified by, *e.g.*, the ship's dimensions, is moving along the route, 'attached' to the point with coordinate changing in time as

$$r(t) = \int_0^t v(s)ds, \quad t \leq t_{\overline{R}}, \tag{8.30}$$

where $t_{\overline{R}}$ is the total time on the route. Note that, owing to the fact that the length of the attraction interval is negligible compared with $\overline{R}$, we do not take into account that the initial and final coordinates of the point are γ_{ta}^- and $\overline{R}-\gamma_{ta}^+$ and not 0 and $\overline{R}$, respectively.

Similar to (8.25) and (8.26), we are now able to construct the *equivalent rate of intersections* $\lambda_{e,m}(r)$ assuming for simplicity, a constant speed $v(t) = v_0$. Thus,

$$E[N_V((r,r+\Delta r),\Delta t)] = \lambda_V(1)\Delta r\Delta t. \tag{8.31}$$

The exact equality in (8.31) is due to the assumptions that our route is a line and that the line process is a stationary (temporal) homogeneous Poisson process. Therefore, the equivalent rate is also a constant, *i.e.*,

$$\lambda_{em} = \Delta t\,\lambda_V(1) = \frac{\gamma_{ta}}{v_0}\lambda_V(1), \tag{8.32}$$

where $\Delta t = \gamma_{ta}/\gamma_0$ is the time needed for the moving attraction interval to pass the interval $(r, r+\Delta r)$, $\Delta r \to 0$. In other words, we count the number of intersections of $(r, r+dr)$ with the lines from the temporal Poisson line process only during this interval of time. In this way, the temporal component of the process is 'eliminated' and its r-parameterization is performed.

As previously, the intersection can lead to an accident. Let the corresponding probability of an accident θ_m also be a constant. Similar to (8.27) and (8.28), the probability of moving along the route V_{ξ_1,ξ_2} without accidents can be obtained in the following simple form:

$$P(\overline{R}) = \exp\{-\theta_m \lambda_{e,m} \overline{R}\} , \tag{8.33}$$

as the corresponding accident rate, due to the assumptions, is a constant.

A partially alternative way of reasoning could be the following. Assume from the start that the point process generated by intersections of a route R_{ξ_1,ξ_2} that is not necessarily a line route with the temporal-planar paths is a non-homogeneous Poisson process (homogeneous in time, for simplicity) with rate $\lambda_R(r,1)$. This can be a reasonable initial assumption. Similar to (8.32), the r-dependent equivalent rate can be obtained asymptotically as $\gamma_{ta} \to 0$, *i.e.*,

$$\lambda_{em}(r) = \frac{\gamma_{ta}}{v_0} \lambda_R(r,1)[1 + o(1)]. \tag{8.34}$$

It is important that, as previously, everything is written in r-parameterization. Marginal cases for (8.34) are obvious. When the speed $v_0 \to 0$, the equivalent rate tends to infinity (time of exposure of a ship to the temporal path process tends to infinity); when $v_0 \to \infty$, the rate tends to zero. Similar to (8.27), Equation (8.33) can be written in an r-dependent way, *i.e.*,

$$P(\overline{R}) = \exp\left\{ - \int_0^{\overline{R}} \theta_m(r) \lambda_{em}(r) dr \right\}. \tag{8.35}$$

Finally, assuming independence of the two causes (the fixed and the moving obstacles) of accidents, the probability of survival on the route is given by the following equation:

$$P(\overline{R}) = \exp\left\{ - \int_0^{R} (\theta_f(r) \lambda_{ef}(r) + \theta_m(r) \lambda_{em}(r)) dr \right\}.$$

As we already mentioned, a direct application of the described approach to the "safety at sea" assessment in terms of absolute values is not so easy, as the corresponding data are very scarce. Therefore, comparing, *e.g.*, probabilities of an accident for different routes can be very helpful. The data on fixed obstacles and on the intensity of navigation from the special navigation charts and sometimes from other sources can be used for constructing the corresponding *relative charts of safety*. The resulting chart can be, *e.g.*, coloured in a manner similar to the charts of sea depths: the larger the depth, the more intensive the colour. For example, the open sea can be characterized by the relative rate 1 (for fixed obstacles), the harbour area by 5, the close-to-shore area by 3, *etc*.

As probabilities of accidents are very small, it is sufficient to consider only the two first terms in the Taylor expansions of exponents in Equations (8.27) and (8.35). Therefore, the failure (accident) rates should just be integrated along different routes and compared afterwards.

Example 8.3 Let θ_f be a constant and assume that there are three possible routes $R_{i,\xi_1,\xi_2}, i = 1,2,3$ connecting ξ_1 on ξ_2 with lengths 1500 km, 1200 km, 1000 km, respectively. Divide the area of navigation into the three zones with different relative rates for the fixed obstacles: 1 (open sea), 3 (intermediate) and 5 (harbour area), respectively. The following table shows the fraction of the length of the i th route belonging to the j th zone (j=1,2,3):

Table 8.1. Relative values for the three navigation zones

j/i	1	2	3
1	0.8	0.8	0.6
2	0.2	0.1	0.2
3	0	0.1	0.2

Thus, the longest route is in the safest zone. In line with this table, the following relative values of integrated hazards H_i (which are, in fact, the corresponding cumulative failure rates) can easily be computed for each route as

$$H_1 = 1.5(0.8 \cdot 1 + 0.2 \cdot 3) = 2.1,$$

$$H_2 = 1.2(0.8 \cdot 1 + 0.1 \cdot 3 + 0.1 \cdot 5) = 1.92,$$

$$H_3 = 1(0.6 \cdot 1 + 0.2 \cdot 3 + 0.2 \cdot 5) = 2.2,$$

giving preference to the second route as the route with the smallest integrated hazard. These relative values should be used only for comparison. It follows from (8.27) and (8.28) and our assumptions that

$$\frac{1 - P(\overline{R}_i)}{1 - P(\overline{R}_j)} = \frac{H_i}{H_j}[1 + o(1)], \quad i, j = 1,2,3, i \neq j.$$

Therefore, the considered approach gives a possibility of finding the best solution in a comparative way.

8.4 Multiple Availability on Demand

8.4.1 Introduction

The main objective of this section is to show how the method of the per demand failure rate works as a fast repair approximation for obtaining some characteristics of repairable systems, which generalize the conventional availability.

Consider a repairable system that starts operating at $t = 0$. A system is perfectly repaired upon failure and the process restarts. Assume that all corresponding time intervals are statistically independent and that the periods of operation and repair are identically distributed exponential random variables.

One of the main reliability characteristics of repairable systems is availability $A(t)$ – the probability of being in the operating state at an arbitrary instant of time t. This probability can be obtained (see, *e.g.*, Rausandt and Hoylandt, 2004) as

$$A(t) = \frac{\mu}{\mu+\lambda} + \frac{\lambda}{\mu+\lambda}\exp\{-(\lambda+\mu)t\}, \tag{8.36}$$

where λ and μ are parameters of exponential distributions of times to failure and to repair, respectively. When $t \to \infty$, the steady state (stationary) availability is obtained from Equation (8.36) as

$$A = \lim_{t\to\infty} A(t) = \frac{\mu}{\mu+\lambda}.$$

The stationary availability A is often used in practice as the main availability characteristic.

It is quite natural to generalize the notion of availability to a setting when a system should be available (be in the operating state) at multiple instants of time, $0 < t_1 < t_2 < \ldots < t_n,\ n \geq 1$.

Definition 8.2. Multiple availability $A_m(t)$ of a repairable system is the probability of the following event: a system is in the operating state at each moment of demand in $[0, t)$.

The unavailability of a system defines the corresponding criterion of failure. Therefore, the Cdf of the time to failure of the described system is $F(t) = 1 - A_m(t)$. As usual, denote the corresponding failure rate by $\lambda(t)$. First, let the instants of demand be deterministic, *i.e.*, $0 < t_1 < t_2 < \ldots t_n < t$. Owing to the memoryless property of the exponential distribution, multiple availability $A_m(t)$ can be obtained for this case as the product of the corresponding availabilities, *i.e.*,

$$A_m(t) = A(t_1)A(t_2 - t_1)\cdots A(t_k - t_{k-1}). \tag{8.37}$$

Demands are usually stochastic; therefore, the corresponding model should be developed. In what follows, we assume for simplicity that the point process of demands is the homogeneous Poisson process with rate λ_d and that it is statistically independent of the alternating renewal process (Ross, 1996), which describes the operation and repair of our system.

There are many applications where multiple availability can be considered as an important probabilistic characteristic of repairable systems. Performance of various information-measuring systems, for instance, which operate continuously and deliver information only on user demand at some instants of time, can be adequately described by multiple availability.

The described criterion of failure in the above definition of multiple availability is rather strong, as a system should be available at each instant of demand. Similar to the shock model of Section 8.3, the following weaker versions of this criterion can be considered:

- Not all demands need to be serviced;
- If a demand occurs when a system is in the state of repair, but the time from the start of repair does not exceed a predetermined value, then this event is not considered a failure.

These assumptions are quite natural. The first describes a situation where the quality of a system's performance can still be considered acceptable if the number of non-serviced demands, *i.e.*, when a system is in a state of repair at the moment of a demand, is less than $k, k \geq 2$. The second assumption can be applied, for instance, to information-measuring systems when the quality of information, measured before the system's last failure, is still acceptable after τ_a units of repair time (information is slowly ageing). This is similar to the setting of Section 8.3.2. We use two methods for deriving the corresponding expressions for multiple availability:

- The method of integral equations to be solved *via* the Laplace transforms;
- An approximate method based on the concept of the per demand failure rate.

We will show that the method of integral equations leads to the exact explicit solutions only for the simplest criterion of failure and for the exponential distributions of times to failure and repair. In other cases, the solution can be obtained *via* the numerical inversion of the Laplace transforms.

On the other hand, simple expressions for multiple availability can be obtained *via* the method of the per demand failure rate on the basis of the following fast repair assumption:

The mean time of repair is much smaller than the mean time to failure and than the mean time between consecutive demands.

The fast repair approximation also makes it possible to derive approximate solutions for the problems that cannot be obtained by the first method, *e.g.*, for the case of arbitrary time-to-failure and time-to-repair distributions.

8.4.2 Simple Criterion of Failure

In accordance with Definition 8.2 and using a straightforward reasoning, the following integral equation with respect to $A_m(t)$ can be obtained (Finkelstein and Zarudnij, 2002):

$$A_m(t) = \exp\{-\lambda_d t\} + \int_0^t \lambda_d \exp\{-\lambda_d t\} A(x) A_m(t-x) dx, \tag{8.38}$$

where the first term on the right-hand side defines the probability of the absence of demands in $[0,t)$, whereas the integrand is a joint probability of the following three events:

- Occurrence of the first demand in $[x, x+dx)$;
- The system is in the operating state at time instant x;
- The system is in the operating state at each moment of demand in $[x,t)$ (the corresponding probability is $A_m(t-x)$).

Equation (8.38) can be easily solved *via* the Laplace transform. Applying this transformation to both sides of (8.38) and using the Laplace transform of $A(t)$, defined by Equation (8.36), the following expression for the Laplace transform of $A_m(t)$ can be obtained after elementary transformations:

$$A_m^*(s) = \frac{s + \lambda_d + \lambda + \mu}{s^2 + s(\lambda_d + \lambda + \mu) + \lambda_d \lambda}. \tag{8.39}$$

Finally, performing the inverse transformation, we get

$$A_m(t) = B_1 \exp\{s_1 t\} + B_2 \exp\{s_2 t\}, \tag{8.40}$$

where the roots of the denominator in (8.39) are

$$s_{1,2} = \frac{-(\lambda_d + \lambda + \mu) \pm \sqrt{(\lambda_d + \lambda + \mu)^2 - 4\lambda_d \lambda}}{2}$$

and $B_1 = \dfrac{s_2}{s_2 - s_1}$, $B_2 = -\dfrac{s_1}{s_2 - s_1}$.

Equation (8.40) is an exact solution of the problem. The following fast repair assumptions will be used to obtain approximations for $A_m(t)$ and the corresponding errors of approximation:

$$\frac{1}{\lambda} >> \frac{1}{\mu}, \quad (\lambda << \mu); \tag{8.41}$$

$$\frac{1}{\lambda_d} >> \frac{1}{\mu}, \quad (\lambda_d << \mu); \tag{8.42}$$

$$t >> \frac{1}{\mu}. \tag{8.43}$$

Assumption (8.41) implies that the mean time to failure is much larger than the mean time of repair, and this is obviously the case in many practical situations, as was also discussed in Section 8.3 (Equation (8.14)). The mean time of repair usually does not exceed several hours, whereas it can be thousands of hours for λ^{-1}. This assumption is important for the method of the current section, as it will be used for approximating availability $A(t)$ (Equation (8.37)) by its stationary characteristic A in the interval between two consecutive demands (and for the corresponding expansions in the series as well). Therefore, this interval should be sufficiently large, which is stated by Assumption (8.42). There exist, of course, systems with a very high intensity of demands (*e.g.*, a database system where queries arrive every few seconds), but here we assume a sufficiently low intensity of demands, which is relevant to many engineering applications. An onboard navigational system is a relevant example. Assumption (8.43) means that the mission time t should be much larger than the mean time of repair.

With the help of these assumptions, an approximate relationship for $A_m(t)$ can be obtained *via* the method of the per demand failure rate in the following form:

$$A_m(t) = \exp\left\{-\int_0^t \lambda(u)du\right\}$$

$$\approx \exp\{-(1-A)\lambda_d t\} = \exp\left\{-\frac{\lambda\lambda_d t}{\lambda+\mu}\right\}, \qquad (8.44)$$

where $\lambda(t)$ is the failure rate, which corresponds to the Cdf $F(t) = 1 - A_m(t)$. Indeed, the demands occur in accordance with the Poisson process with rate λ_d, and at each instant of demand the system is available with approximate probability A. Thus, similar to Equation (8.4),

$$\lambda(t) = \theta\lambda_d,$$

$$\theta = (1-A) = \frac{\lambda}{\lambda+\mu}.$$

These equations were used when deriving (8.44).

What is the source of the error in the approximate Relationship (8.44)? This error results from the fact that the system is actually available on demand with probability $A(t_d)$, where t_d is the time since the previous demand and not with a stationary probability A. Assumptions (8.41) and (8.42) imply that $A(t_d) \approx A$. The error in (8.44) that results from this approximation can be estimated in the following way. When $t >> \mu^{-1}$(Assumption (8.43)) the second term on the right-hand side of the exact formula (8.40) is negligible. The corresponding error δ is defined as

$$\delta = \frac{s_2}{s_2 - s_1}\exp\{s_1 t\} - \exp\left\{-\frac{\lambda\lambda_d t}{\lambda+\mu}\right\}.$$

Using Assumption (8.41) for expanding s_1 and s_2 in series ($\lambda/\mu \to 0$) and Assumption (8.42) for further simplification eventually results in (Finkelstein and Zarudnij, 2006)

$$\delta = \frac{\lambda\lambda_d}{\mu^2}(1+\lambda_d t)(1+o(1)). \qquad (8.45)$$

The general case of arbitrary distributions of the time to failure with mean $\overline{T}$ and of the time to repair with mean $\overline{\tau}$ can also be considered using assumptions similar to Assumptions (8.41)–(8.43). This is because the stationary value of availability A for a general alternating renewal process is equal to $\overline{\tau}/(\overline{T}+\overline{\tau})$ and, therefore, depends only on the corresponding mean values. Relationship (8.44) in this case becomes

$$A_m(t) \equiv \exp\left\{-\int_0^t \lambda(u)du\right\}.$$

$$\approx \exp\{-(1-A)\lambda_d t\} = \exp\left\{-\frac{\bar{\tau}\lambda_d t}{\bar{T}+\bar{\tau}}\right\}.$$

Furthermore, the method of the per demand failure rate can be generalized to the non-homogeneous Poisson process of demands. In this case, as follows from Equation (8.4), $\lambda_d t$ should be substituted by

$$\Lambda_d(t) = \int_0^t \lambda_d(u)du.$$

It is difficult to estimate the error of approximation for the case of arbitrary distributions, as was done in the exponential case. Taking into account the corresponding heuristic reasoning, we can expect that this error will have the same 'structure' as in Equation (8.45), where $1/\mu$ and $1/\lambda$ should be replaced by $\bar{\tau}$ and $\bar{T}$, respectively.

8.4.3 Two Consecutive Non-serviced Demands

The strong criterion of failure given by Definition 8.2 can be naturally relaxed in the following way (Finkelstein and Zarudnij, 2002).

Definition 8.3. The failure of a repairable system that services stochastic demands occurs when a system is in the repair state at two consecutive moments of demand.

In accordance with this definition, multiple availability $A_m^{(2)}(t)$ of a system is the probability of operating without failures in $[0,t)$. As stated earlier, this setting can be quite typical for some information-processing systems. If, for example, a scheduled 'correction' of a navigation system *via* a satellite fails (the system was unavailable), we can still wait for the next correction, but usually not more.

Similar to (8.38), the following integral equation with respect to $A_m^{(2)}(t)$ is obtained:

$$A_m^{(2)}(t) = \exp\{-\lambda_d t\}$$

$$+ \int_0^t \lambda_d \exp\{-\lambda_d x\}\, A(x) A_m^2(t-x)dx$$

$$+ \int_0^t [\lambda_d \exp\{-\lambda_d x\}(1-A(x)) \int_0^{t-x} \lambda_d \exp\{-\lambda_d y\} \tilde{A}(y) A_m^{(2)}(t-x-y)dy]dx\,, \quad (8.46)$$

where $\tilde{A}(t)$ is the availability of the system at t given that at $t=0$ the system was in the repair state, *i.e.*,

$$\tilde{A}(t)=\frac{\mu}{\mu+\lambda}-\frac{\mu}{\mu+\lambda}\exp\{-(\lambda+\mu)t\}\,.$$

The first two terms on the right-hand side of (8.46) have a similar meaning to Equation (8.38), whereas the third term defines the joint probability of the following events:

- Occurrence of the first demand in $[x,x+dx)$;
- The system is in the repair state at x (with probability $1-A(x)$);
- Occurrence of the second demand in $[x+y,x+y+dy)$;
- The system is in the operational state at $x+y$, whereas it was in the repair state at the previous demand (with probability $\tilde{A}(y)$);
- The system operates without failures in $[x+y,t)$ (with probability $A_m^{(2)}(t-x-y)$).

Equation (8.46) can also be solved *via* the Laplace transform. After elementary transformations:

$$A_m^{(2)*}(s)=\frac{(s+\lambda_d)(s+\lambda_d+\lambda+\mu)^2}{s(s+\lambda_d)(s+\lambda_d+\lambda+\mu)^2+s\lambda_d\lambda(s+2\lambda_d+\lambda+\mu)+\lambda_d^2\lambda(\lambda_d+\lambda)}$$

$$=\frac{\mathrm{P}_3(s)}{\mathrm{P}_4(s)}, \tag{8.47}$$

where $P_3(s)$ and $P_4(s)$ denote the corresponding polynomials in the numerator and the denominator, respectively. The inverse transformation results in

$$A_m^{(2)}(t)=\sum_1^4\frac{P_3(s_i)}{P_4'(s_i)}\exp\{s_it\}, \tag{8.48}$$

where $P_4'(s)$ is the derivative of $P_4(s)$ and $s_i, i=1,2,3,4$ are the roots of the denominator in (8.47), *i.e.*,

$$\mathrm{P}_4(s)=\sum_0^4 b_k s^{4-k}=0 \tag{8.49}$$

and b_k are defined as

$$b_0=1,\quad b_1=2\lambda+2\mu+3\lambda_d,\quad b_2=(\lambda_d+\lambda+\mu)(3\lambda_d+\lambda+\mu)+\lambda_d\lambda,$$
$$b_3=\lambda_d[(\lambda_d+\lambda+\mu)^2+\lambda(\lambda_d+\lambda+\mu)+\lambda(\lambda_d+\lambda)],$$
$$b_4=\lambda_d^2\lambda(\lambda_d+\lambda).$$

Equation (8.48) defines the exact solution of the problem. The solution can also be obtained numerically by solving Equation (8.49) and substituting the corresponding roots in (8.48).

As in the previous section, a simple, approximate formula based on the method of the per demand failure rate can also be used. Let Assumptions (8.41)–(8.43) hold. All $b_k, k = 0,1,2,3,4$ in Equation (8.49) are positive, which means that there are no positive roots for this equation. Consider the smallest root in absolute value, s_1. Owing to assumption (8.41):

$$s_1 \approx -\frac{b_4}{b_3} \approx -\frac{\lambda\lambda_d(\lambda+\lambda_d)}{\mu^2}, \quad \frac{\mathrm{P}_3(s_1)}{\mathrm{P}_4'(s_1)} \approx 1.$$

It can also be shown that the absolute values of other roots are much larger than $|s_1|$. Thus, Equation (8.48) can be written as the following fast repair exponential approximation:

$$A_m^2(t) \approx \exp\left\{-\frac{\lambda\lambda_d(\lambda+\lambda_d)}{\mu^2}t\right\}. \tag{8.50}$$

It is difficult to assess the corresponding approximation error directly, as was done in the previous section, because the root s_1 is also defined approximately. On the other hand, the method of the per demand failure rate can be used for obtaining $A_m^{(2)}(t)$. Similar to (8.44),

$$\begin{aligned} A_m^{(2)}(t) &= \exp\left\{-\int_0^t \lambda(u)du\right\} \\ &\approx \exp\left\{-A(1-A)^2\lambda_d t\right\} = \exp\left\{-\frac{\mu\lambda^2\lambda_d t}{(\lambda+\mu)^3}\right\}. \end{aligned} \tag{8.51}$$

Indeed, failure occurs in $[t, t+dt)$ if a demand occurs in this interval (with probability $\lambda_d dt$) and the system is unavailable at this moment of time and at the moment of the previous demand, whereas it was available at the demand prior to the latter one. Owing to the fast repair assumptions (8.41) and (8.42), this probability is approximately equal to $(\mu\lambda/(\lambda+\mu))^3$, as the stationary values of $A(t)$ and $\tilde{A}(t)$ are both equal to $\mu/(\lambda+\mu)$. Taking again into account these assumptions, we observe that Approximations (8.50) and (8.51) are really 'close'.

As in Section 8.4.2, the generalization to arbitrary distributions with finite means is performed, *i.e.*,

$$A_m^2(t) \approx \exp\left\{-A(1-A)^2\lambda_d t\right\} = \exp\left\{-\frac{\lambda_d \bar{T}\bar{\tau}^2 t}{(\bar{T}+\bar{\tau})^3}\right\}. \tag{8.52}$$

8.4.4 Other Weaker Criteria of Failure

The case of not more than N non-serviced demands in $[0,t)$ (not necessarily consecutive) is also considered in a similar manner (Finkelstein and Zarudnij, 2002). The failure in this case is described by the following definition.

Definition 8.4. The failure of a repairable system occurs when more than $N \geq 1$ demands are non-serviced in $[0,t)$.

Denote the corresponding probability of the failure-free operation by $A_{m,N}(t)$. Cumbersome integral equations can be derived (Finkelstein and Zarudnij, 2002) and solved in terms of the corresponding Laplace transforms. The Laplace transform of $A_{m,N}(t)$ should then be inverted using numerical methods. On the other hand, the fast repair approximation, as previously, allows for the simple heuristic approach.

Consider the point process of moments of unavailability on demand of our system. As follows from (8.44), this point process can be approximated by the Poisson process with the rate $(1-A)\lambda_d$. This leads to the following approximate result for arbitrary (not very large) N:

$$A_{m,N}(t) \approx \exp\left\{-\frac{\lambda\lambda_d}{\lambda+\mu}t\right\}\sum_{n=0}^{N}\frac{1}{n!}\left(\frac{\lambda\lambda_d}{\lambda+\mu}t\right)^n, \quad N = 1,2,\dots \, . \tag{8.53}$$

Thus, a rather complicated problem has been immediately solved *via* the Poisson approximation, based on the per demand failure rate $\lambda\lambda_d(\lambda+\mu)^{-1}$. When $N=0$, we arrive at the case of ordinary multiple availability: $A_{m,0}(t) \equiv A_m(t)$.

Another weaker definition of failure is based on the time redundancy concept discussed for a different setting in Section 8.2.2.

Definition 8.5. The failure of a repairable system that services stochastic demands, occurs when the repair action is not completed in time $\tau_a > 0$.

As previously, the corresponding multiple availability $A_{m,\tau}(t)$ is defined as the probability of a system functioning without failures in $[0,t)$. Definition 8.5 means that if a demand occurs when the system is in a state of repair, which is completed within (remaining) time $\tau_a > 0$, then this event is not qualified as a failure of a system. Therefore, the delay τ_a is considered to be acceptable. Note that if $\tau_a = 0$, then $A_{m,\tau}(t) = A_m(t)$.

To obtain a simple approximate formula by means of the method of the per demand failure rate, as in the previous case, consider the Poisson process with rate $(1-A)\lambda_d$, which approximates the point process of the non-serviced demands. In accordance with Equations (8.2) and (8.3), multiplying the rate of this initial process by the probability of 'failure on demand', *i.e.*, $\exp\{-\mu\tau_a\}$, the corresponding failure rate can be obtained (Finkelstein and Zarudnij, 2002) as

$$A_{m,\tau}(t) \approx \exp\{-\lambda_d(1-A)(\exp\{-\mu\tau_a\})t\}$$
$$= \exp\left\{-\frac{\lambda\lambda_d t}{\lambda+\mu}\exp\{-\mu\tau_a\}\right\}.$$

8.5 Acceptable Risk and Thinning of the Poisson Process

In this section, we will consider a simple example of the operation of thinning for the Poisson process of shocks with rate $\lambda_r(t)$ (Finkelstein, 2007c).

Example 8.4 Assume that each shock causes a random loss C_i. Let $C_i, i = 1,2,...$ be i.i.d. random variables with the continuous Cdf $G(c), c \geq 0$. Our interest is in considering the overall consequences of shocks in $[0,t)$.

Divide the c-axis into n regions, *i.e.*,

$$[0,c_1), [c_1,c_2),...,[c_{n-1},\infty).$$

The probability that the loss from a single shock does not exceed the level c_i is $G(c_i)$, and the probability that it is in the region $[c_i,c_j),\ i<j; i,j<n; c_n \equiv \infty$ is

$$p_{i,j} = G(c_j) - G(c_i), \quad p_{i,n} = 1 - G(c_i), \quad p_{i,0} = G(c_i) - G(0) = G(c_i).$$

The first step is to derive the probability $P_j(t)$ that all events that occurred in $(0,t]$ resulted in a loss not exceeding c_i. In accordance with Equation (8.4), this probability can be defined as

$$P_i(t) = \exp\left\{-\int_0^t (1-p_{i,0})\lambda_r(x)dx\right\}. \tag{8.54}$$

Similar to (8.54), the probability that all events resulted in a loss in a range of c_i to c_j is

$$P_{i,j}(t) = \exp\left\{-\int_0^t (1-p_{i,j})\lambda_r(x)dx\right\}.$$

Specifically, for the three regions:

$$P_s(t) = \exp\left\{-\int_0^t (1-p_s)\lambda_r(x)dx\right\}, \quad P_{s,u}(t) = \exp\left\{-\int_0^t (1-p_{s,u})\lambda_r(x)dx\right\} \tag{8.55}$$

$$P_u(t) = \exp\left\{-\int_0^t (1-p_u)\lambda_r(x)dx\right\},$$

where $P_s(t)$ is the probability that all events from the Poisson process in $[0,t)$ result in a 'safe loss'; $P_{s,u}(t)$ denotes the probability that all events result in a loss in $[c_s,c_u)$. Eventually, $P_u(t)$ denotes the supplementary probability that all events result in a loss in the region $[c_u,\infty)$.

The strongest criterion of the corresponding acceptable risk is when all events result in a loss from the first region. It is reasonable to consider a weaker version of this acceptance criterion allowing, for example, not more than $k=1,2,...$ events to result in a loss from the intermediate region $[c_s,c_u)$ (an event in $[c_u,\infty)$ is 'not allowed' at all). For simplicity, let the underlying process be the homogeneous process with rate λ_r. It is clear (Ross, 1996) that this process can be split into three Poisson processes with rates

$$\lambda_r p_s, \lambda_r p_{s,u}, \lambda_r p_u \ .$$

Due to our acceptable risk criterion, the risk in $[0,t)$ is considered unacceptable if at least one event occurs from the process with rate $\lambda_r p_u$ or if more than k events occur from the process with rate $\lambda_r p_{s,u}$. These considerations lead to the following equation for the probability of safe (with acceptable risk) performance:

$$P_{s,k}(t)=\exp\{-\lambda_r p_u t\}\exp\{-\lambda_r p_{s,u} t\}\sum_0^k \frac{(\lambda_r p_{s,u} t)^i}{i!} \ .$$

When there is no intermediate region, $c_u=c_s$ and we arrive at

$$P_{s,0}(t)\equiv P_s(t)=\exp\{-\lambda_r p_u t\}=\exp\{-\lambda_r(1-p_s)t\} \ ,$$

which coincides with the first equation in (8.55).

8.6 Chapter Summary

In this chapter, we have considered several meaningful examples of application of the concept of the per demand failure rate to different reliability problems. A basic feature of all models is an underlying point process of events that can be terminated in some way. Termination usually means the failure of a system or a mission failure. When the underlying process is a homogeneous (or non-homogeneous) Poisson process, the corresponding failure rate can be 'constructed' and, therefore, the probability of termination can usually be obtained in an explicit way. Termination of renewal processes, however, cannot be modelled explicitly, and only bounds and approximations exist for reliability measures of interest.

In Sections 8.2 and 8.4, we consider the weaker criteria of failure when, *e.g.*, not every event from the underlying process can result in the failure of a system or when these events should not be too close in time. The solutions are obtained in terms of the corresponding Laplace transforms, but effective and simple approximate results are derived *via* the method of the per demand failure rate.

In Section 8.3 the developed 1-dimensional approach is applied to obtain the survival probability of an object moving in the plane and encountering moving or (and) fixed obstacles. In the "safety at sea" application terminology, each foundering or collision results in a failure (accident) with a predetermined probability. It is shown that this setting can be reduced to the 1-dimensional setting, which is suitable for applying the method of the per demand failure rate.

9

Failure Rate of Software

9.1 Introduction

This chapter is devoted to software reliability modelling and, specifically, to a discussion of some of the software failure rate models. It should not be considered a comprehensive study of the subject, but rather a brief illustration of the methods and approaches of the previous chapters. In Section 9.2, for instance, we consider several well-known 'empirical' models for software failure rates that can be described in terms of the corresponding stochastic intensity processes defined and studied in Chapters 4 and 5. In Section 9.3, a different approach is presented based on a stochastic model similar to the model used for constructing the failure rate for spatial survival in Section 8.3 (Finkelstein, 1999c). For a more detailed basic treatment of software reliability issues, the reader is referred to, *e.g.*, the books of Musa *et al.* (1987), Xie (1991), Pham (2000) and Singpurwalla and Wilson (1999).

Assessing software reliability is not easy. Perhaps the major difficulty is that we are concerned primarily with design faults, which is a very different situation from that considered by conventional hardware reliability theory. A fault (or bug) refers to a manifestation of a mistake in the code made by a programmer or designer with respect to the specification of the software (Ledoux, 2003). Similar to hardware reliability, software reliability is defined in Singpurwalla and Wilson (1999) as *the probability of failure-free operation of a computer code for a specified mission time in a specified input environment.* Activation of a fault by an input value leads to an incorrect output that is a failure. There are two major causes of randomness in software reliability models, *i.e.*, the unknown 'locations' of bugs and the random nature of input values. Therefore, the stochastic modelling of software reliability can be justified by these factors.

Define a software program as a set of complete machine instructions that executes within a single computer and accomplishes a specific function (Musa *et al.*, 1987). It can formally be described as the following mapping:

$$X \stackrel{G}{\rightarrow} Y ,$$

where X and Y are input and output domains, respectively, and G is a function that maps each $x \in X$ onto a single $y \in Y$. A fault (bug) is defined as a defect of a program that causes one or more values of the input domain to be mapped into incorrect values of the output domain. Denote the set of all faults by X_f.

In real applications, the factors that cause the selection of a particular input value are numerous and complex. The crucial role for the corresponding probabilistic analysis is played by the *operational profile* $p(\overline{x})$ (Pasquini *et al.*, 1996). Assume for simplicity that X is a domain in the Euclidian space $X \subset \Re^m$. The value $p(\overline{x})d\overline{x} \equiv p(x_1, x_2, ..., x_m)dx_1...dx_m$ is interpreted as the probability of choosing an input value in the m-dimensional parallelepiped $[\overline{x}, \overline{x} + d\overline{x}]$. Note that the time, which usually defines a real operational profile, is not a part of the model yet. This case will be considered in Section 9.3. In accordance with the given definitions, the following integral:

$$C_f = \int_{X_f} p(\overline{x})d\overline{x} \tag{9.1}$$

can be viewed as a measure of software reliability as it takes into account the total volume of bugs in a program and probabilities of choosing these faults by a program. As the total volume of 'faulty inputs' is usually much smaller than the volume of the entire X, the following assumption is reasonable:

$$C_f << \int_X p(\overline{x})d\overline{x} = 1.$$

This approach is too general and some other important assumptions should be made and relevant special cases should be discussed. Furthermore, as mentioned, it does not take into account the dependence of the operational profile on time. However, before considering these issues in Section 9.3, we briefly describe several well-known empirical models of software reliability focusing on the corresponding intensity processes.

9.2 Several Empirical Models for Software Reliability

There are numerous models postulating the functions that can define software reliability. In this section we briefly discuss only some straightforward 'historical' models for software reliability that had a substantial impact on further development in this area. Most approaches can roughly be divided into two groups. The first direction deals with modelling the time between successive failures, whereas the second approach considers the numbers of failures up to t.

Assume that a program is set into operation at $t = 0$. After the first failure at random S_1, instantaneous debugging is performed and operation of the program continues up to $S_2 > S_1$, *etc.* As a bug is removed and the program is corrected, it is reasonable to assume that the length of the subsequent cycle is larger (in some suitable stochastic sense) than the length of a previous cycle. For example, the geometric process of Section 4.3.3 for $a < 1$ can be considered as the corresponding sequence of the stochastically increasing cycle durations. Note that debugging can also be imperfect, *i.e.*, new bugs can be 'created' during this operation. In this chapter, for simplicity, we only consider the case of the perfect debugging.

9.2.1 The Jelinski–Moranda Model

This model is probably one of the first meaningful models of software reliability. It has also formed the basis for several other models that have been developed later. Jelinski and Moranda (1972) assume that software contains an unknown number of initial bugs N and that each time software fails, a bug is detected and instantaneously corrected. Each bug has an 'independent input' of size $\lambda > 0$ into the failure rate of the software. Thus, the first cycle is characterized by the failure rate $N\lambda$, the failure rate at the second cycle is $(N-1)\lambda$ and the failure rate at the i th cycle is defined by the number of remaining bugs in the program, *i.e.*, $\lambda(N-i+1)$. The process stops when no bugs are left in the program. As previously, S_i denotes the arrival time of the i th failure with realizations $s_i, i = 1,2,...$. Therefore, the intensity process (stochastic intensity) and the CIF for this process, similar to Equations (4.13) and (4.14), are

$$\lambda_t = \sum_{i \geq 1}^{N} \lambda(N-i+1) I(S_{i-1} \leq t < S_i),\ t \geq 0, \tag{9.2}$$

$$\lambda(t \mid H(t)) = \sum_{i \geq 1}^{N} \lambda(N-i+1) I(s_{i-1} \leq t < s_i),\ t \geq 0, \tag{9.3}$$

where $H(t) = 0 = s_0 \leq s_1 < s_2 < ... < s_{n(t)}$ is the observed history of failures in $[0,t)$ and $S_0 = s_0 = 0$.

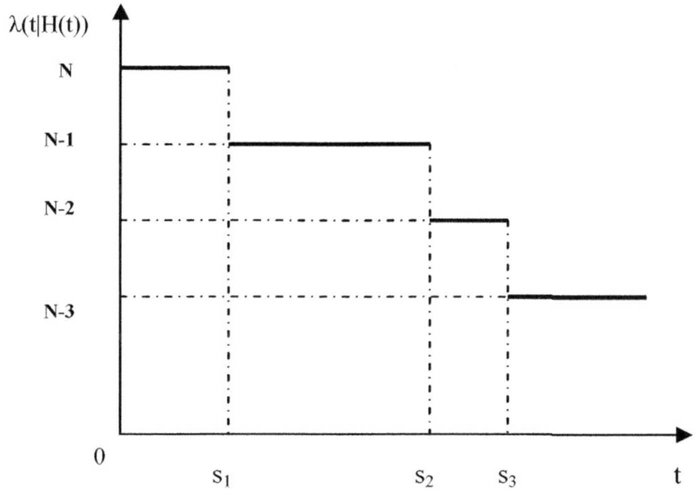

Figure 9.1. The CIF for the Jelinsky–Moranda model

As in Chapter 4, these formulas can be written in a compact way, *i.e.*,

$$\lambda_t = \lambda(N - \widetilde{N}(t)),$$

$$\lambda(t \mid H(t)) = \lambda(N - \widetilde{n}(t)),$$

where $\widetilde{N}(t)$ denotes the random number of the last failure (debugging) before t and $s_{n(t)}$ denotes the corresponding realization. A graph of the possible shape of $\lambda(t \mid H(t)$ is shown in Figure 9.1.

The assumptions underlying the model of Jelinski and Moranda are clearly unrealistic: in reality all bugs do not contribute equally to the failure rate, but as one of the first models, it played a very important role in the development of software reliability. Note that some authors call the intensity process λ_t for software modelling the *concatenated failure rate* (see, *e.g.*, Singpurwalla and Wilson, 1999; Ledoux, 2003).

9.2.2 The Moranda Model

This is also one of the early models. The cycle durations are again distributed exponentially in this model, but it already takes into account the possibility of a different input of different bugs in software reliability. Moranda (1975) suggests a modification of the Jelinski–Moranda model, where the bugs that appear early contribute more to the failure rate than those that appear later. This seems to be a reasonable assumption as early bugs usually represent more serious faults of the program. There can be different ways of modelling this effect analytically. The simplest model is probably given by the corresponding geometrical reduction procedure (Moranda, 1975).

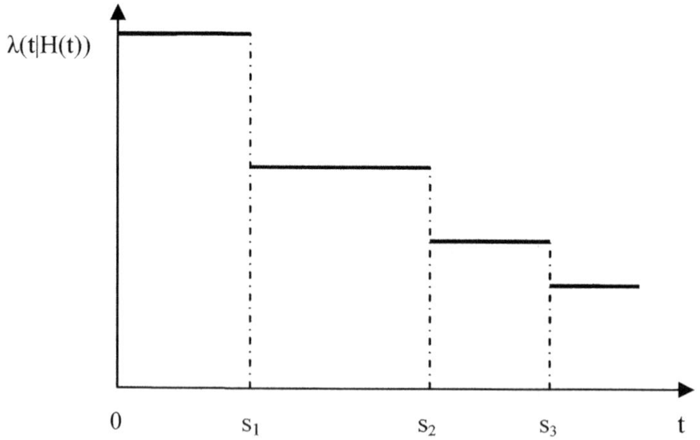

Figure 9.2. The CIF for the Moranda model for $k = 0.5$

The intensity process for this model is defined by the following equation:

$$\lambda_t = \sum_{i \ge 1} \lambda k^{i-1} I(S_{i-1} \le t < S_i), \ t \ge 0, \tag{9.4}$$

where $0 < k < 1$. The exponentially decreasing function $\exp\{-\alpha(k-1)\}$ can also be used for this modelling:

$$\lambda_t = \sum_{i \ge 1} \lambda \exp\{-\alpha(i-1)\} I(S_{i-1} \le t < S_i), \ t \ge 0 .$$

The additional parameter $\alpha > 0$ provides more flexibility for the corresponding statistical inference. Figure 9.2 illustrates this model (compare with Figure 9.1). The important feature of the Moranda model is that there is no assumption on the initial number of bugs in the program.

9.2.3 The Schick and Wolverton Model

This model considers non-exponential interfailure times, which is a significant departure from the previous two models. The intensity process is

$$\lambda_t = \sum_{i \ge 1} \lambda (N - i + 1)(t - S_{i-1}) I(S_{i-1} \le t < S_i), \ t \ge 0 . \tag{9.5}$$

The failure rate at each cycle is proportional not only to the number of remaining bugs, as in the Jelinsky–Moranda model, but to the elapsed time $(t - S_{i-1})$ as well. Therefore, the failure rate is linear at each cycle.

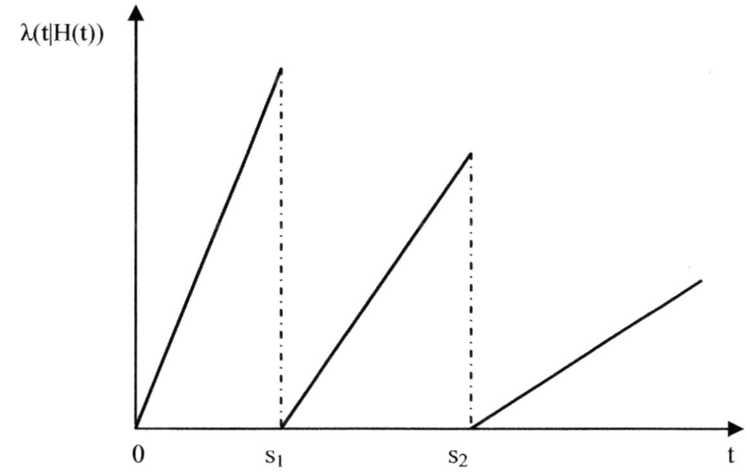

Figure 9.3. The CIF for the Schick and Wolverton model

The Weibull-type generalization can also be considered, *i.e.*,

$$\lambda_t = \sum_{i\geq 1} \lambda(N-i+1)\alpha(t-S_{i-1})^{\beta} I(S_{i-1} \leq t < S_i), \; t \geq 0, \alpha, \beta > 0 .$$

On the other hand, the Moranda model (9.4) can be generalized in a similar way to

$$\lambda_t = \sum_{i\geq 1} \lambda k^{i-1}(t-S_{i-1}) I(S_{i-1} \leq t < S_i), \; t \geq 0 .$$

9.2.4 Models Based on the Number of Failures

The software reliability models of the previous section postulated the form of the corresponding intensity process. Models based on the number of failures usually postulate (explicitly or implicitly) the rate $\lambda_r(t)$ (Section 4.3.1) of the corresponding non-homogeneous Poisson process. Therefore, assume that the software failures occur in accordance with the NHHP with rate $\lambda_r(t)$. If this function is decreasing, then the interarrival times are stochastically increasing with each debugging, and therefore this setting can be described by the reliability growth concept (Ushakov and Harrison, 1994). Obviously, the first choice for the decreasing rate $\lambda_r(t)$ is the decreasing power function, *i.e.*,

$$\lambda_r(t) = at^{-b}, \; a, b > 0 . \tag{9.6}$$

Some programs behave in such a way that the rate, at which failures are observed in software increases initially and then decreases. To accommodate this property, Goel (1985) suggested the following form of the rate:

$$\lambda_r(t) = abct^{c-1} \exp\{-bt^c\}, \quad a, b, c > 0 . \tag{9.7}$$

On the other hand, Musa and Okumoto (1984) postulate the relationship between $\lambda_r(t)$ and the cumulative rate $\Lambda_r(t)$ as

$$\lambda_r(t) = \lambda \exp\{-c\Lambda_r(t)\}, \quad \lambda, c > 0 . \tag{9.8}$$

Thus, the rate at which failures occur exponentially decreases with the expected number of failures. This assumption seems reasonable from a 'physical' point of view. As $\Lambda_r(t)$ is an integral of $\lambda_r(t)$, solving the elementary differential equation results in the following explicit relationships:

$$\lambda_r(t) = \frac{\lambda}{\lambda ct + 1}, \tag{9.9}$$

$$\Lambda_r(t) = \frac{1}{c} \ln(\lambda ct + 1) . \tag{9.10}$$

As $t \to \infty$, Rate (9.9) converges to Rate (9.6) for $b = 1, a = 1/c$.

Another popular model is the model by Goel and Okumoto (1979). These authors argued that $\Lambda_r(t)$ should be bounded, because the expected number of failures over the life of the software is finite (Singpurwalla and Wilson, 1999). Thus, they assume that $\lim_{r\to\infty} \Lambda_r(t) = a > 0$. Note that $\Lambda_r(t)$ in (9.10) is not finite when $t \to \infty$. The crucial assumption in this model, however, is that the expected number of failures in $[t, t+dt)$ is proportional to the product of the expected number of remaining bugs in the software. Therefore,

$$\Lambda_r(t+dt) - \Lambda_r(t) = b(a - \Lambda_r(t))dt + o(dt), \tag{9.11}$$

where $b > 0$ is the fault detection rate that shows the intensity with which the faults are removed. Therefore, another important restrictive assumption is that this rate is constant in time. Relationship (9.11) obviously results in a differential equation with respect to $\Lambda_r(t)$, *i.e.*,

$$\Lambda_r'(t) = b(a - \Lambda_r(t)).$$

Taking into account the boundary conditions (Singpurwalla and Wilson. 1999),

$$\Lambda_r(t) = a(1 - \exp\{-bt\}), \quad \lambda_r(t) = ab\exp\{-bt\}. \tag{9.12}$$

Thus, the asymptotic behaviour of $\Lambda_r(t)$ is more realistic in the Goel and Okumoto model than the infinite limit obtained from Equation (9.10). The exponential decay in $\lambda_r(t)$ also seems to be more likely from general considerations than the more specific shape defined by Equation (9.9).

In accordance with Equation (4.5), the intensity process for the NHPP is deterministic and equal to its rate $\lambda_r(t)$, which can formally be written as

$$\lambda_t = \sum_{i\geq 1} \lambda_r(t) I(S_{i-1} \leq t < S_i) \equiv \lambda_r(t).$$

Most software reliability models are empirical and can be justified (or not) by fitting the failure data. The approach of the next section (Finkelstein, 1999c), by contrast, is theoretical and describes the operation of a program with bugs using some general (although simplified) probabilistic considerations.

9.3 Time-dependent Operational Profile

9.3.1 General Setting

The definition of the operational profile $p(\bar{x}), \bar{x} \in X \subset \Re^m$ was given earlier in Section 9.1. It is clear that the operational profile of software in real usage is time-dependent, and therefore this should be taken into account in software reliability modelling.

Similar to $p(\bar{x})$, define $p(\bar{x},t)$ *via* the probability $p(\bar{x},t)d\bar{x}dt$ of choosing one input value in an infinitesimal domain $(\bar{x},\bar{x}+d\bar{x})\times[t,t+dt)$. Therefore, $\bar{p}(x,t)$ is the density of the corresponding m-dimensional stochastic process.

Let $P_n = \Pr[N = n],\ n = 0,1,2,...$ be the distribution of the number of bugs in X. For each n, denote by $F_n(\bar{x}_{1f},...,\bar{x}_{nf})$ the absolutely continuous joint distribution of the locations (coordinates) of n bugs defined in $X^{(n)} = X\times X....\times X$. Therefore, the corresponding joint density defines the following probability:

$$f_n(\bar{x}_{1f},,...,\bar{x}_{nf})d\bar{x}_{1f}...d\bar{x}_{nf} = \Pr[\bar{X}_{1f} \in (\bar{x}_{1f} + d\bar{x}_{1f}),...,\bar{X}_{nf} \in (\bar{x}_{nf} + d\bar{x}_{nf})],$$

where $\bar{X}_{if}, i = 1,2,...,n$ are the random coordinates of the i th bug.

We are now able to define the Yanoshi density (Daley and Vere-Jones, 1988) $j_n(\bar{x}_{1f},,...,\bar{x}_{nf})$. The product $j_n(\bar{x}_{1f},,...,\bar{x}_{nf})d\bar{x}_{1f}...d\bar{x}_{nf}$ is the probability that there are exactly n bugs, one in each infinitesimal domain $x_{if} \in (\bar{x}_{if} + d\bar{x}_{if})$, $i = 1,2,...,n$. Thus,

$$j_n(\bar{x}_{1f},,...,\bar{x}_{nf}) = P_n\tilde{f}_n(\bar{x}_{1f},,...,\bar{x}_{nf}), \tag{9.13}$$

where

$$\tilde{f}_n(\bar{x}_{1f},,...,\bar{x}_{nf}) = n!\,f_n(\bar{x}_{1f},,...,\bar{x}_{nf}),$$

as there are $n!$ permutations of possible 'positions' of n bugs.

Equation (9.1) can be generalized to the case of the time-dependent operational profile. Assume, first, that n is fixed and that the coordinates of all bugs, $\{\bar{x}_{1f},...,\bar{x}_{nf}\}$, are also deterministic. Then

$$C_f(t,n,\bar{x}_{1f},...,\bar{x}_{nf}) = \int_{X_f} p(\bar{x},t)d\bar{x} = \delta_b \sum_{n=1}^{n} p(\bar{x}_{if},t), \tag{9.14}$$

where δ_b is a measure (volume), which, for simplicity, is the same for each bug. A finite computer representation of an m-dimensional vector owing to machine tolerance stands for an m-dimensional cube. Therefore, the Lebesgue integral in (9.14) is properly defined, and therefore the bug is activated by the corresponding input domain with a measure δ_b. Relationship (9.14) should be understood asymptotically for sufficiently small δ_b, which is obviously the case for software.

The product $C_f(t,n,\bar{x}_{1f},...,\bar{x}_{nf})dt$ defines the probability of choosing a bug as an input value in $[t,t+dt)$ for fixed number of bugs and fixed (deterministic) coordinates. Therefore, $C_f(t,n,\bar{x}_{1f},...,\bar{x}_{nf})$ in this setting can be considered (see the next paragraph) the pdf of the time to the first failure. As the number of bugs N and the coordinates $\{\bar{X}_{1f},...,\bar{X}_{nf}\}$ are random, the corresponding integration should be performed. Thus, taking Equations (9.13) and (9.14) into account, the pdf of the time to the first failure of the software is

$$f_s(t) = \delta_b \sum_{n=1}^{\infty} \int_{X^n} \sum_{i=1}^{n} p(\bar{x}_{if},t) j_n(\bar{x}_{1f},,...,\bar{x}_{nf})\, d\bar{x}_{1f},..., d\bar{x}_{nf} \tag{9.15}$$

and, as usual, the corresponding failure rate is defined as

$$\lambda_s(t) = \frac{f_s(t)}{\int_t^{\infty} f_s(u)du}. \tag{9.16}$$

We are interested in the time to the first failure of the software. The operational profile, however, defines the probability $p(\bar{x},t)d\bar{x}dt$ of choosing a bug that does not take into account the fact that this should be the first bug chosen in $[0,t+dt)$. Therefore, strictly speaking, the corresponding conditional operational profile should substitute $p(\bar{x},t)$ in Equation (9.15). As the total volume of 'faulty inputs' is usually much smaller than the entire volume of X, the impact of this substitution is negligible (Finkelstein, 1999c). With this in mind, we can proceed with the operational profile $p(\bar{x},t)$.

Equations (9.15) and (9.16) describe a general model and additional simplifying assumptions should be imposed for using this model in practice. On the other hand, an alternative approach can be used assuming that the bugs are distributed in X in accordance with the m-dimensional spatial Poisson process. This approach is similar to the spatial survival model of Section 8.3, which was considered for $m = 2$.

Denote by $\lambda_r(\bar{x})$ the rate of the m-dimensional spatial Poisson process defined similar to (8.30). Assume that this rate, which describes the 'density' of bugs in X, is given. Then the failure rate $\lambda_s(t)$ can be constructed directly by generalizing the 2-dimensional approach of Section 8.3 and using the straightforward heuristic reasoning (Finkelstein, 1999c). Therefore,

$$\lambda_s(t) = \delta_b \int_X \lambda_r(\bar{x}) p(\bar{x},t) d\bar{x}[1 + o(1)]. \tag{9.17}$$

Indeed, $\lambda_r(\bar{x})\delta_b[1 + o(1)],\ \delta_b \to 0$ is the probability that the input value chosen by the operational profile $p(\bar{x},t)$ in $[t,t+dt)$ belongs to the bug's area. Another source of approximation is that we assume that a bug was not chosen by the operational profile in $[0,t)$.

9.3.2 Special Cases

Example 9.1 Let the operational profile be uniform in space and time, *i.e.*,

$$p(\bar{x},t) = p.$$

Then Equation (9.17) becomes

$$\lambda_s(t) = p\delta_b \int_X \lambda_r(\bar{x}) d\bar{x} = p\delta_\delta E[N], \tag{9.18}$$

which is a generalization of the Jelinsky–Moranda model to the case of a random number of bugs in a program. After the first debugging, the expected number of remaining bugs is $E[N]-1$, *etc.* Denote $p\delta_b = \lambda$. Then the corresponding intensity process can be defined similar to Equation (9.2) as

$$\lambda_t = \sum_{i\geq 1}^{[E[N]]} \lambda(E[N]-i+1)I(S_{i-1} \leq t < S_i),\ t \geq 0,$$

where the upper index of summation is the integer part of $E[N]$.

Example 9.2 Let the input domain X be one-dimensional, *i.e.*, $X \subset [0,\infty)$. It can be considered as a long 'line' of code. Assume that a program chooses inputs consequently (starting with $x = 0$) moving in one direction. This resembles a process of proofreading in a publishing house. The encountered bug is removed, but the 'reading' starts from the very beginning again (the program restarts, which is often the case in practice). Our interest is in obtaining the expected number of removed bugs in $[0,t)$ for sufficiently large t. This setting is different from the usual renewal-type approach, as the cycles in this process are not identically distributed and not independent (see later). To proceed, some additional assumptions should be made.

Assume that a program starts operating at $t = 0$ and the inputs are 'read' at a constant speed in time, *i.e.*, $v(t) = v$. Therefore, the corresponding operational profile is

$$p(x,t) = \begin{cases} 1, & x = vt, \\ 0, & x \neq vt. \end{cases} \tag{9.19}$$

Denote by $F_i(t), i = 1,2,...$ the Cdf of the ith cycle duration and assume that the distances between the consecutive bugs in a program are i.i.d. random variables. Therefore, an operational profile with a constant speed means that the times between consecutive debugging without restart of the program are identically distributed with the Cdf $F(t) = F_1(t)$, which is a renewal process. For the operation of the program with restart, however, the Cdf of the time to the second bug is the convolution $F^{(1+2)}(t)$, $F^{(1)}(t) \equiv F(t)$; the Cdf of the time to the third bug is $F^{(1+2+3)}(t)$, *etc.* Therefore, the time to the nth failure has the following distribution:

$$L_n(t) = F^{(1+2+...+n)}(t) = F^{\left(\frac{n(n+1)}{2}\right)}(t). \tag{9.20}$$

Specifically, when $F(t) = 1 - \exp\{-\lambda t\}$,

$$L_n(t) = 1 - \exp\{-\lambda t\} \sum_{i=0}^{\frac{n(n+1)}{2}-1} \frac{(\lambda t)^i}{i!}.$$

Equation (9.20), in fact, employs another simplifying assumption that all 'elementary durations' in the convolutions are independent, whereas in reality, *e.g.*, the Cdf of the second cycle is not $F^{(1+2)}(t) = F^{(3)}(t)$ but is defined by the duration, which is a sum of three terms. The first term has the Cdf $F(t)$; the second term has exactly the same duration as the first one and is therefore dependent; the third term has again an independent duration with the Cdf $F(t)$.

Denote, as usual, by $N(t)$ the number of renewals in an ordinary renewal process with the underlying Cdf $F(t)$ and by $N_b(t)$ the number of removed bugs in $[0,t)$ in the described model. It follows from (9.20) that $N_b(t)$ can be obtained from the following stochastic equation:

$$\frac{N_b(t)(N_b(t)+1)}{2} = N(t). \tag{9.21}$$

When $N_b(t) >> 1$ (although we can proceed without this simplifying assumption),

$$N_b(t) = \sqrt{2N(t)}\ . \tag{9.22}$$

Applying the operation of mathematical expectation to both sides of this equation gives

$$E[N_b(t)] = E\left[\sqrt{2N(t)}\right]. \tag{9.23}$$

Although Jensen's inequality cannot be applied to the right-hand side of Equation (9.23), as the square root function is concave, it can be shown using the considerations at the end of Section 4.3.2 that, as $t \to \infty$, the operations of expectation and of the square root can be interchanged in the following sense:

$$E\left[\sqrt{2N(t)}\right] = \sqrt{2}\sqrt{H(t)}[1+o(1)],$$

where $E[N(t)] \equiv H(t)$ is the renewal function for the ordinary renewal process with the governing Cdf $F(t)$.

9.4 Chapter Summary

The aim of this chapter is similar to that of the previous one, *i.e.*, to present some examples of direct failure rate modelling. Most of the initial software reliability models were formulated in the literature in terms of the corresponding failure rates. Therefore, the intensity process approach of Chapter 4 can be illustrated by these models.

The major difficulty in assessing software reliability is that we are dealing primarily with design faults, which is a very different situation from that considered by conventional hardware reliability theory. There are two major causes of randomness in software reliability models: the unknown 'coordinates' of bugs in software and the random nature of input values. Neither of them is easy to model. Stochastic modelling, which takes into account a combination of these sources, can be used, in principle; however, in order to result in something useful in practice, many assumptions should be made. Note that most of the models considered in the literature are based on very strong assumptions.

In Section 9.4 we describe our model (Finkelstein, 1999c), which is based on the concepts of a spatial point process of bugs and an operational profile of software. The combination of these concepts results in a general model for the failure rate of software. If, for example, the operational profile is uniform (homogeneous) in space and time, then this model reduces to the well-known Jelinsky–Moranda model.

10

Demographic and Biological Applications

10.1 Introduction

Up to now, we have implicitly assumed that the lifetimes under consideration are mostly those of engineering (technical) items. Statistical reliability theory usually deals with methods of statistical inference based on lifetime data that describe performance of technical objects. The corresponding distribution functions, parameters of distributions, failure rates and other relevant characteristics are estimated on the basis of available observations (failure times, censored operation intervals, *etc.*). Similar methods are developed in survival analysis and are usually implemented in medical applications. On the other hand, reliability theory possesses the well-developed 'machinery' for stochastic modelling of ageing (deterioration) that eventually leads to failures of technical objects. These methods can be successfully applied to lifespan modelling of humans and other organisms. Thus, not only the final event (*e.g.*, death) can be considered, but the process that results in this event as well. Several simple reliability-based stochastic approaches to the corresponding modelling will be described in what follows. In this chapter, we will not restrict ourselves to discussing the properties of failure (mortality) rates but consider the topic from a broader viewpoint. Note that here we are looking only at some relevant simple models and applications that reflect the research interests of the author in this area and could be helpful to the reader as a source for initial reading.

According to Birren and Renner (1977) "ageing refers to the regular changes that occur in mature genetically representative organisms living under representative environmental conditions as they advance in the chronological age". This definition is meaningful: it emphasizes ageing as a developmental process, specifies the period of maturity through senescence and states that the corresponding sample should be representative. It also focuses on the impact of environment on the ageing process.

The literature on numerous biological theories of ageing is extensive. Various stochastic mortality models are reviewed, for example, in Yashin *et al.* (2000). Most authors agree that the nature of ageing (and, therefore, of death) is associated with "biological wearing" or "wear and tear". Reliability theory possesses well-developed tools for modelling wear in technical systems, and therefore it is natural to apply this technique to biological ageing (Finkelstein, 2005c). Since even the simplest organisms are much more complex than technical systems that are usually considered in reliability analysis, these analogies should not be interpreted too literally and should be regarded as some useful modelling tools. Note that populations of biological organisms, unlike populations of technical devices, evolve in accordance with evolutionary theory.

Various maintenance and repair problems have been intensively studied in reliability theory. The obtained results can definitely be used for modelling mechanisms of maintenance and repair in organisms. However, the notion of reproduction, which is crucial for bio-demography, has not been considered, although notions like stochastic birth and death processes can certainly be useful for the corresponding modelling. Evolutionary theories (Kirkwood, 1997) tend towards a rather controversial view in that all damage, in principle, is repairable and that natural selection can shape the lifetime trajectory of damage and repair, constrained only by the physical limitations of available resources (Steinsaltz and Goldwasser, 2006). However, not all damage in organisms can be reversed: for example, damage to the central nervous system and heart tissue is usually irreversible. In any case, the importance of different repair mechanisms for the survival of organisms is evident, which brings into play stochastic modelling of all 'types' of repair, *i.e.*, perfect, minimal and imperfect repair actions. This topic has been partially studied in reliability theory (Chapter 5), but there are still many open problems.

The future general theory of ageing will probably be built on the basis of unified biological theories that will use stochastic reliability approaches as an important analytical tool. An interesting discussion on general "quality management" of organisms and the pros and cons of exploiting the existing reliability approaches for biological ageing can be found in Steinsaltz and Goldwasser (2006). On the other hand, the mathematical details useful for modelling are discussed in Steinsaltz and Evans (2004).

Vaupel's (2003) conjecture that "after reproduction ceases, the remaining trajectory of life is determined by forces of wear, tear and repair acting on the momentum produced by the Darwinian forces operating earlier in life" resulted in the reliability modelling of Finkelstein and Vaupel (2006). These authors state: "As the force of natural selection diminishes with age, structural reliability concepts can be profitably used in mortality analysis. It means that the design of the structure is more or less fixed at this stage and reliability laws govern its evolution in time. However, it does not mean that these concepts cannot be used for mortality modelling at earlier ages, but in this case they should be combined with the laws of natural selection."

In accordance with a conventional definition, the reliability of a technical object is the probability of performing a designed function under given conditions and in a given interval of time (Rausand and Hoyland, 2004). This definition can be applied for a probabilistic description of a lifespan T of an organisms, where the designed function is understood as being alive. In accordance with Equations (2.31) and (2.32), the main demographic model for the lifetime of humans is the Gompertz (1825) law of mortality, defined by the exponentially increasing mortality rate $\mu(t)$, *i.e.*,

$$\mu(t) = a\exp\{bt\}, \quad a > 0, b > 0 \tag{10.1}$$

and the corresponding distribution function is

$$F(t) = \Pr(T \le t) = 1 - \exp\left\{-\frac{a}{b}[\exp\{bt\} - 1]\right\}. \tag{10.2}$$

In accordance with the conventional notation for demographic literature, the mortality rate (force of mortality), which is equivalent to the failure rate in reliability, is denoted by $\mu(t)$. The Gompertz law has been the main demographic model of human mortality for nearly 200 years. A reasonably good fit (excluding the periods of infant mortality and adolescence) is achieved for numerous human mortality data sets of different countries. A number of attempts were made in the past to justify the exponential form of this empirical model for the human mortality rate by some biological mechanism, but most of these approaches exploited additional assumptions, either explicitly or implicitly equivalent to the desired exponentiality (*e.g.*, Strehler and Mildvan, 1960; Witten, 1985; Koltover, 1997; Gavrilov and Gavrilova, 2001).

Gompertz (1825) gave the following mathematical explanation of his formula. He had assumed that the ability to "resist death" is an inverse function to the mortality rate $\mu(t)$, *i.e.*, $1/\mu(t)$. Furthermore, another assumption stated that the change in this ability is proportional to its value. These assumptions can be described mathematically as

$$d\left(\frac{1}{\mu(t)}\right) = -b\frac{1}{\mu(t)}dt\,.$$

This relationship is equivalent to the following elementary differential equation:

$$\frac{d\mu(t)}{dt} = b\mu(t)$$

with the initial condition $\mu(0) = a$. Therefore, the solution to this equation is given by Equation (10.1).

There can be other popular mortality curves in demographic practice that can fit empirical data. The power law for $\mu(t)$ is also sometimes used in the literature. Another modification is the Makeham (1860) law of mortality, which adds a constant term A to the Gompertz curve (10.1), *i.e.*,

$$\mu(t) = A + a\exp\{bt\}\,.$$

The constant term is believed to account for the so-called baseline mortality, which does not depend on age and usually models 'natural hazards', although many authors do not agree with this explanation. Makeham (1890) derived a mathematical justification of this curve using the corresponding second-order differential equation with respect to $\mu(t)$ (Marshall and Olkin, 2007).

Living organisms reproduce themselves, and this is a crucial distinction between population studies and reliability-based reasoning. Populations of organisms evolve in time, and a special discipline called 'population dynamics', based on methods of specific stochastic processes, deals with this phenomenon. However, mortality rates and some other characteristics can be studied without relying on the methods of population dynamics.

We now turn to the statistical definition of the mortality rate for populations of individuals. As in Section 2.1 (see Equation (2.6)), consider a cohort of N individuals born at $t = 0$ and denote by $N(t)$ the number of those who are alive at time t. Note that, in demography, a cohort is a group of individuals born in the same period of time, generally the same calendar year. Therefore, Definition 2.1 for the failure (mortality) rate holds, and Equation (2.6) in a new notation reads:

$$\mu(t) = \lim_{\Delta t \to 0} \frac{N(t+\Delta t) - N(t)}{N(t)\Delta t}, \quad N(t) \to \infty . \tag{10.3}$$

Cohort measures, describing lifetime random variables, are easily and unambiguously obtained using standard statistical tools. The procedures are the same for engineering and biological items. In the case of humans, however, one must wait approximately 100 years in order for cohort data to be complete. Therefore, many cohort mortality experiments are performed with organisms having short life spans (*e.g.*, medflies, worms and mice). On the other hand, human mortality data sets are usually presented not as cohort data sets but as so-called period data sets. The reason for this is that mortality characteristics change with calendar (chronological) time. Therefore, in addition to the age of an object x (previously denoted by t), we must also consider the calendar time t.

The term *period* means that the data (the number of deaths and the number of survived individuals at each age) are collected for the time period $[t, t + \Delta t)$. Sometimes these data are called *cross-sectional* to emphasise the importance of the calendar variable t. Let, for instance, the numbers of living individuals of ages 0,1,2,... in some population be recorded for 1 January 2000 and the corresponding numbers for each age of those who died in [2000,2001] also be stated. The data are usually organized in the form of life tables, which for some European countries have been in existence already for hundreds of years. In its most basic form, a period life table is a listing of ages and the corresponding probabilities of death within the next year.

Denote by $N(x,t)$ an age-specific population size at time t: the number of individuals of age x. See Keding (1990) and Arthur and Vaupel (1984) for discussion of this quantity. We will call $N(x,t), x \geq 0$ the *population age structure* at time t. Alternatively, $N(x,t)$ is often called the *population density* (Finkelstein, 2005a). Let Δt be our period. Note that in demography the period is usually equal to one year and $N(x,t)$ is also usually defined as the number of individuals whose age at time t is $[x]$, but other units of time can also be used.

Similar to (10.3), define the mortality rate for a population with the age structure $N(x,t), x \geq 0$ as a function of age and time as

$$\mu(x,t) = \lim_{\Delta t \to 0} \frac{N(x+\Delta t, t+\Delta t) - N(x,t)}{N(x,t)\Delta t}, \quad N(x,t) \to \infty . \tag{10.4}$$

Using these definitions, mortality rates and survival probabilities can be estimated from the data. What is the difference between Equations (10.3) and (10.4)?

Imagine that a population is stationary, *i.e.*, the age structure does not depend on the calendar time t. In this case $N(x,t) \equiv N(x)$ and both definitions coincide. The assumption of stationarity, however, is very unrealistic for human populations. Owing to healthcare achievements, improvements in a lifestyle and a decrease in natural hazards (at least in the developed countries), life expectancy is constantly increasing. Oeppen and Vaupel (2002) state, for instance, that female life expectancy in the country with maximum life expectancy (Japan) is increasing every year by approximately 3 months. This trend has been observed already for more than 50 years. Therefore, human populations are definitely non-stationary, and the second argument in $\mu(x,t)$ captures this phenomenon.

Note that, owing to exponential representation (2.5), the univariate mortality rate $\mu(x)$ uniquely defines the corresponding Cdf and, therefore, completely characterizes the lifetime random variable T. Unlike this cohort setting, the lifetime random variable for the period setting cannot be unambiguously defined only *via* $\mu(x,t)$ without additional simplifying assumptions. This is the main complication, which should be taken into account when analysing mortality data. However, most of the practical demographic methods pay no attention to this important phenomenon. We will consider this topic in more detail in Section 10.6.

Let X_t denote a random age at time t of an individual chosen at random (with equal chances) from a population of size

$$N(t) = \int_0^\infty N(u,t)du .$$

Therefore, we interpret X_t as a random age in a population with an age structure $N(x,t),\ x \geq 0$. Let

$$f(x,t) = \frac{N(x,t)}{\int_0^\infty N(x,t)dx} \tag{10.5}$$

and

$$F(x,t) = \Pr[X_t \leq x] = \frac{\int_0^x N(u,t)du}{\int_0^\infty N(u,t)du} \tag{10.6}$$

be the pdf and the Cdf of X_t, respectively. The latter can be equivalently interpreted as the proportion of individuals in our population whose age does not exceed x. It is obvious that the described notion of a random age is relevant only for a period setting, as we count 'lives' in a period $[t, t+\Delta t)$ for different ages. In the cohort setting, however, the age of all individuals is the same.

Remark 10.1 Equations (10.5) and (10.6) define, in fact, the estimates of the pdf and the Cdf, respectively (observed period values). In order to avoid possible confusion, we assume as in (10.3) and (10.4) that the population size tends to infinity.

Remark 10.2 When a population is stationary, Equation (10.6) becomes

$$\Pr[X \le x] = \frac{\int_0^x N(u)du}{\int_0^\infty N(u)du},$$

where $\Pr[X \le x]$ is the probability that the age (and not the age at death, as when defining the 'usual' cohort Cdf of a lifetime random variable) of an individual is less than or equal to x. Note that, as a population is stationary in this example, the cohort and the period settings can be considered equivalent. The corresponding probability density function is defined by

$$\frac{d}{dx}\Pr[X \le x] = \frac{N(x)}{\int_0^\infty N(u)du}.$$

We will continue with further studies of mortality in the non-stationary populations in Section 10.6, but first we will discuss several models of mortality and ageing that can be described in terms of a usual cohort setting.

10.2 Unobserved Overall Resource

Following Finkelstein (2003b), we assume that an organism at birth ($t = 0$) acquires an overall unobserved random resource R, described by the Cdf $F_0(r)$ *i.e.*, $F_0(r) = \Pr[R \le r]$ and the corresponding mortality (failure) rate $\mu_0(r)$. We also assume that the process of an organism's ageing is described by an increasing, differentiable and deterministic (for simplicity) cumulative function $W(t)$ ($W(0) = 0$) to be called wear. The wear increment in $[t, t + dt)$ is defined as $w(t) + o(dt)$. Additionally, let $W(t) \to \infty$ as $t \to \infty$. Under these assumptions, we formally arrive at the accelerated life model (see also Section 5.2.1 and Chapter 6), *i.e.*,

$$\Pr[T \le t] \equiv F(t) = F_0(W(t)) \equiv \Pr[R \le W(t)], \quad (10.7)$$

where

$$W(t) = \int_0^t w(u)du; \quad w(t) > 0, t \in [0, \infty).$$

The corresponding mortality rate $f(t) / \bar{F}(t)$ is obtained from (10.7) as

$$\mu(t) = w(t)\mu_0(W(t)). \quad (10.8)$$

These formulas are similar to Equations (5.1) and (5.2) of Chapter 5, but we obtain them now in a different way. Note that death (failure) occurs when the wear $W(t)$ reaches the boundary R. Substitution of the deterministic wear $W(t)$ in (10.7) by the increasing stochastic process $W_t, t \geq 0$ leads to the following general relationship (Finkelstein, 2003b):

$$F(t) = \Pr[T \leq t] = \Pr[R \leq W_t] = E[F_0(W_t)], \tag{10.9}$$

where the expectation is obtained with respect to $W_t, t \geq 0$. As the mortality rate is a conditional characteristic, it cannot be obtained from (10.8) as a simple expectation: $\mu(t) = E[w_t \mu_0(W_t)]$ and, similar to Equations (3.5) and (3.6), the corresponding conditioning should be performed, *i.e.*,

$$\mu(t) = E[w_t \mu_0(W_t) \mid T > t], \tag{10.10}$$

where w_t denotes the stochastic rate of diffusion: $dW_t \equiv w_t dt$.

A good candidate for $W_t, t \geq 0$ is the standardized gamma process, which, according to Definition 5.9, has stationary independent increments and $W_t - W_s$ ($t > s$) has a gamma density with scale parameter 1 and shape parameter $(t - s)$. The Wiener process with drift can also sometimes be used for modelling, although its realizations are not monotone. An assumption of 'strict' monotonicity is usually natural for the modelling of wear. This process is defined (Ross, 1996) in the following way.

Definition 10.1. The Wiener process (Brownian motion) with drift is the stochastic process $W_t, t \geq 0$, $W_0 = 0$, with stationary independent increments. Its values at $\forall t > 0$ are normally distributed with mean at (a is a drift coefficient) and variance t.

The wear in $[t, t+h)$ can also be defined in a natural way as the following increment (Lemoine and Wenocur, 1985; Wenocur, 1989; Singpurwalla, 1995):

$$W_{t+\Delta t} - W_t = a(W_t)\varepsilon(\Delta t) + b(W_t)\Delta t, \quad \forall t \in [0, \infty), \tag{10.11}$$

where $\varepsilon(\Delta t)$ is a random variable with a positive support and finite first two moments and $a(\cdot)$ and $b(\cdot)$ are continuous positive functions of their arguments. Letting $\Delta t \to 0$, we arrive at the continuous version of (10.11) in the form of Ito's stochastic differential equation (Singpurwalla, 1995), *i.e.*,

$$dW_t = a(W_t)d\eta_t + b(W_t)dt,$$

where $\eta_t, t \geq 0$ is, for example, a gamma process if $\varepsilon(\Delta t)$ has a gamma density with scale parameter 1 and shape parameter Δt. Integrating this equation with the initial condition $W_0 = 0$ results in

$$W_t(t) = \int_0^t a(W_u)d\eta(u) + \int_0^t b(W_u)du. \tag{10.12}$$

The following two examples are really meaningful and probably deserve to be presented in separate sections.

Example 10.1 As a specific case of an unobserved resource model, consider now a discrete resource $R = N$ with the Cdf $F_0(n) \equiv P(N \le n)$. The following simple reliability interpretation is meaningful. Let N be the random number of initially (at $t = 0$) operating independent and identically distributed components with constant failure rates λ. Assume that these components form a parallel system, which, according to Gavrilov and Gavrilova (2001), models the lifetime of an organism (generalization to the series-parallel structure is straightforward). In each realization $N = n, n \ge 1$, our degradation process of pure death $W_t, t \ge 0$ in this setting is just the number of failed components. When this number reaches n, the death of an organism occurs. The transition rates of the corresponding Markov chain are $n\lambda, (n-1)\lambda, (n-2)\lambda,$ *etc.* Denote by $\mu_n(t)$ the mortality rate, which describes T_n –the time to death for the fixed $N = n, n = 1,2,...$ ($n = 0$ is excluded, as there should be at least one operating component at $t = 0$). It is shown in Gavrilov and Gavrilova (2001) that as $t \to 0$, this mortality rate tends to an increasing power function (the Weibull law), which is a remarkable fact. On the other hand, for random N, similar to (10.9), the mortality rate is given as the following conditional expectation with respect to N:

$$\mu(t) = E[\mu_N(t) \mid T > t]. \tag{10.13}$$

Therefore, similar to the continuous case, $\mu(t)$ is a conditional expectation (on the condition that the system is operable at t) of a random mortality rate $\mu_N(t)$. Note that, for small t, this operation can approximately result in the unconditional expectation

$$\mu(t) \approx E[\mu_N(t)] = \sum_{n=1}^{\infty} P_n \mu_n(t), \tag{10.14}$$

where $P_n \equiv \Pr[N = n]$, but the limiting transition, as $t \to 0$, should be performed carefully in this case. As $t \to \infty$, we observe the following mortality plateau (Finkelstein and Vaupel, 2006):

$$\mu(t) \to \lambda. \tag{10.15}$$

This is due to the fact that the conditional probability that only one component with the failure rate λ is operating tends to 1 as $t \to \infty$ (on the condition that the system is operating).

Assume now that N is Poisson distributed with parameter η. Taking into account that the system should be operating at $t = 0$,

$$P_n = \frac{\exp\{-\eta\}\eta^n}{n!(1 - \exp\{-\eta\})}, \quad n = 1,2,\ldots.$$

It can be shown *via* direct integration and using the discrete versions of Equations (3.4)–(3-6) that the time to death in our simplified model has the following Cdf (Steinsaltz and Evans, 2004):

$$F(t) = \Pr[T \le t] = \frac{1 - \exp\{-\eta \exp\{-\lambda t\}\}}{1 - \exp\{-\eta\}}. \quad (10.16)$$

The corresponding mortality rate is

$$\mu(t) = \frac{F'(t)}{1 - F(t)} = \frac{\eta\lambda \exp\{-\lambda t\}}{\exp\{\eta \exp\{-\lambda t\}\} - 1}. \quad (10.17)$$

Performing, as $t \to \infty$, the limiting transition in (10.17), we also arrive at the mortality plateau (10.15).

In fact, the mortality rate given by Equation (10.17) is far from the exponentially increasing Gompertz law (10.1). The Gompertz law can erroneously follow from Approximation (10.14) if this approximation is used formally, without considering a proper conditioning in (10.13), as in Gavrilov and Gavrilova (2001). The relevant discussion can be found in Steinsaltz and Evans (2004).

Example 10.2 We will now combine the resource model (10.7)–(10-9) with the shock model of Section 8.1. We assume that the i th shock causes our system's failure with probability $\theta(t)$, and with the complementary probability $1-\theta(t)$ it only increases the accumulated wear by a random amount W_i. Assume that these random variables are i.i.d. ($W_i = W,\ i = 1,2,...$) and that they are characterized by the density $f(w)$ and the moment generating function $M_W(t)$, *i.e.*,

$$M_W(t) = E[\exp\{tW_i\}] = \int_0^\infty \exp\{tW\} f(w)dw.$$

Failure occurs when the accumulated wear reaches the initial resource R. Other important assumptions from the computational point of view are: the Cdf of R is exponential with the failure rate μ_0 and the process of shocks is the nonhomogeneous Poisson process with rate $\nu(t)$. After cumbersome technical derivations (Cha and Finkelstein, 2008), the following equation for the mortality (failure) rate can be obtained:

$$\mu(t) = (1 - M_W(-\mu_0)(1-\theta(t)))\nu(t).$$

It is clear that when $W = 0$ (this means that shocks do not increase wear), this formula reduces to (8.5). If W follows the exponential distribution with mean m, then the corresponding mortality (failure) rate can be derived explicitly as (Cha and Finkelstein, 2008)

$$\mu(t) = \left(1 - \frac{1-\theta(t)}{\mu_0 m + 1}\right)\nu(t).$$

10.3 Mortality Model with Anti-ageing

Following contemporary biological views, assume that there exist two processes: ageing and anti-ageing (regeneration), to be modelled by stochastic processes of wear and anti-wear, respectively (Finkelstein, 2003b). Denote the resulting stochastic process with independent increments by W_t^ρ. Assume that the process of anti-wear decreases each increment of wear. For example, Equation (10.11) is generalized in this case to

$$\begin{aligned} W_{t+\Delta t}^\rho - W_t^\rho &= a(W_t)\varepsilon(\Delta t) + b(W_t)\Delta t - \rho(t)[a(W_t)\varepsilon(\Delta t) + b(W_t)\Delta t] \\ &= (1-\rho(t))[a(W_t)\varepsilon(\Delta t) + b(W_t)\Delta t], \quad \forall t \in [0,\infty), \end{aligned} \tag{10.18}$$

where $\rho(t)$, $0 \le \rho(t) \le 1$, is a decreasing function (the case of a decreasing stochastic process $\rho_t, t \ge 0$, which is independent of the process of wear W_t, can be considered as well). Assume also that $\rho(t) \to 0$ as $t \to \infty$, which means that the anti-ageing mechanism deteriorates with age. Therefore, this function describes the ability of an organism to decrease its wear in each increment.

Similar to the previous section, we will model biological ageing by the process W_t^ρ. Ageing for humans actually starts at the age of maturity, *i.e.*, 25 to 30 years. This means that $\rho(t)$ is very close or equal to 1 up to this age. The described combined process of wear and anti-wear can be defined directly *via* the rate of diffusion w_t, *i.e.*,

$$w_t^\rho = (1-\rho(t))w_t \,. \tag{10.19}$$

We will use this convenient definition in what follows. This means that the rate of diffusion is smaller due to the anti-ageing mechanism by the time-dependent factor $(1-\rho(t))$. Thus, the formulas of the previous section can be written with the obvious substitution of w_t by w_t^ρ. Equation (10.9), for example, becomes

$$F(t) = E\left[F_0\left(\int_0^t (1-\rho(u))w_u du\right)\right] \tag{10.20}$$

and Equation (10.10) is modified to

$$\mu(t) = E[w_t^\rho \mu_0(W_t^\rho) \mid T > t] = (1-\rho(t))E[w_t \mu_0(W_t^\rho) \mid T > t] \,. \tag{10.21}$$

Specifically, when the mortality rate $\mu_0(t) = \mu_0$ is a constant, Equation (10.21) simplifies to

$$\mu(t) = \mu_0(1-\rho(t))E[w_t \mid T > t] \,.$$

Equations (10.20) and (10.21) imply that

$$E\left[F_0\left(\int_0^t (1-\rho(u))w_u du\right)\right]$$
$$= 1 - \exp\left\{-\int_0^t (1-\rho(u))E[w_u \mu_0(W_u^{\rho}) \mid T > u]du\right\}.$$

Consider now the survival function $\bar{F}(t \mid x)$, which describes the corresponding remaining lifetime, *i.e.*,

$$\bar{F}(t \mid x) = \frac{\bar{F}(t+x)}{\bar{F}(x)}$$
$$= \exp\left\{-\int_x^{x+t} (1-\rho(u))E[w_u \mu_0(W_u^{\rho}) \mid T > u]du\right\}.$$

As $\rho(t) \to 0$ for $t \to \infty$, the following asymptotic relationship holds:

$$\bar{F}(t \mid x) = \exp\left\{-\int_x^{x+t} E[w_u \mu_0(W_u^{\rho}) \mid T > u]du\right\}(1+o(1)). \qquad (10.22)$$

It follows from Equation (10.22) that when $x \to \infty$, the remaining lifetime still depends on the initial distribution $F_0(r)$. Thus, the influence of the initial resource R is not fading out, as intuition would probably suggest.

The model to be considered further is defined by the triple $\{R, w_t, \rho_t\}$ (Finkelstein, 2003b). Assume that the human lifetime is programmed genetically at birth by the triple $\{R, \hat{w}_t, \hat{\rho}_t\}$, where $\hat{w}_t$ and $\hat{\rho}_t$ are 'stochastic programs'. Realizations of these stochastic programs $\hat{w}(t)$ and $\hat{\rho}(t)$ (as well as r_0) are embedded individually at birth. Therefore, $\hat{w}(t), \hat{\rho}(t)$ describe the 'designed' trajectories for individuals in some baseline time scale, whereas realizations $w(t), \rho(t)$ describe what is happening in the real life of an individual. Given a realization $r_0, w(t), \rho(t)$, the time to death t_d in our model is uniquely defined from the following equation:

$$r_0 = \int_0^{t_d} (1-\rho(u)w(u)du. \qquad (10.23)$$

We can assume the following genetic interpretation of the triple. Different (not related) individuals have stochastically independent triples. A reasonable assumption is that the parents and their offspring have dependent triples. Thus, *e.g.*, $F_0(r)$ should be understood as some averaged, marginal distribution whereas the conditional distribution should be defined by the corresponding history (information on the parents and grandparents, for instance). Identical twins or the outcomes of cloning are genetically identical and must exhibit the maximal extent of dependence between their triples. Therefore, it can be supposed for simplicity that they are embedded at birth with identical realizations $r_0, \hat{w}(t), \hat{\rho}(t)$. Does this mean that actual realizations $w(t)$ and $\rho(t)$, and consequently the time of death t_d, will be the same? An obvious answer is negative, because, for example,

- Realization of these programs in real time can be influenced by external factors effecting the 'designed at birth' baseline time scale;
- The programs can have errors (bugs). These errors can be embedded at birth or acquired during the lifetime.

A number of biological theories agree that errors in the processes of repair, replication and transcription of DNA are responsible for ageing (Cunningham and Brookbank, 1988). The role of genetics at the 'global level' is illustrated, *e.g.*, by the classical studies on the age at death of 58 sets of identical twins in Bankand and Jarvic (1978). The corresponding intrapair mean difference in age of death is about 3 years, and it is 6 years for non-identical twins of the same sex. Also, humans whose parents and grandparents lived long live on average six years longer than those whose parents and grandparents died before the age of 50 (Cunningham and Brookbank, 1988).

In the context of our triple mode, the following interesting question arises: what is more important in defining the time of death, the initial resource R or the process of anti-wear defined by $\rho(t)$? The answer obviously depends on the shape of $\rho(t)$, and this is illustrated by the following example.

Example 10.3 Consider two marginal cases. Let $\rho(t)$ decreases to 0 very sharply. Then, for sufficiently large t

$$\int_0^t (1-\rho(u))w(u)du \approx \int_0^t w(u)du\,, \tag{10.24}$$

implying that, if r_0 is not too small, then $t_d \approx t_d^0$, where t_d^0 denotes the time of death in Model (10.23) when $\rho(t) \equiv 0$ (there is no anti-ageing).

On the other hand, let $\rho(t)$ be a step function for some $\tilde{t} > 0$, *i.e.*,

$$\rho(t) = \begin{cases} 1, & 0 \le t < \tilde{t}, \\ 0, & t \ge \tilde{t}. \end{cases}$$

Then

$$\int_0^t (1-\rho(u))w(u)du = \int_{\tilde{t}}^t w(u)du\,,\ t \ge \tilde{t}\,.$$

This means that $t_d = \tilde{t} + \tilde{t}_d$, where $\tilde{t}_d$ is obtained from Equation (10.23). Note that the lower limit of integration in (10.23) is substituted by $\tilde{t}$ and the upper limit is substituted by $\tilde{t} + \tilde{t}_d$, *i.e.*,

$$r_0 = \int_{\tilde{t}}^{\tilde{t}+\tilde{t}_d} (1-\rho(u)w(u)du\,.$$

Assume that $\tilde{t} >> \tilde{t}_d$, which implies that $t_d \approx \tilde{t}$. This assumption means that $\tilde{t}$ is sufficiently large and the wear is sufficiently 'intensive'. Therefore, the anti-wear process is more important in defining t_d than r_0 (given it is not too large) in this marginal case.

The shape of $\rho(t)$ can be rather close to the step function (10.24). For humans it is very close to 1 up to 25 to 30 years, decreases rather slowly up to middle age and then decreases more substantially up to 70 to 80 years. Eventually it drops sharply. This shape can be considered as a baseline for $\rho(t)$.

We do not need special biological evidence to prove a lifetime dependence on environment. There can be different ways to model the impact of environment. In the context of our triple model, assume that ageing and anti-ageing processes depend on some overall environmental (lifestyle) scalar parameter l, which (for simplicity) does not depend on t, *i.e.*, $w_t(l), \rho_t(l)$. It should be understood, however, that quantifying l is a very difficult task. Therefore, it is reasonable to use only some general qualitative considerations and simple clarifying examples. For instance, different lifestyles of humans can be ordered by the value of parameter l. Let l_g stand for a 'good lifestyle' and l_b for a 'bad lifestyle' and $l_g < l_b$. Therefore, ageing is more intensive and anti-ageing is less intensive for a 'bad lifestyle', *i.e.*,

$$\rho(t, l_g) > \rho(t, l_b), \ \forall t \in (0, \infty),$$

$$w(t, l_g) < w(t, l_b), \ \forall t \in (0, \infty).$$

It is reasonable to use the accelerated life model (ALM) defined by Equation (5.2) for this kind of modelling. Assume that the scale transformation function for the case under consideration is linear. Therefore, generalizing Equation (10.19) in realizations leads to

$$w(t, \rho, l) \equiv (1 - \rho(lt))w(lt). \tag{10.25}$$

The case $l = 1$ corresponds to the baseline process $(1 - \rho_t)w_t$. The ALM describes the change in 'biological time' for our model owing to the environmental influence. This interpretation is close to our virtual age-based reasoning of Chapter 5. Let $0 < l_g < 1 < l_b$ and $t_d(l)$ denote the time of death in a realization with the scale parameter l. Similar to (10.23), and changing the variable of integration to $y = lu$, the following equation is obtained:

$$r_0 = \int_0^{t_d(l)} (1 - \rho(lu)w(lu)du = \frac{1}{l} \int_0^{lt_d(l)} (1 - \rho(y)w(y)dy. \tag{10.26}$$

If the difference in lifestyles $l_b - l_g$ is sufficiently large, the difference $t_d(l_g) - t_d(l_b)$ can also be large. This is clearly seen, because owing to our assumptions, the integrand on the right-hand side of (10.26) is an increasing function. In this way, we can compare the impacts of r_0 and l on the time of death. The following example illustrates this reasoning.

Example 10.4 Assume that the dependence on l in $w(t,l)$ can be ignored and consider the setting of Example 10.3. It can be shown (Finkelstein, 2003b) that the difference in the corresponding lifetimes in this case is

$$t_d(l_g)-t_d(l_b)=\tilde{t}\left(\frac{l_b-l_g}{l_b l_g}\right)+(\tilde{t}_d(l_g)-\tilde{t}_d(l_b)). \qquad (10.27)$$

Under the same assumptions as in Example 10.3, the second summand on the right hand side of (10.27) can be ignored. Therefore, if $\tilde{t}$ is sufficiently large and l_b-l_g is not too small, we can say that the impact of lifestyle is decisive. This example prompts a more general conjecture: the influence of the embedded (genetic) parameters is 'damped' by the impact of environment at least for sufficiently old individuals.

10.4 Mortality Rate and Lifesaving

Mortality of humans in developed countries is declining with time, which is a consequence of improving conditions of life. By "conditions of life" or mortality conditions we mean the whole range of factors, with healthcare quality being the major one. Numerous advances in healthcare have resulted in saving lives (lifesaving) of humans, where previously these lives were lost. Therefore, life expectancy at birth and other characteristics are improving. Oeppen and Vaupel (2002) state, for instance, that female life expectancy in the country with the maximum life expectancy (currently Japan) is increasing every year by approximately three months. This trend has been observed already for more than 50 years.

The Gompertz law of human mortality (10.1) usually gives a reasonable fit to the real demographic data for ages beyond 30. Assume that this model is used for fitting mortality data in a developed country at some calendar time t_0 (*e.g.*, $t_0=1950$). As in Equation (10.4), denote the corresponding period mortality rate by $\mu(x,t_0)$, where x is the age at death. Bongaarts and Feeney (2002) show that the mortality rate in contemporary populations with a high level of life expectancy tends to improve over time by a similar factor at all adult ages, which defines the Gompertz shift model:

$$\mu(x,t)=\theta(t)\mu(x,t_0), \quad t>t_0, \theta(t_0)=1, \qquad (10.28)$$

where the function $\theta(t)$ is decreasing with time t and does not depend on age x. This model was verified using contemporary data for different developed countries and the corresponding values for $\theta(t)$ were obtained (see also *e.g.*, Oopen and Vaupel, 2002). Equations (10.1) and (10.28) also show that the logarithms of mortality rates at different time instants are practically parallel. Note that (10.28) can be also obviously interpreted as the proportional hazards (PH) model (Section 7.3).

The relevant natural example of the described lifesaving is the convergence of mortality rates of 'old cohorts' after the reunification of East and West Germany at $t_0=1990$ (Vaupel *et al.*, 2003). Mortality rates in East and West Germany differed noticeably before the reunification, and the East German rates had improved to the level of those of the West shortly thereafter. This is a consequence of a direct (better healthcare) and of an indirect (better environment eliminates some causes of death) lifesaving. It is worth noting that the older the cohorts were, the more pronounced this effect was, as the quality of the healthcare is more important for older subpopulations. In what follows we will describe probabilistically the simplified cohort version of lifesaving.

Let $\mu(t)$, as previously, denote the cohort mortality rate for some population. Suppose that for some reason (*e.g.*, better healthcare), $\mu(t)$ is reduced to a new level $\mu_r(t)$ to be modelled by a function $\theta(t), 0 < \theta(t) \leq 1, \forall t \geq 0$ as

$$\mu_r(t) = \theta(t)\mu(t). \tag{10.29}$$

We see that Relationship (10.29) (in a slightly different notation) is the cohort version of the Gompertz shift model (10.28). The following useful reasoning can give a reasonable justification of (10.29) in terms of lifesaving. Assume that each life, characterized by the initial mortality rate $\mu(t)$, is saved (cured) with probability $1-\theta(t)$ (or, equivalently, a proportion of individuals who would have died are now resuscitated and given another chance). Those who are saved experience a minimal repair (Section 4.3.1). The number of resuscitations (repairs) is unlimited. Under these assumptions, it was proved analytically in Vaupel and Yashin (1987) that the described lifesaving procedure results in the mortality rate given by Equation (10.29). As a corollary to this result, a point process of saved lives is the NHPP with rate

$$\mu_s(t) = (1-\theta(t))\mu(t).$$

A result similar to (10.29) was obtained for different reliability-related settings by Brown and Proschan (1983) (for $\theta(t) \equiv \theta$), Block *et al.* (1985) and Finkelstein (1999a), where an object (organism) subject to the non-homogeneous Poisson process of shocks (*e.g.*, diseases) with rate $\mu(t)$ was considered. It was assumed that a shock, affecting an object at time $t \in (0,\infty)$, independently of the previous shocks, causes a failure (death) with probability $\theta(t)$ and is harmless to an object with a complementary probability $1-\theta(t)$. Then the mortality (failure) rate is given by Equation (10.29) (see Section 8.2 for details). We will proceed with applications of this model in the next section.

It is important for demographic practice to define the lifesaving ratio $R_\theta(t)$ in terms of the mean remaining lifetime as improvements in medical, socio-economic and environmental conditions usually have a more substantial effect on older people. In accordance with Equation (2.7), this ratio can be defined as

$$R_\theta(t) = \frac{\int_0^\infty \exp\left\{-\int_t^{t+x} \theta(u)\mu(u)du\right\}dx}{\int_0^\infty \exp\left\{-\int_t^{t+x} \mu(u)du\right\}dx}.$$

Example 10.5 Let $\mu(t) \equiv \mu$ and $\theta(t)$ is a step function (young people are perfectly cured, whereas old people are not cured at all), *i.e.*,

$$\theta(t) = \begin{cases} 0, & 0 \le t < t_0, \\ \theta, & t \ge t_0. \end{cases}$$

Then

$$R_\theta(t) = \begin{cases} \mu(t_0 - t) + \theta^{-1}, & 0 \le t < t_0, \\ \theta^{-1}, & t \ge t_0. \end{cases}$$

This function decreases in t for $0 < t < t_0$ and then equals θ^{-1} for $t \ge t_0$.

10.5 The Strehler–Mildvan Model and Generalizations

In this section, we will justify and generalize the application of the shock model (8.5) (see also Example 10.2).

As in Section 10.2 (Equation (10.9)), consider the first passage-type setting but with an additional feature of 'killing' events (Singpurwalla, 1995; Aven and Jensen, 1999). Let $W_t, t \ge 0$ denote an increasing stochastic process of damage accumulation and let $R(t)$ be a function that defines the corresponding boundary. Death occurs when W_t exceeds $R(t)$ for the first time. Let, as previously, $W(t)$ denote the increasing realization of this process. The time-independent case $R(t) = R$ (initial resource) was considered in Section 10.2.

Let $P_t, t \ge 0$ be a point process of external instantaneous harmful events (external stresses or demands for energy) with rate $\nu(t)$. Following reliability terminology, we will call these events "shocks". As previously, assume that each shock results in death with probability $\theta(t)$ and is 'survived' with the complementary probability $1 - \theta(t)$. This can be now interpreted in the following more detailed way: each shock has a random magnitude $Y_i = Y, i = 1,2,...$ with a common distribution function $G(y)$. Death at age t occurs when this magnitude exceeds $R(t) - W(t)$. Therefore, the function $\theta(t)$ that was previously unspecified has now the following clear probabilistic meaning:

$$\theta(t) = \Pr[Y > R(t) - W(t)] = 1 - G(R(t) - W(t)). \tag{10.30}$$

We also assume for simplicity that a shock is the only cause of death. The corresponding generalization to the case when death also occurs when $W(t)$ reaches the boundary $R(t)$ can be performed as well (Finkelstein, 2007d). In the original Strehler–Mildvan (1960) model, which was widely applied to human mortality data (see Riggs and Millecchia, 1992; Riggs and Hobbs, 1998, among others), our $R(t) - W(t)$ means the remaining vitality at time t. It was also supposed in this model that this function linearly decreases with age, which can be a reasonable assumption, as some biological markers of human ageing can behave linearly (Nakamura *et al.*, 1998). But an important, unjustified assumption was that the distribution function $G(y)$ is exponential (Yashin *et al.*, 2000). The combination of linearity of $R(t)-W(t)$ and of exponentiality of $G(y)$ results in the exponential form of the corresponding mortality rate, and therefore cannot be considered a justification of the empirical Gompertz law of human mortality. Arbeev *et al.* (2005) consider modification of this model and apply it to modelling human cancer incidence rates. They assume that $R(t)-W(t)$ is decreasing exponentially. Our approach, based on Equation (10.29), does not need additional assumptions on $G(y)$ and $R(t)-W(t)$. Note that Equation (10.29) was obtained under the crucial assumption that the point process of shocks is the NHPP. Therefore, the corresponding survival function is similar to (8.5), *i.e.*,

$$\bar{F}(t)=\exp\left\{-\int_0^t \theta(u)\nu(u)du\right\}.$$

Unfortunately, Strehler and Mildvan (1960) did not make this crucial assumption.

Equation (10.29) states that the resulting mortality rate is the simple product of the rate of the Poisson process and of the probability $\theta(t)$. Therefore, its shape can be easily analysed. When $R(t)-W(t)$ decreases, the probability $\theta(t)$ increases with age, which is in line with the accumulation of degradation reasoning. If, additionally, the rate of harmful events $\nu(t)$ is not decreasing, or not decreasing faster than $\theta(t)$ is increasing, the resulting mortality rate $\mu(t)$ is also increasing. The following possible scenarios can result in a *decreasing* mortality rate $\mu(t)$ (other cases can also be considered as in Finkelstein, 2007d):

- $\theta(t)$ is decreasing, as the boundary function $R(t)$ is increasing faster than $W(t)$: additional vitality is additively 'earned' by an organism with age. Let, for instance, $W(t)=wt,\ R(t)=bt;\ 0<w<b$. Then

  $$\theta(t)=\Pr[Y>R(t)-W(t)]=1-G((b-w)t)$$

  is decreasing in t;
- The rate of initial harmful events $\nu(t)$ is decreasing. This assumption can be quite realistic, *e.g.*, for human populations in developed countries when the exposure to stresses of different kinds decreases at advanced ages.

Thus, the case of 'negative ageing' can still formally occur within the framework of the suggested generalized Streller–Mildvan model. In the next section we will show how in some instances the 'unnatural' mortality rate can be transformed.

10.6 'Quality-of-life Transformation'

We have briefly discussed several of the simplest ageing classes of distributions in Chapter 3. Although it is a common perception in the biological and demographic sciences that the shape of the mortality rate alone is sufficient for defining ageing properties of organisms, this is not true. In fact, the accumulated damage, which is responsible for the age-related changes, combined with other factors, eventually determines the shape of the mortality rate. Specifically, the additive degradation models can often (but not always; see the previous section) result in an increasing mortality rate (Sumita and Shanthikumar, 1985). It seems intuitively unnatural that a degradable object can be characterized by a decreasing mortality rate. Therefore, a regularization procedure will now be suggested, which can eventually result in the increasing 'mortality' rate for a supplementary lifetime random variable (Finkelstein, 2007d).

Denote by $q(t) \le 1$ a quality of life index at age t. The function $q(t)$ defines a weight that is given to the unit increment of life at age t. Humans at advanced ages usually have restrictions of various kinds showing a substantial deterioration in vitality and functions that decrease the quality of life at this stage. Although formally vitality and 'functioning' decrease at all adult ages, the noticeable decline in the corresponding quality of life due to these processes occurs usually only at relatively advanced ages.

These considerations are somehow similar to the starting point of the *Quality Adjusted Life Years* (QALYs) approach (see, *e.g.*, Humnik *et al.*, 2001), but our goal is different. The QALYs approach is focused on solving individual healthcare decision problems, when, for instance, an operation with probability p can add a number of quality years ($q = 1$) but can also result in death ($q = 0$) with probability $1 - p$, whereas without the operation a patient lives with a lower quality of life, *i.e.*, $q < 1$. Our interest is not in a specific deterioration in abilities of individuals with concrete health problems, but rather in modelling a general trend, which shows the decline in quality of life as a manifestation of senescence. Therefore, we will assume that $q(t) = 1,\ t \in [0, t_s)$ and that this function monotonically decreases for $t \ge t_s$, where t_s is the starting age of senescence: a noticeable decline in 'abilities and possibilities'.

Let, as previously, T be a lifetime random variable with the Cdf $F(t)$ and the mortality rate $\mu(t)$. Denote by $Q(T)$ a 'weighted lifetime': a random variable weighted in accordance with the quality of life function $q(t)$, *i.e.*,

$$Q(T) = \int_0^T q(u)du\,, \tag{10.31}$$

where the function $q(t)$ should be such that $Q(\infty) = \infty$.

When $q(t) \equiv 1$, the lifetimes are equal: $Q(T) = T$. Thus, $Q(T)$ in an 'integrated way' already reflects not only the length of life but its quality as well. The distribution function of $Q(T)$ is derived easily *via* the generic Cdf $F(t)$ as

$$\Pr[Q(T) \le t] = \Pr[T \le Q^{-1}(t)] = F(Q^{-1}(t))\,, \tag{10.32}$$

where $Q^{-1}(t)$ is the inverse function to $Q(t)$, which exists and increases as the function $Q(t)$ increases. In accordance with the definition, the corresponding mortality rate $\mu_q(t)$ is

$$\mu_q(t) = \frac{d(F(Q^{-1}(t))}{dt\,(1 - F(Q^{-1}(t))}\,.$$

$$= \frac{d(F(Q^{-1}(t)))d(Q^{-1}(t))}{d(Q^{-1}(t))dt} = \frac{d(Q^{-1}(t))}{dt}\mu(Q^{-1}(t))\,. \tag{10.33}$$

Our intention now is to show that, for example, in the case of the ultimately decreasing mortality rate $\mu(t)$, which is usually qualified as negative senescence, the function $\mu_q(t)$ can still increase, which is somehow more intuitively acceptable for models with degradation. Note that negative senescence is not just a theoretical concept, as it can be encountered in nature (certain plants and fish have the constant or decreasing mortality rates). It is natural to model $q(t)$ as a decreasing power function for large t. A generalization to the regularly varying functions (Bingham *et al.*, 1987) is rather straightforward. Let $q(t) \propto t^{-\alpha}, 0 < \alpha < 1$. By this notation we mean proportionality. The case $\alpha = 1$ will be considered separately, whereas the range $\alpha > 1$ is not allowed, as the function $Q(t)$ should take the value of infinity at $t = \infty$. Under these assumptions

$$Q(t) \propto t^{-\alpha+1} = t^{\frac{k}{n}}, k < n; \quad Q^{-1}(t) \propto t^{\frac{n}{k}}.$$

It follows from (10.33) that, for a constant mortality rate $\mu(t)$, the rate $\mu_q(t)$ is already increasing and $\mu_q(t) \propto t^{n/k-1}$. It is easy to see that it will still be increasing even for decreasing mortality rate $\mu(t) \propto t^{-\beta}$, if $0 < \beta < 1 - k/n$. Thus, under some reasonable assumptions, a regularization procedure has been performed resulting in the increasing rate $\mu_q(t)$. The following example deals with the case $\alpha = 1$.

Example 10.6 Let $F(t) = 1 - \exp\{-\mu t\}$ and

$$q(t) = \begin{cases} 1, & t \le t_s, \\ \dfrac{k}{(t - t_s) + k}, & t > t_s. \end{cases}$$

where $k > 0$, which means that $q(t) \infty k/t$ for sufficiently large t. Therefore,

$$Q(t) = \begin{cases} t, & t \le t_s, \\ t_s + k\left[\ln\left(\dfrac{t - t_s}{k} + 1\right)\right], & t > t_s. \end{cases} \tag{10.34}$$

It is easy to see that the inverse function $Q^{-1}(t)$ is linear in $[0, t_s]$ and is exponentially increasing for $t > t_s$. It follows from Equations (10.33) and (10.34) that $\mu_q(t)$ is also increasing for $t > t_s$ and is constant in $[0, t_s]$. Thus, $\mu_q(t)$ already has the desired non-decreasing shape.

10.7 Stochastic Ordering for Mortality Rates

We continue now describing the properties of age-specific mortality rates $\mu(x, t)$, where x is the age at death and t is the corresponding calendar (chronological) time. We combine here methods and approaches of modern mathematical demography (Kefytz and Casewell, 2005) with the corresponding reliability-related reasoning.

Equations (10.4)–(10.6) define the mortality rate $\mu(x,t)$ and the age X_t for a population at the calendar time t with an age structure $N(x,t)$. Now let $N(x,t), x \geq 0$ and $N^*(x,t), x \geq 0$, be age structures for two populations with random ages X_t and X_t^*, respectively (Finkelstein, 2005a). The corresponding definitions are given in Section 10.1. Specific types of these age structures will be considered later, but now we are interested in the stochastic comparison of X_t and X_t^* for a fixed t. In accordance with Definition 3.4 (Equation (3.40)), we say that the age X_t^* defined by the age structure $N^*(x,t), x \geq 0$ is stochastically larger than the age X_t defined by the age structure $N^*(x,t), x \geq 0$ and write

$$X_t^* \geq_{st} X_t \tag{10.35}$$

if the corresponding age distribution functions are ordered as

$$F^*(x,t) \leq F(x,t), \ \forall x > 0 . \tag{10.36}$$

As follows from (10.6), Inequality (10.36) is equivalent to

$$\frac{\int_0^x N^*(u,t)du}{\int_0^\infty N^*(u,t)du} \leq \frac{\int_0^x N(u,t)du}{\int_0^\infty N(u,t)du}; \forall x > 0 , \tag{10.37}$$

and the age structure $N^*(x,t), x \geq 0$ gives larger probabilities to ages beyond x, than $N(x,t), x \geq 0$.

Stochastic comparison of populations at different time instants can also be of interest. The following inequality:

$$X_{t_2} \geq_{st} X_{t_1}; t_2 > t_1$$

means that the population with the age structure $N(x,t_2), x \geq 0$ is stochastically older than the population with the age structure $N(x,t_1), x \geq 0$, which certainly is the case in practice (under reasonable restrictions on fertility and migration), because mortality rates (at least in the developed countries) are declining with t. If this inequality holds for all ordered t_1 and t_2 in some interval of time, we say that the population is ageing in this interval of time.

10.7.1 Specific Population Modelling

We have already stated in Section 10.1 that, whereas the period mortality rate $\mu(x,t)$ is properly defined by Equation (10.4), the corresponding lifetime random variable for the period setting cannot be unambiguously defined only *via* $\mu(x,t)$. Additional simplifying assumptions should be employed, and this is what is usually done in applications. On the other hand, we know, that in accordance with exponential representation (2.5), the failure rate $\lambda(t)$ always defines the corresponding absolutely continuous distribution function for the cohort setting.

Consider a population that is closed to migration and experiences a constant birth rate B_0 annually. These simplifying assumptions are very natural and allow for detailed mathematical modelling. The age structure in this case can be defined *via* the corresponding cohort survival function, *i.e.*,

$$N(x,t) = B_0 \exp\left\{-\int_0^x \mu(u,t-x+u)du\right\}$$

$$\equiv B_0\, l_c(x,t-x)\,, \tag{10.38}$$

where $l_c(x,t-x)$ denotes the life table survival probability of a cohort of age x born at time $t-x$ and $\mu(u,t-x+u)$ is the mortality rate for this cohort. Therefore, the lifetime random variable is defined for a cohort of age x *via* the corresponding Cdf. Equation (10.38) and some of the forthcoming considerations can be generalized to the case of time-dependent birth rates, but for simplicity we assume that B_0 is a constant. On the other hand, all generalizations that consider migration are usually extremely difficult.

Let $N(x,t), x \geq 0$ be the same population age structure as in (10.38). As in Bongaarts and Feeney (2002), we now artificially 'freeze' the mortality conditions at time t in the following way:

$$N(x,t) = B_0 \exp\left\{-\int_0^x \mu^*(u,t)du\right\}. \tag{10.39}$$

The function $\mu^*(x,t)$ can be now interpreted as the mortality rate for a stationary population with the age structure $N(x,t), x \geq 0$. Therefore, the corresponding lifetime random variable can also be defined *via* $\mu^*(x,t)$ in the usual way using the exponential representation for the Cdf. Note that, although the integrals (and therefore the corresponding survival functions) in Equations (10.38) and (10.39) are obviously equal, the integrands are not equal. On the other hand, the exponential representation *via* the mortality rate $\mu(x,t)$ for the same age structure reads (Preston and Coale, 1982)

$$N(x,t) = B_0 \exp\left\{-\int_0^x \mu(u,t)du\right\}\exp\left\{-\int_0^x I(u,t)du\right\}, \tag{10.40}$$

where $I(u,t)$ is the intensity of a population growth, *i.e.*,

$$I(x,t) = \frac{\partial N(x,t)/\partial t}{N(x,t)}.$$

It can be seen from Equations (10.38)–(10.40) that

$$\int_0^x I(u,t)du = \int_0^x (\mu(u,t-x+u) - \mu(u,t))du$$

and

$$I(x,t) = \mu^*(x,t) - \mu(x,t)$$

in this specific case (see also Arthur and Vaupel, 1984). Equation (10.40) can formally be transformed into

$$N(x,t) = B_0 \exp\left\{ -\int_0^x \mu(u,t)D(u,t)du \right\}, \tag{10.41}$$

where

$$D(x,t) = 1 + \frac{I(x,t)}{\mu(x,t)} \tag{10.42}$$

is a distortion factor for the case of mortality that is changing in time. If, for example, a population is growing, $D(x,t) > 1$. Under additional assumptions (see later), Bongaarts and Feeney (2002) show that $D(x,t)$ does not depend on age x and they develop methods for calculating the corresponding bias for life expectancy.

Consider now a hypothetical population (also closed to migration and with a constant birth rate B^*) and define a new hypothetical age structure $N^*(x,t), x \geq 0$ *via* the mortality rate $\mu(x,t)$ as

$$N^*(x,t) = B^* \exp\left\{ -\int_0^x \mu(u,t)du \right\}. \tag{10.43}$$

Therefore, $\mu(x,t)$ can also be interpreted as the mortality rate for a stationary population with the age structure $N^*(x,t), x \geq 0$. Equations (10.38)–(10.43) will be used for comparing the Cdfs of X_t^* and X_t and also for comparing different definitions of life expectancy. To proceed with these comparisons we need a useful and simple lemma (Finkelstein, 2005a).

Lemma 10.1. Let $f(x)$ and $g(x)$ be continuous functions such that $g(x)$ is decreasing and the integral of $f(x)$ in $[0,\infty)$ is finite.

Then

$$\frac{\int_0^x f(u)g(u)du}{\int_0^\infty f(u)g(u)du} > \frac{\int_0^x f(u)du}{\int_0^\infty f(u)du}, \quad \forall x > 0.$$

Proof. Applying the mean value theorem:

$$\frac{\int_0^x f(u)g(u)du}{\int_0^\infty f(u)g(u)du} = \frac{\int_0^x f(u)g(u)du}{\int_0^x f(u)g(u)du + \int_x^\infty f(u)g(u)}$$

$$= \frac{g(0,x)\int_0^x f(u)du}{g(0,x)\int_0^x f(u)du + g(x,\infty)\int_x^\infty f(u)du} > \frac{\int_0^x f(u)du}{\int_0^\infty f(u)du},$$

where $g(0,x)$ and $g(0,\infty)$ are the corresponding mean values, which exist due to our assumptions. As $g(x)$ is decreasing, $g(0,x) > g(0,\infty)$, and therefore the inequality follows. ■

The following result (Finkelstein, 2005a) shows that random ages X_t^* and X_t are ordered as in Inequalities (10.35) and (10.36), which define the usual stochastic ordering.

Theorem 10. 1. Let the mortality rate $\mu(x,t)$ decrease in calendar time t. Assume that population age structures $N(x,t), x \geq 0$ and $N^*(x,t), x \geq 0$ are given by Equations (10.39) and (10.43), respectively.

Then Ordering (10.35) holds.

Proof. In accordance with Inequality (10.36) and Equations (10.40) and (10.43), we must show that

$$\frac{\int_0^x \exp\left\{-\int_0^y \mu(u,t)du\right\}dy}{\int_0^\infty \exp\left\{-\int_0^y \mu(u,t)du\right\}dy} < \frac{\int_0^x \exp\left\{-\int_0^y \mu(u,t)du\right\}\exp\left\{-\int_0^y I(u,t)du\right\}dy}{\int_0^\infty \exp\left\{-\int_0^y \mu(u,t)du\right\}\exp\left\{-\int_0^y I(u,t)du\right\}dy}.$$

As the mortality rate decreases with time, the population is growing. Therefore, $I(u,t) > 0$ and the corresponding exponential function in the integrand is monotonically decreasing with y. Therefore, the result immediately follows from Lemma 10.1 after noting that

$$\int_0^\infty \exp\left\{-\int_0^y \mu(u,t)du\right\}dy < \infty.$$

■

Under the foregoing assumptions, this result can be interpreted as follows:

A random age X_t in the observed population is stochastically smaller than a random age X_t^* in a hypothetical population constructed via the current mortality rate $\mu(x,t)$ at time t.

Lemma 10.2. Let the mortality rate $\mu(x,t)$ decrease in time t. Assume that population age structures $N(x,t), x \geq 0$ and $N^*(x,t), x \geq 0$ are given by Equations (10.40) and (10.43), respectively.

Then

$$\frac{\int_0^x \mu(u,t)N^*(u,t)du}{\int_0^\infty \mu(u,t)N^*(u,t)du} < \frac{\int_0^x \mu(u,t)N(u,t)du}{\int_0^\infty \mu(u,t)N(u,t)du}; \forall x > 0 .$$

Proof. Substituting Relationships (10.40) and (10.43) into this inequality:

$$\frac{\int_0^x \mu(y,t)\exp\left\{-\int_0^y \mu(u,t)du\right\}dy}{\int_0^\infty \mu(y,t)\exp\left\{-\int_0^y \mu(u,t)du\right\}dy}$$

$$< \frac{\int_0^x \mu(y,t)\exp\left\{-\int_0^y \mu(u,t)du\right\}\exp\left\{-\int_0^y I(u,t)du\right\}dy}{\int_0^\infty \mu(y,t)\exp\left\{-\int_0^y \mu(u,t)du\right\}\exp\left\{-\int_0^y I(u,t)du\right\}dy},$$

which is proved using exactly the same argument as when proving Theorem 10.1 after noting that

$$\int_0^\infty \mu(y,t)\exp\left\{-\int_0^y \mu(u,t)du\right\}dy = 1 < \infty . \qquad \blacksquare$$

We will use the supplementary results of this section for ordering life expectancies for different populations. But first, various definitions of life expectancies should be given (Finkelstein, 2005a).

10.7.2 Definitions of Life Expectancy

10.7.2.1 Period Life Expectancy at Birth

Denote by $e(0,t)$ the period life expectancy at birth at calendar time t. It is usually defined in demography as the average age at death that would be observed for a group of individuals (a cohort) who experience over the course of their lives the age-specific mortality rates $\mu(x,t)$ observed during the time period (for the fixed time t), *i.e.*,

$$e(0,t) = \int_0^\infty l_c(x,t)dx\,, \tag{10.44}$$

where $l_c(x,t)$ is the life table survival probability, defined for the corresponding cohort (compare with $l_c(x,t-x)$ in Equation (10.38)) as

$$l_c(x,t) = \exp\left\{-\int_0^x \mu(u,t)du\right\}, \tag{10.45}$$

which corresponds to exponential representation (2.5) in demographic notation. The period life expectancy at age x is the analogue of the mean remaining lifetime (2.7) in reliability theory, *i.e.*,

$$e(x,t) = \frac{\int_x^\infty l_c(u,t)du}{l_c(x,t)}\,.$$

The period life expectancy $e(0,t)$ is the main index that is convenient for calculation and is used in practical demography. It works perfectly for stationary populations. However, when mortality rates are changing with time, as follows from Equations (10.41) and (10.44)–(10.45), it leads to distortions in estimating the 'true' life expectancy. To calculate this distortion, Bongaarts and Feeney (2002) have suggested the following model.

10.7.2.2 Life Expectancy at Current Mortality Conditions

Definition (10.44) is based only on age-specific mortality rates obtained without any direct information on the population age structure, which is reasonable only for stationary populations.

The life expectancy at current mortality conditions is defined by Bongaarts and Feeney (2002) for the specific 'frozen' population, described by Equation (10.39) as

$$e^*(0,t) = \int_0^\infty l^*(x,t)dx\,, \tag{10.46}$$

where $l^*(x,t)$ is the survival function, defined *via* the mortality rate $\mu^*(x,t)$ similar to Equation (10.45).

Alternatively, the life expectancy at current mortality conditions can be written by means of the observed mortality rate as the following integral of the survival function in Equation (10.38):

$$e^*(0,t) = \int_0^\infty \left(\exp\left\{ -\int_0^x \mu(u,t-x+u)du \right\} \right) dx$$
$$= \int_0^\infty l_c(x,t-x)dx .$$

This index is often called the *cross-sectional average life* (CAL) in advanced demographic studies.

10.7.2.3 Average Age at Death

Another period measure, which is used for calculating life expectancy, at least under the assumption of a birth rate that is constant in time, is the life expectancy at death:

$$A(t) = \frac{\int_0^\infty x\mu(x,t)N(x,t)dx}{\int_0^\infty \mu(x,t)N(x,t)dx} . \tag{10.47}$$

In other words, it is the average age at death at current mortality rates. This measure already reflects population characteristics and can be estimated from the period life table data, although the calculations are not so easy compared with conventional calculations of $e(0,t)$.

Under additional assumptions (see the next section), Bongaarts and Feeney (2002) showed that $e^*(0,t) = A(t)$. Note that we do not need these additional assumptions for comparing $e(0,t)$ with $A(t)$. It can easily be seen that all three measures are equal for stationary populations, *i.e.*,

$$e(0,t) = e^*(0,t) = A(t) .$$

When mortality is changing with time, however, they are different and the use of conventional $e(0,t)$ should be justified.

10.7.2.4 Mean Population Age

The mean age of individuals in a population with the age structure $N(x,t),\ x \geq 0$ can be defined as

$$e(t) = \frac{\int_0^\infty xN(x,t)du}{\int_0^\infty N(x,t)du} .$$

It is clear that this is not a very good indicator of life expectancy, but it can be used as a helpful additional characteristic. As the human mortality rate $\mu(x,t)$ increases in age x, the following comparison of $e(t)$ with the average age at death $A(t)$ is straightforward:

$$\frac{\int_0^\infty x\mu(x,t)N(x,t)dx}{\int_0^\infty \mu(x,t)N(x,t)dx} > \frac{\int_0^\infty xN(x,t)dx}{\int_0^\infty N(x,t)dx}. \tag{10.48}$$

To prove this inequality, it is sufficient to consider the 'modified' population age structure $\mu(x,t)N(x,t),\ x \geq 0$. Under the assumption of the mortality rate $\mu(x,t)$ that increases with age, this structure gives larger probabilities to ages beyond x than the age structure $N(x,t),\ x \geq 0$, which results in an inequality similar to Inequality (10.37) and, finally, in Inequality (10.48). Note that human mortality is described by a mortality rate that increases with age x, as defined by the Gompertz law (10.1) and the Gompertz shift model (10.28).

10.7.3 Comparison of Life Expectancies

10.7.3.1 Comparison of $e(0,t)$ *with* $e^*(0,t)$

As previously, we will make this comparison for a population that experiences no migration and a constant annual birth rate. As the population is growing (the mortality rate is decreasing in calendar time t),

$$\mu^*(x,t) - \mu(x,t) > 0;\ \forall x \geq 0, \tag{10.49}$$

which obviously leads to the corresponding ordering of life expectancies (see Equations (10.44) and (10.46)) and to a distortion $\Delta(t)$:

$$\Delta(t) \equiv e(0,t) - e^*(0,t) > 0. \tag{10.50}$$

This is a general result for the population of the defined type, which can also be formulated as

$$\Delta(t) = \int_0^\infty \left(\exp\left\{ -\int_0^x \mu(u,t)du \right\} \right) dx - \int_0^\infty \left(\exp\left\{ -\int_0^x \mu(u,t-x+u)du \right\} \right) dx.$$

Bongaarts and Feeney (2002) make additional assumptions for estimating this distortion that they call the *tempo bias*. They assume that changes in the population age structure $N(x,t),\ x \geq 0$ owing to mortality decline with time t are modelled as the *age-independent* shift $s(t)$ to the larger ages, *i.e.*,

$$N(x,t) = \begin{cases} B, & x < s(t), \\ N(x - s(t),0), & x \geq s(t). \end{cases} \tag{10.51}$$

Note that Equation (10.51) leads to the same shift in mortality rates. Formally, this is a rather stringent assumption, but assuming the Gompertz law for mortality curves with the fixed t, we immediately arrive at the Gompertz shift model (10.28), as the exponential function 'converts shifts into multipliers'. It was also proved by these authors that

$$\mu(x,t) = \left(1 - \frac{de^*(0,t)}{dt}\right)\mu^*(x,t). \tag{10.52}$$

Equation (10.52) shows that when the life expectancy $e^*(0,t)$ is increasing, the observed mortality rate $\mu(x,t)$ is smaller than $\mu^*(x,t)$. Using numerical procedures, Bongaarts and Feeney (2002) obtained the values of $e^*(0,t)$ and the corresponding tempo bias Δt. It turned out that the average tempo bias, *e.g.*, for females in France, Japan, Sweden and the USA for the period from 1980 to 1995 is rather large: 2.3 years, 3.3 years, 1.6 years and 1.6 years, respectively. However, a question still remains: is $e^*(0,t)$, defined for a specific population under the stringent conditions, the best candidate for the 'true' life expectancy?

10.7.3.2 Comparison of $e(0,t)$ *with* $A(t)$

The following theorem (Finkelstein, 2005a) is a direct consequence of Equations (10.43) and (10.47).

Theorem 10.2 Let $N^*(x,t), x \ge 0$ be an age structure for a hypothetical population defined by equation (10.43).

Then the average age at death for this population $A^*(t)$ is equal to the conventional life expectancy $e(0,t)$:

$$A^*(t) \equiv \frac{\int_0^\infty x\mu(x,t)N^*(x,t)dx}{\int_0^\infty \mu(x,t)N^*(x,t)dx} = e(0,t). \tag{10.53}$$

Theorem 10.3 Let the mortality rate $\mu(x,t)$ decrease in calendar time t. Assume that the population age structures $N(x,t)$ and $N^*(x,t)$ are given by Equations (10.39) and (10.43), respectively.

Then the conventional life expectancy $e(0,t)$ is larger than the average age at death (10.47):

$$e(0,t) - A(t) > 0. \tag{10.54}$$

Proof. In accordance with Theorem 10.2 and Equation (10.47), we must prove that

$$\frac{\int_0^\infty x\mu(x,t)N^*(x,t)dx}{\int_0^\infty \mu(x,t)N^*(x,t)dx} - \frac{\int_0^\infty x\mu(x,t)N(x,t)dx}{\int_0^\infty \mu(x,t)N(x,t)dx} > 0. \tag{10.55}$$

As the ordering of survival functions leads to the same ordering of the corresponding mean values, Inequality (10.55) immediately follows from Lemma 10.2 (the sign of inequality of this lemma will be opposite for survival functions). Note that for proving Inequality (10.54) we do not need additional proportionality assumptions. ■

10.7.3.3. Comparison with a Hypothetical Cohort

The following alternative comparison with a hypothetical cohort can also be helpful.

Let M denote the maximum age in the life table, *e.g.*, $M = 110$ years. The age structure $N(x,t), x \geq 0$ means that $B_0(t-x)$ individuals were born at $t-x$ and $N(x,t)$ of whom had survived to t. Let us shift the 'life trajectories' of survivors backwards by $M-x$ units of time. This means that the whole population with size $N(t)$ will be born at $t-M$ and the cohort for this whole population can be considered. As mortality rates are declining with t,

$$\mu(x,t-M+x) > \mu(x,t).$$

This inequality also means that $e(0,t) > e_s(0,t)$, where $e_s(0,t)$ denotes the life expectancy of the described shifted cohort.

10.7.4 Further Inequalities

In this section, only the case of stationary populations will be considered. Denote by $l_c(x)$ the life table survival probability for some stationary population, which corresponds to the general time-dependent Equation (10.45). In accordance with Remark 10.2, the pdf of the age of an individual chosen at random (with an equal chance) from a population of size $N(t)$ is

$$f_a(x) = \frac{l_c(x)}{\int_0^\infty l_c(u)du}. \tag{10.56}$$

Define the mean life expectancy E by averaging the corresponding stationary life expectancy at x (Section 10.7.2.1)

$$e(x) = \frac{\int_x^\infty l_c(u)du}{l_c(x)}$$

with respect to pdf (10.56), *i.e.*,

$$E = \frac{\int_0^\infty e(x) l_c(x) dx}{\int_0^\infty l_c(x) dx} = \frac{\int_0^\infty \left(\int_x^\infty l_c(u) du \right) dx}{\int_0^\infty l_c(u) du} . \tag{10.57}$$

Thus, E is the average time to death for an individual chosen randomly (with an equal chance) from the whole population at some fixed time t.

Remark 10.3 A different type of averaging can be considered in a cohort setting. Assume as in Keyfitz and Casewell (2005) that death deprives an individual of the remainder of his life expectancy (see also Example 4.1). Thus death, which had occurred in $[x, x+dx]$, deprived an individual of $e(x)$ years, *i.e.*, his life expectancy at x. Let, as usual, $f(t)$ denote the lifetime pdf. The average life deprivation at death is therefore

$$\widetilde{E} = \int_0^\infty f(u) e(u) du = \int_0^\infty e(u) \mu(u) l_c(u) du$$
$$= \int_0^\infty \left(\mu(x) \left(\int_x^\infty l_c(u) du \right) \right) dx .$$

A better statistical interpretation is, however, the one involving the notion of 'another chance of life' or lifesaving as defined in Section 10.4. The corresponding reliability interpretation of this operation is: a single 'minimal repair at death'.

Consider now two stationary populations with survival functions $l_{c1}(x)$ and $l_{c2}(x)$, respectively. Let

$$l_{c1}(x) > l_{c2}(x), \forall x > 0 , \tag{10.58}$$

which can be interpreted as the (usual) stochastic ordering between the corresponding lifetime random variables.

Theorem 10.4. Let Ordering (10.58) for two stationary populations hold.
Then

$$E_1 > E_2 . \tag{10.59}$$

Proof. We will outline a sketch of this proof that can be made mathematically strict in an obvious, although cumbersome, way. Let

$$l_{c1}(x) = l_{c2}(x) + \delta(x) ,$$

where $x_0 \in (0,\infty)$ and the continuous function $\delta(x)$ is 0 outside the interval $[x_0-\varepsilon, x_0+\varepsilon]$; $\delta(x_0-\varepsilon)=\delta(x_0+\varepsilon)=0$. Assume that ε is sufficiently small and that the area

$$\Delta = \int_0^\infty \delta(x)dx = \int_{x_0-\varepsilon}^{x_0+\varepsilon} \delta(x)dx$$

is also sufficiently small. Assume that $\delta(x)$ does not change the monotonicity of $l_{c2}(x)$ and therefore, the function $l_{c1}(x)$ is a survival function. Transformation of E_1 results in the following:

$$E_1 = \frac{\int_0^\infty \left(\int_x^\infty l_{c1}(u)du\right)dx}{\int_0^\infty l_{c1}(u)du} = \frac{\int_0^{x_0+\varepsilon}\left(\int_x^\infty l_{c1}(u)du\right)dx + \int_{x_0+\varepsilon}^\infty\left(\int_x^\infty l_{c1}(u)du\right)dx}{\int_0^\infty l_{c1}(u)du}$$

$$= \frac{\int_0^{x_0+\varepsilon}\left(\int_x^\infty l_{c2}(u)du + \Delta\right)dx + \int_{x_0+\varepsilon}^\infty\left(\int_x^\infty l_{c2}(u)du\right)dx}{\int_0^\infty l_{c2}(u)du + \Delta}$$

$$= \frac{\Delta(x_0+\varepsilon) + \int_0^\infty\left(\int_x^\infty l_{c2}(u)du\right)dx}{\int_0^\infty l_{c2}(u)du + \Delta} = \frac{\dfrac{\Delta(x_0+\varepsilon)}{\int_0^\infty l_{c2}(u)du} + E_2}{1 + \dfrac{\Delta}{\int_0^\infty l_{c2}(u)du}}$$

$$= \left(E_2 + \frac{\Delta(x_0+\varepsilon)}{\int_0^\infty l_{c2}(u)du} - \frac{\Delta}{\int_0^\infty l_{c2}(u)du}\right)(1+o(1)),$$

which holds asymptotically for sufficiently small Δ. On the other hand, the difference in the last line is positive for $(x_0+\varepsilon) > 1$, and therefore Inequality (10.59) holds in this case.

Assume that our survival functions differ outside the initial small interval $[0,1)$. Thus, using a sequence of properly arranged infinitesimal steps of the described type, we can 'transform' any survival function $l_{c2}(x)$ into the survival function $l_{c1}(t)$. It can be shown under reasonable assumptions that this small (compared with $[1,\infty)$) initial interval will not 'spoil' the described procedure.

■

We will now construct a counterexample showing that a weaker assumption than (10.58) does not result in Ordering (10.59). Assume that life expectancies at birth for two populations are ordered as

$$e_1(0) > e_2(0). \tag{10.60}$$

As life expectancy is an integral of the corresponding survival function, Inequality (10.60) follows from Inequality (10.58). Let the graphs of $l_{c1}(x)$ and $l_{c2}(x)$ cross only once at x_c in such a way that $l_{c1}(x) < l_{c2}(x)$ for $x \in (0, x_c)$ and $l_{c1}(x) > l_{c2}(x)$ for $x \in (x_c, \infty)$. Assume first that the corresponding life expectancies are equal, *i.e.*,

$$e_1(0) = \int_0^\infty l_{c1}(x)dx = \int_0^\infty l_{c2}(x)dx = e_2(0). \tag{10.61}$$

Considering areas under these survival curves it is easy to derive taking into account Equations (10.57) and (10.61) that $E_1 > E_2$, as

$$\int_x^\infty l_{c1}(x)dx > \int_x^\infty l_{c2}(x)dx, \forall x > 0. \tag{10.62}$$

We will now use the following *variation principle*. Transform 'slightly' the curve $l_{c1}(x)$ in $x \in (0, x_c)$ (not changing monotonicity and its values at 0 and x_c) in such a way that its values are smaller than those of $l_{c1}(x)$ in this interval. It follows from Equation (10.61) that

$$e_2(0) = \tilde{e}_1(0) + \varepsilon, \tag{10.63}$$

where $\tilde{e}_1(0)$ is the life expectancy that corresponds to the new survival curve and $\varepsilon > 0$ is a sufficiently small quantity. Equation (10.63) means that, in contrast to Assumption (10.60), the inequality $\tilde{e}_1(0) < e_2(0)$ holds. However, as ε can be made as small as we wish, Inequality (10.62) is not violated (excluding an initial interval that can be made arbitrarily small) and the mean life expectancies defined by Equation (10.57) are ordered as $\tilde{E}_1 > E_2$. Therefore, an ordering of life expectancies at birth (which is weaker than (10.58)) does not imply the same direction in ordering of the mean life expectancies. This reasoning can also be made mathematically strict, but the idea and the result are obvious.

Remark 10.4 The author is grateful to Professor Joshua Goldstein for the setting of Section 10.7.4.

10.8 Tail of Longevity

In this section, we will briefly consider (see Finkelstein and Vaupel, 2006 for the full version) a practical demographic modification of the remaining lifetime concept in a cohort setting to finite but large populations. Another important feature of this approach is that the suggested characteristic is defined *via* the two distributions. The first one is the ordinary distribution of a lifespan, whereas the second one is the distribution of a lifetime of the last survivor in a population.

Consider a stationary population of a sufficiently large size N. As usual, denote by X the random age at death and by ω_N the random maximum age at death (the age at last death) in this population. It is challenging to define the tail of longevity as some remaining *potential* lifetime, taking into account the maximum lifetime variable ω_N.

Denote by $\tau(\omega_N, q)$ the q-quantile for the distribution of ω_N, *i.e.*,

$$\Pr[\omega_N \leq \tau(\omega_N, q)] = q$$

and by $\tau(q_0)$ the q_0-quantile for the distribution of X, *i.e.*,

$$\Pr[X \leq \tau(q_0)] = q_0 .$$

Vaupel (2003) defines the *tail of longevity* as the difference

$$TL(q, q_0) \equiv \tau(\omega_N, q) - \tau(q_0) \tag{10.64}$$

and *the relative tail of longevity* as

$$RTL(q, q_0) \equiv \frac{\tau(\omega_N, q)}{\tau(q_0)} - 1 . \tag{10.65}$$

Our main focus is on the relative tail. Relative measures are necessary for adequate comparisons of tails in different populations.

Vaupel (2003) considered specific values of quantiles: $q = 0.5$ and $q_0 = 0.9$. The latter value marks the left endpoint of the post-reproductive zone for some organisms, where the force of natural selection is no longer active. The median of the maximal lifespan distribution $\tau(\omega_N, 0.5)$ is just a reasonable choice for a quantile of this distribution. Not that formally we do not rely on specific values of q and q_0, as the only reasonable restriction is that the corresponding quantiles should be properly ordered ($\tau(\omega_N, q) > \tau(q_0)$), which is obviously the case in reality.

In accordance with (2.5), the Cdf of the age at death X is

$$F(t) = 1 - \exp\left(- \int_0^t \mu(u) du \right), \tag{10.66}$$

where $\mu(t)$ is the corresponding mortality rate. Let

$$S(t) = N \exp\left(- \int_0^t \mu(u) du \right) \tag{10.67}$$

be the expected number of members of the population who will survive at t, starting with initial value $S(0) = N$.

In line with general considerations on the distribution of the maximum of N i.i.d. random variables, Thatcher (1999) showed that the Cdf of ω_N, for large N, can be defined as

$$\begin{aligned} F_N(t) &\equiv \Pr[\omega_N \le t] = (F(t))^N \\ &= \left(1 - \frac{S(t)}{N}\right)^N \approx \exp(-S(t)) \\ &= \exp\left(-N \exp\left(-\int_0^t \mu(u)du\right)\right). \end{aligned} \tag{10.68}$$

Using Equation (10.68), the quantile $\tau(\omega_N, q)$ is obtained from

$$S(\tau(\omega_N, q)) = -\ln q . \tag{10.69}$$

Therefore, taking into account Equation (10.67),

$$\int_0^{\tau(\omega_N, q)} \mu(u)du \approx \ln N - \ln(-\ln q) . \tag{10.70}$$

The second term on the right in Equation (10.70) is of minor importance, as N is large and we are not interested in the 'too high quantiles' when studying the maximal value distributions. For large enough N, Relationship (10.70) can be practically considered an equality, and this will be assumed in what follows.

Doubling the sample size N will only slightly increase $\tau(\omega_N, q)$ for sufficiently large N. The increase from N to N^2 or N^3 gives a substantial increase, depending on the shape of the mortality rate: it is smaller for increasing failure rates and larger for constant and decreasing failure rates. This result follows from Equation (10.70).

Our goal is to compare $\tau(\omega_N, q)$ with the quantile $\tau(q_0)$ obtained from (10.67), *i.e.*, $F(\tau(q_0)) = q_0$. Note that the quantile $\tau(q_0)$, chosen as 0.9, defines the starting point of old age (Vaupel, 2003). Formally, however, we are not very concerned with the concrete values of q_0 and q as we only need the following ordering: $\tau(q_0) < \tau(\omega_N, q)$.

Redundancy is the main tool in designing reliable technical structures. The idea that redundant structures constitute a plausible lifetime model seems very attractive, as the extremely high 'reliability of humans' is likely to exist in nature only with the help of redundancy on different levels. In what follows, we will prove the conjecture of Vaupel (2003) that redundancy decreases the relative tail of longevity. In Finkelstein and Vaupel (2006) it was also proved that heterogeneity in a population increases the tail of longevity.

Consider the case of loaded redundancy when n i.i.d. components operate in parallel. The case of unloaded redundancy (standby) is considered in a similar way. Mortality rates of the simplest redundant structures of identical components with constant mortality rates, operating in parallel, were analysed by Gavrilov and Gavrilova (1991, 2002). These authors show that for sufficiently small t, the mortality rate of the fixed parallel structure (loaded redundancy) approximately follows the power law, and the mortality rate of a structure with a random number of initially operating components approximately follows the Gompertz law (see Example 10.1 for more details). The Cdf of the time to death (failure) of the described system is

$$F_n(t) = (F(t))^n, \ n = 1,2,...$$

and the corresponding quantile $\tau(n,q_0); \tau(1,q_0) \equiv \tau(q_0)$ is obtained from equation $F_n(\tau(n,q_0)) = q_0$ or, equivalently,

$$F(\tau(n,q_0)) = q_0^{\frac{1}{n}}. \tag{10.71}$$

This means that the effect of redundancy of this type changes the baseline level q_0 to $q_0^{1/n}$. For reasonable parameter values, this usually leads to a substantial increase in the quantile.

What about the maximal lifespan quantile? The only difference from the baseline $\tau(\omega_N,q)$ is the size of the sample, which is now nN, because the maximum value is observed at the failure of the last of the nN components. Therefore, Equation (10.70) for obtaining $\tau(\omega_N,q)$ becomes the following equation:

$$\int_0^{\tau(\omega_{nN},q)} \mu(u)du = \ln N + \ln n - \ln(-\ln q) \tag{10.72}$$

for obtaining $\tau(\omega_{nN},q)$. Usually, n is small with respect to N (although this is probably not the case for the molecular or genetic level). As $N \to \infty$, the term $\ln n$ is negligible. Therefore,

$$\tau(\omega_{nN},q)/\tau(\omega_N,q) \to 1 \text{ as } N \to \infty. \tag{10.73}$$

Theorem 10.5. Let the sample size N be sufficiently large.

Then the relative tail of longevity for a system with a loaded redundancy structure is smaller than the relative tail of longevity for a non-redundant system, *i.e.*,

$$RTL(n,q,q_0) < RTL(q,q_0), \ n = 2,3,... .$$

Proof. It follows from (10.73) that for large enough N

$$\frac{\tau(n,q_0)}{\tau(q_0)} > \frac{\tau(\omega_{nN},q)}{\tau(\omega_N,q)} \tag{10.74}$$

and, in accordance with the definition of the relative tail of longevity,

$$\frac{RTL(n,q,q_0)+1}{RTL(q,q_0)+1} = \frac{\tau(\omega_{nN},q)\tau(q_0)}{\tau(n,q_0)\tau(\omega_N,q)} < 1,$$

which results in the statement of this theorem. ■

Similar, for large enough N

$$\frac{\tau(\omega_{(n+1)N},q)\tau(n,q_0)}{\tau(n+1,q_0)\tau(\omega_{nN},q)}<1,\ \ n=1,2,...,\ \ \tau(1,q_0)\equiv\tau(q_0), \tag{10.75}$$

which means that the relative tail of longevity decreases with n.

Inequalities (10.74) and (10.75) hold for parameters values that are reasonable in demographic practice and show that the loaded redundancy decreases the relative tail of longevity.

Example 10.7 Consider the exponential case $\mu(t)=\mu$ and let the level of redundancy be $n=2$. From Equation (10.72) it follows that

$$\tau(\omega_{2N},q)=\frac{\ln N+\ln 2-\ln(-\ln q)}{\mu}.$$

Using Equation (10.70):

$$\frac{\tau(\omega_{2N},q)}{\tau(\omega_N,q)}=\frac{\ln N+\ln 2-\ln(-\ln q)}{\ln N-\ln(-\ln q)}\approx 1+\frac{\ln 2}{\ln N}. \tag{10.76}$$

From Equation (10.71):

$$\tau(2,q_0)=\frac{-\ln(1-\sqrt{q_0})}{\mu}.$$

Therefore,

$$\frac{\tau(2,q_0)}{\tau(q_0)}=\frac{\ln(1-\sqrt{q_0})}{\ln(1-q_0)}. \tag{10.77}$$

If, for example, $N=10^6$, $q_0=0.9$, $q=0.5$, then Relationship (10.76) gives the value of 1.05, whereas Relationship (10.77) gives the value of 1.3. The corresponding values for relative tails of longevity are

$$RTL(q,q_0)\equiv\frac{\tau(\omega_N,q)}{\tau(q_0)}-1=\frac{\ln N-\ln(-\ln q)}{-\ln(1-q_0)}-1\approx 5.15,$$

$$RTL(2,q,q_0)\equiv\frac{\tau(\omega_{2N},q)}{\tau(2,q_0)}-1=\frac{\ln N+\ln 2-\ln(-\ln q)}{-\ln(1-\sqrt{q_0})}-1\approx 3.95,$$

which shows a decrease of approximately 30% in the relative tail of longevity owing to redundancy.

10.9 Chapter Summary

Reliability theory possesses a well-developed methodology for stochastic modelling of ageing and failures in technical objects. Therefore, it is reasonable to apply these methods to lifespan modelling of humans and other organisms. Some simple stochastic models and approaches are discussed in this chapter. Among reliability-oriented books, as far as we know, this topic is considered for the first time.

We revise the original Strehler–Mildvan model, which was widely applied to human mortality data, and show that from mathematical point of view it is valid only under the assumption of the Poisson property of the point process of shocks (demands for energy). We also show that the operation of thinning of the Poisson process described in Chapter 8 can be used for a probabilistic explanation of the lifesaving procedure, which results in the decrease in mortality rates of contemporary human populations in developed countries.

We apply the concept of stochastic ordering to stochastic comparisons of different populations. An important specific feature of this modelling, as compared with a traditional reliability one, is that the mortality rate in demographic studies of non-stationary populations is not only a function of age, but of calendar time as well. In this case, the mortality rate does not uniquely define the corresponding lifetime distribution and the main exponential representation of reliability and survival analysis (Equation (2.5)) is not valid without additional simplifying assumptions.

Finally, in the last section, the tail of longevity for human populations is discussed. This notion is somehow close to the notion of the mean remaining lifetime, but the corresponding definition is based on two population distributions: on the 'ordinary' lifetime distribution and on the distribution of time to death of the last survivor. The tail of longevity is the difference between the properly chosen quantiles of these distributions and it describes the potential remaining lifetime for a population as a whole.

References

Aalen OO (1988). Heterogeneity in survival analysis. *Statistics in Medicine*, 7, 1121-1137.

Aalen OO, Gjessing HK (2001). Understanding the shape of the hazard rate: A process point of view. *Statistical Science*, 16, 11-22.

Abdel-Hameed M (1975). A gamma wear process. *IEEE Transactions on Reliability*, 24, 152-153.

Abdel-Hameed M (1987). Inspection and maintenance policies of devices subject to deterioration. *Advances in Applied Probability*, 19, 917-931.

Andersen PK, Borgan O, Gill RD, Keiding N (1993). *Statistical Models Based on Counting Processes*. Springer-Verlag, New York.

Anderson JE, Louis TA (1995). Survival analysis using a scale change random effects model. *Journal of the American Statistical Association*, 90, 669-679.

Arbeev K, Ukraintseva S, Arbeeva L, Yashin AI (2005). Mathematical models for human cancer incidents rates. *Demographic Research*, 12, 237-232.

Arjas E, Norros I (1989). Change of life distribution via a hazard transformation: an inequality with application to minimal repair. *Mathematics of Operations Research*, 14, 355-361.

Arnold BC, Zahedi H (1988). On multivariate mean residual life functions. *Journal of Multivariate Analysis*, 25, 1-9.

Arthur WB, Vaupel JW (1984). Some general relationships in population dynamics. *Population Index*, 50, 214-226.

Ascher H, Feingold H (1984). *Repairable Systems Reliability*. Marcel Dekker, New York.

Aven T, Jensen U (1999). *Stochastic Models in Reliability*. Springer-Verlag, New York.

Badia FG, Berrade MD, Campos CA, Navascues MA (2001). On the behaviour of aging characteristics in mixed populations. *Probability in the Engineering and Informational Sciences*, 15, 83-94.

Bagdonavicius V, Nikulin M (2002). *Accelerated Life Models. Modelling and Statistical Analysis*. Chapman & Hall.

Bai DS, Yun WJ (1986). An age replacement policy with minimal repair cost limit. *IEEE Transactions on Reliability*, 31, 452-459.

Bankand L, Jarvik LS (1978). A longitudinal study of aging human twins. In: Schneider EL (ed.) *Genetics of Ageing*. Plenum Press, New York, 231-244.

Barlow RE (1985). A Bayes explanation of an apparent failure rate paradox. *IEEE Transactions on Reliability*, 34, 107-108.

Barlow RE, Hunter LC (1960). Optimal preventive maintenance policies. *Operations Research*, 8, 90-100.

Barlow RE, Proschan F (1965). *Mathematical Theory of Reliability*. Wiley, New York.
Barlow RE, Proschan F (1975). *Statistical Theory of Reliability and Life Testing*. Holt, Renerhart & Winston, New York.
Bassan B, Kochar S, Spizzichino F (2002). Some bivariate notions of IFR and DMRL and related properties. *Journal of Applied Probability*, 39, 533-544.
Basu AP (1971). Bivariate failure rate. *Journal of American Statistical Association*, 66, 103-104.
Baxter LA, Kijima M, Tortorella M (1996). A point process model for the reliability of the maintained system subject to general repair. *Stochastic Models*, 12, 37-65.
Beard RE (1959). Note on some mathematical mortality models. In: Woolstenholme GEW, O'Connor M (eds.) *The Lifespan of Animals*. Little, Brown and Company, Boston, 302-311.
Beard RE (1971). Some aspects of theories of mortality, cause of death analysis, forecasting and stochastic processes. In: Brass W (ed.) *Biological Aspects of Demography*. Taylor & Francis, London, 57-68.
Bedford T, Cooke R (2001). *Probabilistic Risk Analysis. Foundations and Methods*. Cambridge University Press.
Beichelt FE (1981). A generalized block-replacement policy. *IEEE Transactions on Reliability*, 30, 171-173.
Beichelt FE, Fatti LP (2002). *Stochastic Processes and their Applications*, Taylor and Francis, London.
Beichelt FE, Fischer K (1980). General failure model applied to preventive maintenance policies. *IEEE Transactions on Reliability*, 29, 39-41.
Bergman B (1978). Optimal replacement under a general failure model. *Advances in Applied Probability*, 10, 431-451.
Bergman B (1980). On the optimality of stationary replacement strategies. *Journal of Applied Probability*, 17, 178-186.
Bergman B (1985). Reliability theory and its applications. *Scandinavian Journal of Statistics*, 12, 1-41.
Bingham NH, Goldie CM, Teugels JL (1987). *Regular Variation*. Cambridge University Press.
Birren JE, Renner VJ (1977). *Handbook of Psychology of Aging*. Van Nostrand, New York.
Block HW, Joe H (1997). Tail behaviour of the failure rate functions of mixtures. *Lifetime Data Analysis*, 3, 269-288.
Block HW, Savits TH (1980). Multivariate increasing failure rate distributions. *Annals of Probability*, 8, 793-801.
Block HW, Savits TH (1997). Burn-in. *Statistical Science*, 12, 1-19.
Block HW, Borges W, Savits TH (1985). Age-dependent minimal repair. *Journal of Applied Probability*, 22, 370-386.
Block HW, Li Y, Savits TH (2001). Behaviour of failure rates of mixtures and systems. In: Hayakawa Y, Irony T, Xie M (eds.) *System and Bayesian Reliability: Essays in Honour of Professor R.E. Barlow*. Series on Quality, Reliability and Engineering Statistics, World Scientific Press, Singapore, 269-288.
Block HW, Li Y, Savits TH (2003). Initial and final behaviour of failure rate functions for mixtures and functions. *Journal of Applied Probability*, 40, 721-740.
Block HW, Mi J, Savits TH (1993). Burn-in and mixed populations. *Journal of Applied Probability*, 30, 692-702.
Block HW, Savits TH, Singh H (1998). The reversed hazard rate function. *Probability in the Engineering and Informational Sciences*, 12, 69-90.
Boland PJ, El-Neweihi E (1998). Statistical and information based minimal repair for k out of n systems. *Journal of Applied Probability*, 35, 731-740.

Bongaarts J, Feeney G (2002). How long do we live? *Population and Development Review*, 28, 13-29.

Bradley D, Gupta RC (2003). Limiting behaviour of the mean residual life. *Annals of the Institute of Statistical Mathematics*, 55, 217-226.

Brindley EC, Thompson WA (1972). Dependence and aging aspects of multivariate survival. *Journal of American Statistical Association*, 67, 822-830.

Brown M, Proschan F (1983). Imperfect repair. *Journal of Applied Probability*, 20, 851-862.

Brown J, Mahoney J, Sivazlian B (1983). Hysteresis repair in discounted replacement problems. *IIE Transactions*, 15, 156-65.

Calabra R, Pulcini G (1987). On the asymptotic behaviour of the mean residual life function. *Reliability Engineering*, 19, 165-170.

Carey JR, Liedo P, Vaupel JW (1992). Slowing of mortality rates at older ages of medfly cohorts. *Science*, 258, 457-461.

Cha JH (2000). On a better burn-in procedure. *Journal of Applied Probability*, 37, 1099-1103.

Cha JH (2006). An extended model for optimal burn-in procedures. *IEEE Transactions on Reliability*, 55, 189-1999.

Cha JH, Finkelstein MS (2008). On a model with shocks and initial resource. Technical Report 293, University of Free State.

Cha JH, Mi J (2008). Some probability functions in reliability and their applications. *Naval Research Logistics*, DOI 10.1002/nav.

Chan J, Shaw L (1993). Modelling repairable systems with failure rates that depend on age and maintenance. *IEEE Transactions on Reliability*, 42, 566–571.

Chandra NK, Roy D (2001). Some results on reversed hazard rate. *Probability in the Engineering and Informational Sciences*, 15, 95-102.

Clarotti CA, Spizzichino F (1990). Bayes burn-in and decision procedures. *Probability in Engineering and Informational Sciences*, 4, 437-445.

Clayton DG (1978). A model of association in bivariate life tables and its application in epidemiological studies of familial tendency in chronic disease incidence. Biometrika, 65, 141-151.

Clayton D, Cusick J (1985). Multivariate generalizations of the proportional hazards model. *Journal of the Royal Statistical Society*, 148, 82-117.

Cook RD, Johnson ME (1986). Generalized Burr-Pareto-logistics distributions with applications to a uranium exploration data set, *Technometrics*, 28, 123-131.

Cox DR (1972). Regression models and life tables (with discussion). *Journal of the Royal Statistical Society*, 34, 187-202.

Cox DR, Isham V (1980). *Point Processes*. University Press, Cambridge.

Cox DR, Oakes D (1984). *Analysis of Survival Data*. Chapman and Hall, London.

Crowder MJ (2001). *Classical Competing Risks*. Chapman and Hall.

Cunningham JW, Brookbank JW (1988). *Gerontology*. Harper and Row, New York.

Dagpunar JS (1997). Renewal-type equations for a general repair process. *Quality and Reliability Engineering International*, 13, 235-245.

Daley DJ, Vere-Jones D (1988). *An Introduction to the Theory of Point Processes*. Springer –Verlag, New York.

David HA, Moeschberger ML (1978). *The Theory of Competing Risks*. Griffin.

Denton FT, Spencer BG (1999). How old is old? Revising the definition based on life table criteria. *Mathematical Population Studies*, 7, 147-159.

Di Crescenzo A (2007). A Parrondo paradox in reliability theory. *The Mathematical Scientist*, 32, 17-22.

Dorado C, Hollander M, Sethuraman J (1997). Nonparametric estimation for a general repair model. *Annals of Statistics*, 25, 1140-1160.

Doyen L, Gaudoin O (2004). Classes of imperfect repair models based on reduction of failure intensity or virtual age. *Reliability Engineering and System Safety*, 84, 45-56.

Doyen L, Gaudoin O (2006). Imperfect maintenance in the generalized competing risks framework. *Journal of Applied Probability*, 43, 835-839.

Dufresne F, Gerber HU, Shiu ESW (1991). Risk theory with the gamma process. *ASTIN Bulletin*, 21, 177–192.

Ellingwood BR, Mori Y (1993). Probabilistic methods for condition assessment and life prediction of concrete structures in nuclear power plants. *Nuclear Engineering Design*, 142, 155–66.

Esaulova V (2006). Failure rates modelling for heterogeneous populations. PhD Dissertation, University of Magdeburg.

Feller W (1968). *An Introduction to Probability Theory*, Vol. I, 2nd edn. John Wiley & Sons, New York.

Feller W (1971). *An Introduction to Probability Theory*, Vol. II, 2nd edn. John Wiley & Sons, New York.

Finkelstein (1989). Perfect, minimal and imperfect repair (In Russian). *Reliability and Quality Control*, 3, 17-21.

Finkelstein MS (1992a). Some notes on two types of minimal repair. *Advance in Applied Probability*, 24, 226-228.

Finkelstein MS (1992b). A restoration process with dependent cycles (translation from Russian). *Automat. Remote Control*, 53, 7, 1115-1120.

Finkelstein MS (1993). A scale model of general repair. *Microelectronics and Reliability*, 33, 41-46.

Finkelstein MS (997). A concealed age of distribution functions and the problem of general repair. *Journal of Statistical Planning and Inference*, 65, 315-321.

Finkelstein MS (1998). A point process stochastic model with application to safety at sea. *Reliability Engineering and System Safety*, 60, 227-234.

Finkelstein MS (1999a). Wearing-out components in variable environment. *Reliability Engineering and System Safety*, 66, N3, 235-242.

Finkelstein MS (1999b). Multiple availability on stochastic demand. *IEEE Transactions on Reliability*, 48, 19-25.

Finkelstein MS (1999c). *Point Processes for Safety and Reliability* (in Russian). Elektropribor, St. Petersburg.

Finkelstein MS (1999d). A point process model for software reliability. *Reliability Engineering and System Safety*, 63, 67-71.

Finkelstein MS (2000). Modelling a process of non-ideal repair. In: Limnios N, Nikulin M (eds.) *Recent Advances in Reliability Theory*. Birkhauser, Boston, 41-53.

Finkelstein MS (2001). How does our n-component system perform? *IEEE Transactions on Reliability*, 50, 414-418.

Finkelstein MS (2002a). On the shape of the mean residual life function. *Applied Stochastic Models in Business and Industry*, 18, 135-146.

Finkelstein MS (2002b). On the reversed hazard rate. *Reliability Engineering and System Safety*, 78, 71-75.

Finkelstein MS. (2003a). Simple bounds for terminating Poisson and renewal processes. *Journal of Statistical Planning and Inference*, 113, 541-548.

Finkelstein MS (2003b). A model of biological aging and the shape of the observed failure rate. *Lifetime Data Analysis*, 9, 93-109.

Finkelstein MS (2003c). A model for spatial survival. *Statistics and Probability Letters*, 62, 371-378.

Finkelstein MS (2003d). On one class of bivariate distributions. *Statistics and Probability Letters*, 65, 1-6.

Finkelstein MS (2004a). Alternative time scales for systems with random usage. *IEEE Transactions on Reliability*, 53, 261-265.
Finkelstein MS (2004b). On the exponential formula of reliability and survival analysis. *IEEE Transactions on Reliability*, 53, 265-269.
Finkelstein MS (2004c). Minimal repair in heterogeneous populations. *Journal of Applied Probability*, 41, 281-286.
Finkelstein MS (2004d). On the multivariate analogue of exponential formula. *South African Statistical Journal*, 38, 47-57.
Finkelstein MS (2005a). On stochastic comparison of population densities and life expectancies. *Demographic research*, 13/6, 144-161.
Finkelstein MS (2005b). Lifesaving explains mortality decline with time. *Mathematical Biosciences*, 196, 187-197.
Finkelstein MS (2005c). On some reliability approaches to human aging. *The International Journal of Reliability, Quality and Safety Engineering*, 12, 337-346.
Finkelstein MS (2007a). On statistical and information-based virtual age of degrading systems. *Reliability Engineering and System Safety*, 92, 676-682.
Finkelstein MS (2007b). On some ageing properties of general repair processes. *Journal of Applied Probability*, 44, 506-513.
Finkelstein MS (2007c). Shocks in homogeneous and heterogeneous populations. *Reliability Engineering and System Safety*, 92, 569-575.
Finkelstein MS (2007d). Ageing: Damage accumulation versus increasing mortality rate. *Mathematical Biosciences*, 207, 104-112.
Finkelstein MS, Esaulova V (2001). Modelling a failure rate for the mixture of distribution functions. Probability in Engineering and Informational Sciences, 15, 383-400.
Finkelstein MS, Esaulova V (2005). On the weak IFR ageing of bivariate lifetime distributions. *Applied Stochastic Models in Business and Industry*, 21, 265-272.
Finkelstein MS, Esaulova V (2006a). Asymptotic behaviour of a general class of mixture failure rates. *Advances in Applied Probability*, 38, 244-262.
Finkelstein MS, Esaulova V (2006b). Mixture failure rate ordering. *Communications in Statistics, Theory and Methods*, 35, 1943-1955.
Finkelstein MS, Esaulova V (2008). On asymptotic failure rates in bivariate frailty competing risks model. *Statistics and Probability Letters*, doi:1016/j.spl.2007.11.012.
Finkelstein MS, Vaupel JW (2006). The relative tail of longevity and the mean remaining lifetime. *Demographic Research*, 14/6, 111-138.
Finkelstein MS, Zarudnij VI (2002). Laplace transform methods and fast repair approximations for multiple availability and its generalizations. *IEEE Transactions on Reliability*, 51, 168-177.
Fleming TR, Harrington DP (1991). *Counting Processes and Survival Analysis*. John Wiley, New York.
Gaver DP (1963). Random hazards in reliability problems. *Technometrics*, 5, 211-226.
Gavrilov NA, Gavrilova NS (1991). *The Biology of Life Span: A Quantitative Approach*. Harwood Academic Publishers.
Gavrilov NA, Gavrilova NS (2001). The reliability theory of ageing and longevity. *Journal of Theoretical Biology*, 213, 527-545.
Gnedenko BV, Belyaev Yu.K, Solovyev AD (1964). *Mathematical Methods in Reliability Theory* (in Russian). Nauka, Moscow. English translation: (1969). Academic Press, New York.
Goel AL (1985). Software reliability models: assumptions, limitations and applicability. *IEEE Transactions on Software Engineering*, SE-11, 1411-1423.
Goel AL, Okumoto K (1979). Time-dependent error detection rate model for software reliability and other performance measures. *IEEE Transactions on Reliability*, 28, 206-211.

Govil KK, Aggrawal KK (1983). Mean residual life function for normal, gamma and lognormal densities. *Reliability Engineering*, 5, 47-51.

Gompertz B (1825). On the nature of the function expressive of the law of human mortality and on a new mode of determining the value of life contingencies. *Philosophical Transactions of the Royal Society*, 115, 513-585.

Grabski F (2003). The reliability of an object with a semi-Markov failure rate. *Applied Mathematics and Computation*, 135, 1-16.

Guess F, Proschan F (1988). Mean residual life: Theory and applications. In: Krishnaiah PR, Rao CR (eds.) *Handbook of Statistics*. Elsevier, Amsterdam, 9, 215-224.

Gumbel EJ (1960). Bivariate exponential distribution. *Journal of American Statistical Association*, 55, 698-707.

Guo R, Love CE (2002). Statistical analysis of an age model for imperfectly repaired systems. *Quality and Reliability Engineering International*, 8, 133-146.

Gupta RC (2001). Nonmonotonic failure rates and mean residual life functions. In: Hayakawa Y, Irony T, Xie M (eds.) *System and Bayesian Reliability: Essays in Honour of Professor R.E. Barlow*. Series on Quality, Reliability and Engineering Statistics, World Scientific Press, Singapore, 147-163.

Gupta RC (2003). On some association measures in bivariate distributions and their relationships. *Journal of Statistical Planning and Inference*, 117, 83-98.

Gupta RC, Akman HO (1995). Mean residual life functions for certain types of nonmonotonic aging. *Communications in Statistics-Stochastic Models*, 11, 219-225.

Gupta RC, Gupta PL (1996). Ageing characteristics of the Weibull mixtures. *Probability in the Engineering and Informational Sciences*, 10, 591-600.

Gupta RC, Kirmani SNUA (2000). Residual coefficient of variation and some characterization results. *Journal of Statistical Planning and Inference*, 91, 23-31.

Gupta RC, Warren R (2001). Determination of change points of nonmonotonic failure rates. *Communication in Statistics-Theory and Methods*, 30, 1903-1920.

Gupta RD, Nanda AK (2001). Some results on reversed hazard rate ordering. *Communications in Statistics-Theory and Methods*, 30, 2447-2457.

Gurland J, Sethuraman J (1995). How pooling failure data may reverse increasing failure rate. *Journal of American Statistical Association*, 90, 1416-1423.

Gut A, Husler J (2005). Realistic variation of shock models. *Statistics and Probability Letters*, 74, 187-204.

Harmer G.P, Abbott D (1999). Parrondo's paradox. *Statistical Science*, 14, 206-213.

Hollander M, Sethuraman J (2002). Nonparametric inference for imperfect repair models. *The Indian Journal of Statistics*, 64, 693-706.

Hougaard P (2000). *Analysis of Multivariate Survival data*. Springer-Verlag, New York.

Hunink M, Glaszion P, Siegel J, Piskin J, Elstein A, Weinstein, M (2001). *Decision Making in Health and Medicine: Integrating Evidence and Values*. Cambridge University Press, New York.

Jelinski Z, Moranda PB (1972). Software reliability research. In: Freiberger W (ed.) *Statistical Methods for the Evaluation of Computer System Performance*. Academic Press, 465-484.

Jiang R, Murphy DNP (1995). Modelling failure data by mixture of two Weibull distributions. *IEEE Transactions on Reliability*, 44, 477-488.

Johnson NL, Kotz S (1975). A vector multivariate hazard rate. *Journal of Multivariate Analysis*, 5, 53-66.

Johnson NL, Kotz S, Balakrishnan N (1994). *Continuous Univriate Distributions*, Vol. 1. Wiley, New York.

Johnson NL, Kotz S, Balakrishnan N (1995). *Continuous Univriate Distributions*, Vol. 2. Wiley, New York.

Kahle W (2006). Optimal maintenance policies in incomplete repair models. *Reliability Engineering and System Safety*, 92, 569-575.

Kahle W, Love CE. (2003). Modelling the influence of maintenance actions. In: Lindquist BH, Doksum KA (eds.) *Mathematical and Statistical Methods in Reliability. Series on Quality, Reliability and Engineering Statistics*. World Scientific, Singapore, 387–400.

Kahle W, Wendt H (2004). On accumulative damage process and resulting first passage times. *Applied Stochastic Models in Business and Industry*, 20, 17-27.

Kalashnikov V (1997). *Geometric Sums: Bounds for Rare Events with Applications*. Kluwer Academic Publishers.

Kaminskij MP, Krivtsov V (1988). A Monte Carlo approach to repairable system reliability analysis. In: Mosleh A (ed.) *Probabilistic Safety Assessment and Management*, Springer-Verlag, New York.

Kaminskij MP, Krivtsov V (2006). G-renewal process in warranty data analysis. *Reliability: Theory and Applications*, 1, 29-34.

Kalbfleisch JD, Prentice RL (1980). *The Statistical Analyses of Failure Time Data*. John Wiley & Sons, New York.

Karlin S (1968). *Total Positivity*. Stanforfd University Press, Palo Alto, California.

Kebir Y (1991). On hazard rate processes. *Naval Research Logistics*, 38, 865-877.

Keiding N (1990). Statistical inference in the Lexis diagram. *Philosophical Transactions of the Royal Society of London*, A 332, 487-509.

Keyfitz N, Casewell N (2005). *Applied Mathematical Demography*. Springer.

Kijima M (1989). Some results for repairable systems with general repair. *Journal of Applied Probability*, 26, 89-102.

Kijima M (1992). Replacement policies of a shock model with imperfect preventive maintenance. *European Journal of Operational Research*, 57, 100-110.

Kijima M, Sumita U (1986). A useful generalization of renewal theory: counting processes governed by non-negative Markovian increments. *Journal of Applied Probability*, 23, 72-78.

Kijima M, Morimura H, Suzuki Y (1988). Periodical replacement problem without assuming minimal repair. *European Journal of Operational Research*, 37, 194-203.

Kirkwood TB (1997). The origins of human aging. *Philosophical Transactions of the Royal Society of London*, B, 352, 1765-1772.

Klemera P, Doubal P (2005). A new approach to the concept and computation of biological age. *Mechanisms of Aging and Development*, 127, 240-248.

Koltover VK (1997). Reliability concept as a trend in biophysics of aging. *Journal of Theoretical Biology*, 184, 157-163.

Kotz S, Balakrishnan N, Johnson NL (2000). *Continuous Multivariate Distributions*. V 1, 2nd edition, Wiley, New York.

Lai CD, Xie M (2006). *Stochastic Ageing and Dependence for Reliability*. Springer.

Lam Y (1988a). A note on the optimal replacement problem. *Advances in Applied Probability*, 20, 479-482.

Lam Y (1988b). Geometric process and the replacement problem. *Acta Mathematicae Applicatae Sinica*, 4, 366-382.

Lam Y (1996). Analysis of a two-component series system with a geometric process model. *Naval Research Logistics*, 43, 491-502.

Lam Y (1997). The rate of occurrence of failures. *Journal of Applied Probability*, 34, 234-247.

Lam Y, Zhang YL, Zheng YH (2002). A geometric process equivalent model for a multistate degenerative system. *European Journal of Operational Research*, 142, 21-29.

Last G, Szekli R (1998). Asymptotic and monotonicity properties of some repairable systems. *Advances in Applied Probability*, 30, 1089-1110.

Ledoux J (2003). Software reliability modelling. In: Pham H (ed.) *Handbook of Reliability Engineering*. Springer, London, 213-234.

Lehmann EL (1966). Some concepts of dependence. *Annals of Mathematical Statistics*, 37, 1137-1153.

Lemoine AJ, Wenocur ML (1985). On failure modelling. *Naval Research Logistics Quarterly*, 32, 479-508.

Li Y (2005). Asymptotic baseline of the hazard rate function of mixtures. *Journal of Applied Probability*, 42, 892-901.

Lindqvist B (1999). Repairable systems with general repair. In: Schueller G, Kafka P (eds.) *Safety and Reliability. Proceedings of the European Conference on Safety and Reliability, ESREL-99*. Balkema, 43-48.

Lindqvist B (2006). On statistical modelling and analysis of repairable systems. *Statistical Science*, 21, 532-551.

Lindqvist B, Elvebakk G, Heggland K (2003). The trend-renewal process for statistical analysis of repairable systems. *Technometrics*, 45, 31-44.

Lynch JD (1999). On conditions for mixtures of increasing failure rate distributions to have an increasing failure rate. *Probability in the Engineering and Informational Sciences*, 13, 33-36.

Lynn NJ, Singpurwalla ND (1997). Comment: "Burn-in" makes us feel good, *Statistical Science*, 12, 13-19.

Makeham WM (1860). On the law of mortality and the construction of annuity tables. *The Assurance Magazine and Journal of the Institute of Actuaries*, 8, 301-310.

Makeham WM (1897). On further developments of Gompertz's law. *The Assurance Magazine and Journal of the Institute of Actuaries*, 28, 152-159.

Mardia KV (1970). *Families of Bivariate Distributions*. Charles Griffin and Company Limited, London.

Marshall AW, Olkin I (2007). *Life Distributions*. Springer.

Meeker WQ, Escobar LA (1998). *Statistical Methods for Reliability Data*. John Wiley, New York.

Meilijson I (1972). Limiting properties of the mean residual lifetime function. *Annals of Mathematical Statistics*, 43, 354-357.

von Mises R (1964) *Selected Papers of Richard von Mises*. American Mathematical Society, Providence, Rhode Island.

Mi J (1994). Burn-in and maintenance policies. *Advances in Applied Probability*, 26, 207-221.

Mi J (1995). Bathtub failure rate and upside-down bathtub mean residual life. *IEEE Transactions on Reliability*, 44, 388-391.

Mi J (1998). A new explanation of decreasing failure rate of a mixture of distributions. *IEEE Transactions on Reliability*, 47, 460-462.

Mi J (1999). Age-smooth properties of mixture models. *Statistics and Probability Letters*, 43, 225-236.

Moranda PB (1975). Prediction of software reliability and its applications. In: *Proceedings of the Annual Reliability and Maintainability Symposium*. Washington DC, 327-332.

Murthy DNP, Xie M, Jiang R (2003). *Weibull Models*. Wiley, New York.

Musa JD, Ianino A, Okumoto K (1987). *Software Reliability: Measurement, Prediction, Application*. McGraw-Hill, New York.

Musa JD, Okumoto K (1984). A logarithmic Poisson execution time model for software reliability measurement. In: *Proceedings of the Seventh International Conference on Software Engineering*. Orlando, FL, 230-237.

Nakamura E, Lane M, Roth G, Ingram D (1998). A strategy for identifying biomarkers of ageing. *Experimental Gerontology*, 33, 42-46.

Nachlas JA (2005). *Reliability Engineering. Probabilistic Models and Maintenance Methods*. Taylor & Francis, Boca Raton.
Navarro J (2008). Likelihood ratio ordering of order statistics, mixtures and systems. *Journal of Statistical Planning and Inference*, 138, 1242-1257.
Navarro J (2008). Characterizations using the bivariate failure rate function. *Statistics and Probability Letters*, doi:10.1016/j.spl.2007.12.004.
Navarro J, Ruiz JM (2004). How to obtain bathtub-shaped failure rate models from normal mixtures. *Probability in the Engineering and Informational Sciences*, 18, 511-531.
Nelsen RB (2006). *Introduction to Copulas*. Springer.
Nelson W (1982). *Applied Life Data Analysis*. Wiley, New York.
Nelson W (1990). *Accelerated Testing, Statistical Models, Test Plans and Data Analysis*. Wiley, New York.
Nicolai RP (2008). *Maintenance Models for Systems Subject to Measurable Deterioration*. Thela Thesis, Rotterdam.
Nicolai RP, Dekker R, van Noortwijk JM (2007). A comparison of models for measurable deterioration: an application to coatings and steel structures. *Reliability Engineering and System Safety*, 92, 1635-1650.
van Noortwijk JM, van der Weide JAM, Kallen MJ, Pandey MD (2007). Gamma process and peaks over the threshold distributions for time-dependent reliability. *Reliability Engineering and System Safety*, 92, 1651-1658.
Oakes D. (1989). Bivariate survival models induced by frailties. *Journal of American Statistical Association*, 84, 487-493.
Oakes D, Dasu T (1990). A note on residual life. *Biometrica*, 77, 409-410.
Oeppen J, Vaupel JW (2002). Broken limits to life expectancy. *Science*, 296, 1029-1031.
Pasquini A, Crespo AN, Matrella P (1996). Sensitivity of reliability growth models to operational profile errors *vs*. testing accuracy. *IEEE Transactions on Reliability*, 45, 531-540.
Patra K, Dey D (2002). A general class of change point and change curve modelling for lifetime data. *Annals of the Institute of Statistical Mathematics*, 54, 517-530.
Pham H (2000). *Software Reliability*. Springer-Verlag, Singapore.
Preston SH, Coale AJ (1982). Age structure, growth, attrition and accession: a new synthesis. *Population Index*, 48, 217-259.
Pulchini, G (2003). Mechanical reliability and maintenance models. In: Pham H (ed.) *Handbook of Reliability Engineering*. Springer, London, 327-348.
Rausand M, Hoylandt A (2004). *System Reliability Theory: Models and Statistical Methods*, 2nd edn. John Wiley, New York.
Riggs JE, Hobbs GR (1998). Non-random sequence of slope-intercept estimates in longitudinal Gompertzian analysis suggests biological relevance. *Mechanisms of Aging and Development*, 100, 269-275.
Riggs JE, Millecchia RJ (1992). Using the Gompertz-Strehler Model of Ageing and Mortality to Explain Mortality Trends in Industrialized Countries. *Mechanisms of Aging and Development*, 65, 217-228.
Ross S (1996). *Stochastic Processes*. John Wiley, New York.
Schweizer B, Sklar A (1983). *Probabilistic Metric Spaces*. Elsevier Science Publishing, New York.
Sedjakin, N. (1966). On one physical principle of reliability theory (in Russian). *Techn. Kibernetika*, 1966, 3, 80-82.
Shaked M (1981). Bounds on the distance of a mixture from its parental distribution. *Journal of Applied Probability*, 18, 853-863.
Shaked M, Shanthikumar J (2007). *Stochastic Orders*. Springer.
Shaked M, Spizzichino F (2001). Mixtures and monotonicity of failure rate functions. In: Balakrishnan N, Rao CR (eds.) *Advances in Reliability*. Elsevier Science, Amsterdam, 20, 185-197.

Shin I, Lim T, Lie C (1996). Estimating parameters of intensity function and maintenance effect for repairable unit. *Reliability Engineering and System Safety*, 54, 1–10.
Singpurwalla ND (1995). Survival in dynamic environment, *Statistical Science*, 10, 86-103.
Singpurwalla ND, Wilson SP (1995). The exponential formula of reliability and survival analysis: does it always hold? *Lifetime Data Analysis*, 1, 187-194.
Singpurwalla ND, Wilson SP (1999). *Statistical Methods in Software Engineering*. Springer, New York.
Singpurwalla ND, Youngren MA (1993). Multivariate distributions induced by dynamic environments. *Scandinavian Journal of Statistics*, 20, 251-261.
Spizzichino F (1992). Reliability decision problems under conditions of ageing. In: Bernardo J, Berger J, Dawid AP, Smith AFM (eds.) *Bayesian Statistics*. Clarendon Press, Oxford, 4, 803-811.
Spizzichino F (2001). *Subjective Probability Models for Lifetimes*. Chapman and Hall, Boca Raton, Florida.
Stadje W, Zuckerman D (1991). Optimal maintenance strategies for repairable systems with general degree of repair. *Journal of Applied Probability*, 28, 384-396.
Steinsaltz D, Evans S (2004). Markov mortality models: implications of quasistationarity and varying initial distributions. *Theoretical Population Biology*, 65, 319-337.
Steinsaltz D, Goldwasser L (2006) Ageing and total quality management: extending the reliability metaphor for longevity. *Evolutionary Ecology Research*, 8, 1445-1459.
Strehler BL, Mildvan AS (1960). General theory of mortality and aging. *Science*, 132, 14-21.
Sumita U, Shanthikumar J (1985). A class of correlated cumulative shocks models. *Advances in Applied Probability*, 17, 347-366.
Sweet AL (1990). On the hazard rate of the lognormal distribution. *IEEE Transactions on Reliability*, 39, 325-328.
Thatcher RE (1999). The long-term pattern of adult mortality and the highest attained age. *Journal of the Royal Statistical Society*, 162, 5-43.
Thomson WA (1988). *Point Process Models with Applications to Safety and Reliability*. Chapman and Hall, New York.
Tsiatis A (1975). A nonidentifiability aspect of the problem of competing risks. *Proceedings of the National Academy of Sciences of the United States of America*, 72, 20-22.
Uematsu K, Nishida T (1987). One-unit system with a failure rate depending upon the degree of repair. *Mathematica Japonica*, 32, 139-147.
Ushakov IA, Harrison RA (1994). *Handbook of Reliability Engineering*. John Wiley & Sons, New York.
Vaupel JW (2003). Post-Darwinian longevity. In: Carey JK, Tuljapurkar S (eds). *Life Span. Evolutionary, Ecological and Demographic Perspectives*. A Supplement to vol.29: *Population and Development Review*, 127-151.
Vaupel JW, Yashin AI (1985). Heterogeneity ruses: some surprising effects of selection on population dynamics. *The American Statistician*. 39, 176-185.
Vaupel JW, Yashin AI (1987). Repeated resuscitation: how life saving alters life tables. *Demography*, 4, 123-135.
Vaupel JW, Carey JR, Christensen K (2003). It is never too late. *Science*, 301, 1679-1681.
Vaupel JW, Manton KG, Stallard E (1979). The impact of heterogeneity in individual frailty on the dynamics of mortality. *Demography*, 16, 439-454.
Wang JL Muller HG, Capra WB (1998). Analyses of the oldest-old mortality: lifetables revisited. *The Annals of Statistics*, 26, 126-163.
Wang H, Pham H (2006). *Reliability and Optimal Maintenance*. Springer, London.
Wenocur ML (1989). A reliability model based on the gamma process and its analytic theory. *Advances in Applied Probability*, 21, 899-918.

Witten M (1985). A return to time, cells, systems, and ageing: III. Gompertzian models of biological ageing and some possible roles for critical elements. *Mechanics of Ageing and Development*, 32, 141–177.

Xie M (1991a). On the solution of renewal-type integral equations, *Communications in Statistics*, 18, 281–293.

Xie M (1991b). *Software Reliability Modelling*. World Scientific, UK.

Yashin AI, Arjas A (1988). A note on random intensities and conditional survival functions. *Journal of Applied Probability*, 25, 630-635.

Yashin AI, Iachine IA (1999). Dependent hazards in multivariate survival problems. *Journal of Multivariate Analysis*, 71, 241-261.

Yashin AI, Manton KG (1997). Effects of unobserved and partially observed covariate processes on system failure: a review of models and estimation strategies. *Statistical Science*, 12, 20-34.

Yashin AI., Iachin, IA, Begun AS (2000). Mortality modelling: a review. *Mathematical Population Studies*, 8, 305-332.

Zahedi H (1991). Proportional mean remaining life model. *Journal of Statistical Planning and Inference*, 29, 221-228.

Zarunij VI (1973). *Reliability of the Marine Navigational Systems* (in Russian). Sudostroenie, Leningrad.

Index

accelerated life model (ALM), 96, 98, 99, 108, 126, 140, 171, 176, 182, 249
acceptable risk, 222, 223
age
- correspondence, 95, 96, 98, 105, 131
- process, 104, 111, 116, 119, 120
- reduction models, 102, 107, 125, 127
- specific population size, 241

ageing, 1, 3, 21, 23, 29, 30, 44, 63, 66, 94, 97, 109, 111, 113, 115, 119, 131, 136, 168, 171, 199, 215, 237, 238, 242, 246, 248, 249, 253, 256, 273
- bivariate, 59, 60
- negative, 61
- positive, 61

anti-ageing, 246, 248, 249
as good as new, 67, 70, 88, 92
asymptotic failure rate, 195
attraction interval, 211, 212
availability, 86, 204, 213, 214, 217
- multiple, 214, 215, 218, 221
- steady state (stationary), 214

bathtub (BT), 31, 32, 34, 35, 139, 142
biological age, 5, 94, 130, 235, 238, 246
bivariate risks, 52

calendar time, 79, 102, 103, 112, 113, 240, 241, 250, 256, 259, 260, 263, 264
CIF, 71, 72, 75, 78, 199, 200, 227, 228
cohort, 240, 242, 250, 251, 257, 260, 265, 266, 268
competing risks, 45, 59, 60, 63, 66, 172, 185
convolution, 21, 74, 77, 200, 203
copula, 63, 64
cross-sectional average life, 262
cycle, 67, 69, 77, 79, 80, 82, 86, 88, 93, 102, 108, 110, 112, 115, 118, 120, 125, 127

debugging, 226, 227, 230, 233, 234
DFR, 29, 30, 31, 60, 61, 63, 66, 136, 139, 142, 169
DFRA, 29
degradation, 67, 85, 95, 97, 98, 101, 103, 104, 106, 127, 128, 244, 253, 255
demography, 238, 240, 256, 260, 261
distribution
- absolutely continuous, 9, 10, 12, 16, 40, 43, 44, 46, 47, 52, 55, 58, 59, 61, 63, 65, 74, 95, 107, 175, 188, 232, 257
- baseline, 137, 145, 171, 182, 183, 196
- Clayton bivariate, 56, 57
- conditional, 108, 125, 134, 247, 255
- Dirichlet, 188
- Durling–Pareto bivariate, 65
- Erlangian, 21, 22, 102
- exponential, 19, 20, 22, 23, 29, 31, 37, 83, 91, 102, 105, 142, 166, 201, 205, 206, 214, 215, 228, 245, 252
- Farlie-Gumbel-Morgenstern, 64
- gamma, 20, 21, 22, 50, 136, 144, 145, 148, 158, 161, 178, 180, 182, 188, 243
- Gompertz, 28, 239, 245, 250, 251, 253, 263, 264, 271
- Gumbel bivariate, 56, 57, 188

inverse Gaussian, 27, 29, 31, 33
lifetime, 11, 19, 40, 44
lognormal, 25, 26, 29, 31, 33, 34
Makeham, 28
marginal, 55, 57, 63, 64
Marshall–Olkin bivariate, 57, 188
mixing, 144, 145, 146, 148, 152, 155, 157, 158, 165, 169, 172, 176, 178, 180, 182, 183, 187, 188, 190, 196
Pareto, 24, 25, 177, 182
stable, 149, 162, 163
truncated normal, 25, 26, 29, 33
Weibull, 23, 24, 26, 31, 32, 37, 127, 139, 143, 152, 161, 165, 173, 177, 181, 182, 188, 229, 244
DLR, 152, 153
DMRL, 29, 30, 60, 61

equivalent rate, 209, 211, 212
environment, 82, 92, 94, 95, 97, 105, 106, 131, 237, 249, 250
evolutionary theory, 238
exponential family, 147, 148
exponential representation
bivariate, 51, 54, 56, 57, 60, 65
unconditional, 47
conditional, 46, 47
marginal, 47
univariate, 12, 14, 15, 17, 40, 43, 44, 133, 144, 159, 241, 257, 261, 273

failure rate, 10-19, 22, 24-28, 30, 39, 68, 69, 73, 75, 77, 79-85, 90-92, 94, 97, 101, 102, 106, 109, 111, 115, 117, 124-128, 238, 239, 244, 245, 257, 270
bending down property, 150, 152
bivariate, 52, 54, 57
concatenated, 228
cumulative, 17, 101, 127, 175, 177, 182
mixture, 4, 6, 133, 135, 136, 137, 139, 141, 142, 144, 147, 149, 151, 154, 155, 158, 160, 165, 168, 172, 173, 175, 177, 179, 181, 185
per demand, 198, 199, 204, 206, 207, 210, 213, 215, 217, 218, 220, 223
reduction of, 95, 125, 126, 128
software, 225-235
frailty, 136, 137, 149, 159, 160, 162, 163, 172, 184, 188
bivariate, 185, 189
gamma, 149, 161, 162
multiplicative, 157, 159
multivariate, 172, 184
shared, 187

gamma process, 71, 128, 129, 243
geometric process, 76, 77, 80, 102, 125, 126
Glaser's theorem, 32, 33
Gompertz shift model, 250, 251, 263, 264

heterogeneous populations, 5, 89, 92, 135, 137, 159, 163, 168, 169

IFR, 29, 30, 31, 44, 60, 61, 63, 66, 114, 116, 118, 121, 136, 139, 142, 144, 169, 171
IFRA, 29
IMRL, 29, 30, 60, 61
intensity process, 71, 72, 75, 78, 79, 80, 85, 90, 91, 92, 110, 111, 112, 113, 114, 126, 127, 133, 225, 226, 227, 229, 230, 231, 234, 235
inter-arrival time, 73, 84, 88, 108, 111, 112, 115, 122
inverse problem, 145, 147, 149
Ito's equation, 243

Jensen's inequality, 48, 201

Laplace transform, 109, 145, 147, 149, 162
L'Hopital's rule, 37, 118
life deprivation, 266
life expectancy, 241, 250, 258, 260, 261, 264, 265, 268
period, 260, 261
lifesaving, 250, 251, 253, 266, 268, 273
line process, 210, 212
longevity, 268, 269

maintenance, 68, 70, 79, 81, 95, 113, 122, 124, 139, 238
mean remaining lifetime (MRL), 9, 11, 13, 14, 16, 18, 29, 22, 26, 28, 29, 34, 35, 39, 43, 44, 54, 83, 93, 101, 104, 105, 107, 116, 127, 166, 167, 251, 261, 273
information-based, 103
mixture, 167
mixture
bounds, 159

continuous, 135
discrete, 135, 137, 138
limiting behaviour, 171, 172, 185, 187
ordering, 149
mixture model
additive, 141, 144, 145, 178
multiplicative, 141, 142, 144, 146, 147, 157, 159, 164, 165, 168, 173, 179, 179, 182, 185, 189. 190
mortality plateau, 119, 244, 245

mortality rate, 1, 3, 7, 12, 28, 83, 97, 119, 137, 159, 238, 239, 249, 243, 246, 250, 253, 254, 256, 258, 261, 262, 264, 268, 273
period, 256

NBU, 29, 30
NWU, 29, 30

obstacles
with fixed coordinates, 208
with moving coordinates, 210
ordering
failure rate, 149, 150, 153, 154, 156
likelihood ratio, 152, 153, 155, 157, 169
stochastic, 61, 137, 150, 151, 153, 156, 157, 255, 259, 266, 268, 273
operational profile of software, 226, 231, 234, 236

parallel system, 17, 22, 23, 41, 44, 83, 244
point process, 67, 70, 71, 74, 80, 81, 85, 92
Poisson process, 6, 7, 51, 72, 73, 76, 82, 91, 103, 109, 114, 115, 129, 130, 197, 198, 200, 202, 204, 206, 253, 273
compound, 129, 130
memoryless property, 200, 214
nonhomogeneous (NHPP), 5, 72, 73, 90, 93, 109, 110, 129, 198, 200, 202, 209, 218, 246, 251, 253
spatial, 207
thinning of, 197, 222
population
age structure, 240, 241, 256, 257, 259, 261, 263, 265
density, 240
mean age, 262
stationary, 247, 257, 258, 261, 262, 265, 269, 273
probability of termination, 197, 223
proportional hazards (PH), 143, 183, 251

quality management, 238
quality of life, 253-255
quality of repair function, 116

random
change point, 165
environment, 45
failure rate, 47, 51, 68, 69, 133, 134, 168
scale, 164
variable, 9, 10, 15, 23, 42, 46, 48, 49, 52, 56, 59, 61, 67, 70, 73, 76, 85, 87, 102, 108, 114, 118, 122, 129, 133, 139, 150, 153, 157, 161, 165, 169, 177, 184, 202, 206, 209, 214, 222, 235, 240, 254, 257, 270
usage, 164, 165, 169
rate of
diffusion, 243, 246
point process, 71
redundancy
structural, 205
time, 205
regime, 95, 96, 98, 99, 101, 105, 107, 108
regularly varying functions, 255
renewal density, 74, 76, 88
renewal equations, 75, 109, 123
renewal process, 67, 68, 73, 74, 76, 77, 79, 81, 84, 86, 92, 93, 107, 108, 111, 124, 125, 128, 131, 197, 200, 201, 214, 217, 223, 234, 235
g-, 107, 108, 110
modulated, 79, 80
quasi-, 77
terminating, 200
repair, 4, 5, 67, 68, 70, 77, 79, 80, 82, 84, 89, 92, 93, 111, 119, 126, 128, 131, 134, 205, 206, 214, 216, 238, 266
fast, 204, 205, 206, 215, 216, 220, 221
general, 4, 5, 67
imperfect, 4, 5, 67, 68, 70, 75, 77, 80, 81, 88, 92, 93, 94, 102, 108, 110, 113, 115, 120, 122, 124, 126, 127, 128, 130, 131, 238
instantaneous, 57, 70, 73, 79, 81, 93, 107, 108, 111, 112, 124, 125

minimal, 5, 67, 68, 81, 93, 94, 104, 105, 109, 111, 115, 119, 121, 123, 125, 128, 131, 251, 266
perfect, 67, 70, 76, 84, 88, 94, 108, 114, 123, 128
repair process, 6, 68, 113, 116, 119, 128, 131
repairable system, 67, 68, 72, 81, 83, 85, 86, 92, 102, 107, 119, 204, 205, 213, 214, 218, 221
reproduction, 238
reversed failure rate, 9, 39, 41, 42, 44
risk, 11, 65, 172, 223
analysis, 14, 29, 135, 168, 176, 197, 198
conditional, 46

safety at sea, 197, 207, 208, 210, 212, 223
scale parameter, 21, 24, 50, 126, 128, 183, 243, 249
Sedjakin principle, 101
series system, 106, 192
shape of
failure rate, 10, 19, 26, 28, 29, 31, 43
mortality rate, 253, 270
MRL function, 28
shape parameter, 21, 25, 50, 126, 128, 183, 243, 249
shocks, 163, 199, 201, 207, 222, 245, 251, 252, 273
fatal, 201
non-fatal, 201
software reliability, 225, 226, 228, 230, 235
Jelimski–Moranda model, 227, 228
Moranda model, 228
Schick and Wolverton model, 229
spares, 109
spatial survival, 207
stability of an age process, 119, 121, 122
asymptotic, 119
stopping time, 47
Strehler–Mildvan model, 252, 273
survival function, 14, 17, 19, 3, 25, 28, 41, 47, 52, 54, 55, 58, 60, 64, 81, 106, 114, 116, 119, 125, 126, 134, 138, 144, 166, 168, 188, 202, 205, 247, 253, 257, 361, 265, 267, 272

unobserved resource model, 242, 244
upside-down bathtub (UBT), 31, 32, 34, 35, 139
ultimately decreasing (increasing), 150, 171, 174, 179, 184, 193, 255

variance, 19, 20, 25, 27, 137, 141, 143, 1149, 157, 159, 169, 243, 249
virtual age, 93, 95, 101, 102, 105. 107, 249
information-based, 95, 102, 103, 107
recalculated, 98, 100, 105, 107, 108
statistical, 95, 97, 98, 100, 105, 106
vitality, 252, 254

waiting time, 42, 44
mean, 43, 44
"weakest populations dying out first" principle, 150, 152, 157, 169, 172
wear, 3, 29, 31, 68, 70, 95, 119, 127, 128, 237, 238, 242, 243, 245, 246, 249
cumulative, 127
Wiener process, 129, 243

Yanoshi density, 252

GPSR Compliance

The European Union's (EU) General Product Safety Regulation (GPSR) is a set of rules that requires consumer products to be safe and our obligations to ensure this.

If you have any concerns about our products, you can contact us on ProductSafety@springernature.com

In case Publisher is established outside the EU, the EU authorized representative is:

Springer Nature Customer Service Center GmbH
Europaplatz 3
69115 Heidelberg, Germany

Batch number: 10755053

Printed by Printforce, the Netherlands